Studies in Classification, Data Analysis, and Knowledge Organization

More information about this series at http://www.springer.com/series/1564

Nadja Bauer · Katja Ickstadt ·
Karsten Lübke · Gero Szepannek ·
Heike Trautmann · Maurizio Vichi
Editors

Applications in Statistical Computing

From Music Data Analysis to Industrial
Quality Improvement

 Springer

Editors
Nadja Bauer
Department of Computer Science
Dortmund University of Applied
Sciences and Arts
Dortmund, Germany

Karsten Lübke
Institute for Empirical Research
and Statistics
FOM University of Applied Sciences
Essen, Germany

Heike Trautmann
Department of Information Systems
University of Münster
Münster, Germany

Katja Ickstadt
Faculty of Statistics
TU Dortmund University
Dortmund, Germany

Gero Szepannek
School of Business Studies
HOST University of Applied
Sciences Stralsund
Stralsund, Germany

Maurizio Vichi
Department of Statistical Sciences
Sapienza University of Rome
Rome, Italy

ISSN 1431-8814 ISSN 2198-3321 (electronic)
Studies in Classification, Data Analysis, and Knowledge Organization
ISBN 978-3-030-25146-8 ISBN 978-3-030-25147-5 (eBook)
https://doi.org/10.1007/978-3-030-25147-5

This Springer imprint is published by the registered company Springer Nature Switzerland AG
The registered company address is: Gewerbestrasse 11, 6330 Cham, Switzerland

Published with permission of TU Dortmund/Jürgen Huhn

Preface

This volume on *Applications in Statistical Computing—From Music Data Analysis to Industrial Quality Improvement* is dedicated to our teacher, mentor, colleague, and friend Claus Weihs. With his retirement approaching, we thought it more than appropriate to honor his professional and scientific career with a Festschrift, and here are some reasons why.

Statisticians are often seen as "reacting" researchers; reacting to new, interesting data that, in turn, require new statistical methodology for a successful analysis. Nowadays, many of them find themselves reacting to the rapid development of the field of data science. Claus is not one of them. He has been proactive throughout his scientific career. In fact, as a statistician he has been one of the designers of the field of data science in Germany. In addition, he made important contributions to many fields of applications ranging from econometrics to music.

Claus started his scientific career in Bonn, Germany, where he studied mathematics with a minor in computer science. Afterward, he received his Ph.D. in Bonn with a dissertation on *Auswirkungen von Fehlern in den Daten auf Parameterschätzungen und Prognosen*, a topic already on the interface between mathematics, computations, and applications.

In 1985, he began to work as a statistical consultant for Ciba-Geigy in Basel, Switzerland. There, he developed statistical models and software for industrial applications. With this expertise, he was perfectly suited for the newly designed chair for "Computational Statistics" at TU Dortmund University that he accepted in 1994, well before the boom of what is now called "Data Science".

Since then Claus has been shaping the field of computational statistics with contributions to many fields at the interface of statistics and data science and to various applications. For example, he participated in the collaborative research centers SFB 475 on "Reduction of Complexity in Multivariate Data Structures" and SFB 823 on "Statistical Modelling of Nonlinear Dynamic Processes". Initialized by the advent hymn "Tochter Zion", he developed statistical methods for music data analysis, a field also covered in this Festschrift.

Moreover, very early on, in 2002, he initiated the Bachelor's degree program "Datenanalyse und Datenmangement" and the Master's program in "Datenwissenschaft" in a joint collaboration with the Faculties of Mathematics and Computer Science at TU Dortmund University. Both, the BA and the MA program, are now called "Data Science".

The community of statisticians and data scientists, TU Dortmund University, especially its Faculty of Statistics, and we as editors, personally, profited in many more ways from Claus' creativity and visionary nature. With this Festschrift, we try to cover some of the many facets of his work.

The volume consists of six parts: Methodological Developments in Data Science, Computational Statistics, Perspectives on Statistics and Data Science, Statistics in Econometric Applications, Statistics in Industrial Applications, and Statistics in Music Applications.

We would like to thank all contributing authors for their support and cooperation in developing this special issue. As the selection of authors naturally is biased toward our own memories and experiences, we apologize to all colleagues whom we did not contact. We hope you will enjoy reading this Festschrift nevertheless. Our special thanks go to Martina Bihn and Veronika Rosteck from Springer, and to Claus' wife Heidrun Gömpel, for their encouragement and their help all along the way including the editing of this volume. Our biggest thanks go to our teacher, mentor, colleague, and friend Claus! We all wish you the very best for your retirement! We are sure statistics and data science will still be part of your future. Stay proactive and enjoy spending more time on singing and jazz, and with family and friends!

Dortmund, Germany   Nadja Bauer
October 2019   Katja Ickstadt
                                                                                        Karsten Lübke
                                                                                       Gero Szepannek
                                                                                      Heike Trautmann
                                                                                       Maurizio Vichi

Contents

Part VI   Statistics in Music Applications

Part I
Methodological Developments in Data Science

Chapter 1
Aviation Data Analysis by Linear Programming in Airline Network Revenue Management

Wolfgang Gaul and Christoph Winkler

Abstract Aviation data comprise, e.g., bookings and cancellations by consumers as well as no show situations before departure, aircraft-type assignments to flight legs and overbooking-decisions to avoid empty seats in airplanes. Here deterministic linear programming (DLP) is a widely used approach to process this kind of data in an area called airline network revenue management for which adaptions of a basic DLP-model to overbooking situations as well as the offering of flexible products are known. We combine these concepts in a model which simultaneously allows the incorporation of overbooking-decisions and the offering of specific as well as flexible products. Additionally, we further extend this integrated formulation to allow the treatment of different booking-classes and aircraft-type assignment considerations. We present characteristics of the new approach, which uses the overlapping science directions of data analysis and operations research, point out differences to already known results in airline network revenue management, describe an example which illustrates how the different aspects can be considered, and indicate the advantages of our model in view of various data settings.

1.1 Motivation

An important part of the scientific literature in airline revenue management tackles network problems, which means that a connection of an airline between two cities may consist of two or more flight legs with additional transfers on the ground (see, e.g., Talluri and van Ryzin 1998, 1999; Bertsimas and Popescu 2003; Bertsimas and de Boer 2005; Adelman 2007; Klein 2007; van Ryzin and Vulcano 2008; Liu and van Ryzin 2008; Topaloglu 2008; Erdelyi and Topaloglu 2009; Topaloglu 2009a, b;

W. Gaul (✉) · C. Winkler
Karlsruhe Institute of Technology (KIT), Kaiserstraße 12, 76128 Karlsruhe, Germany
e-mail: wolfgang.gaul@kit.edu
URL: https://www.wolfgang-gaul.de

C. Winkler
e-mail: christoph.winkler12@web.de

© Springer Nature Switzerland AG 2019
N. Bauer et al. (eds.), *Applications in Statistical Computing*,
Studies in Classification, Data Analysis, and Knowledge Organization,
https://doi.org/10.1007/978-3-030-25147-5_1

3

Kunnumkal and Topaloglu 2010; Erdelyi and Topaloglu 2011; Kunnumkal et al. 2012; Lapp and Weatherford 2014; Vossen and Zhang 2015; Barz and Gartner 2016; Goensch 2017).

Another aspect in airline revenue management is overbooking, the acceptance of booking requests, that exceed the physical capacities of available resources. In order to avoid empty seats, overbooking is practiced to compensate cancellations during the booking-process and no-shows just before departure. While overbooking using a single resource was investigated since the middle of the past century (see, e.g., Beckmann 1958; Subramanian et al. 1999; Karaesmen and van Ryzin 2004b; Amaruchkul and Sae-Lim 2011; Lan et al. 2011), in the last few years several contributions concerning overbooking in an airline network appeared (see e.g., Bertsimas and Popescu 2003; El-Haber and El-Taha 2004; Karaesmen and van Ryzin 2004a, b; Gosavi et al. 2007; Erdelyi and Topaloglu 2009, 2010; Wannakrairot and Phumchusri 2016).

Additionally, products with different properties (or various restrictions) have been designed to meet preferences and the willingness to pay of consumer segments as well as the interests of potential sellers. One striking concept is the distinction between specific and flexible products, introduced in revenue management by Gallego and Phillips (2004). While a specific product is an offering with exactly predefined conditions (e.g., in aviation, restrictions concerning the route of a flight, booking-class, utilization time, etc.), a flexible product is only characterized by some fundamental conditions. The seller can assign a flexible offering to one of several specific products that fulfill the constraints by which the flexible product is described. In Gallego et al. (2004), Petrick et al. (2010, 2012), Koch et al. (2017) this concept was applied to the more complex network-case.

As a solution of sophisticated (network) revenue management problems using dynamic programming fails because of the high dimensionality of the state-variables (see Talluri and van Ryzin 2004 for a detailed discussion), deterministic linear programming (DLP) has been applied where the term "deterministic" means that demand for products is substituted by expected values. Adaptions of DLP concerning overbooking (see, e.g., Bertsimas and Popescu 2003) and the incorporation of flexible products (see, e.g., Gallego et al. 2004) were developed (and are described in the appendix). We propose an extension by combining these two adaptations in a single approach in which, additionally, class-dependent bookings and aircraft-type assignments to flight legs based on fluctuations in demand are considered as a further contribution where data analysis and operations research as two overlapping sciences (see, e.g., Gaul 2007) are applied to find best solutions in underlying data.

Notation needed for the description of the new approach is introduced in Sect. 1.2 together with selected model features. In Sect. 1.3, an example with data from an airline network environment is provided. Finally, a conclusion is presented in Sect. 1.4. An appendix is used to explain how model formulations concerning adaptions of DLP to overbooking, the offering of specific as well as flexible products, and the combination of both aspects can be described.

1.2 Notation and Model Description

1.2.1 Notation

With $|M|$ as notation of the cardinality of a set M consider an airline that operates a set of aircraft-types $\tilde{T} = \{1, \ldots, |\tilde{T}|\}$ to cover a set of flight legs $H = \{1, \ldots, |H|\}$ which induces costs $cost_{h\tau}$ if type $\tau \in \tilde{T}$ is assigned to flight leg $h \in H$. A set of classes $K = \{1, \ldots, |K|\}$ indicates physically available classes of seats in the aircraft-types (e.g., first, business, economy) and sometimes additional booking-classes in which different price settings are charged due to buying restrictions and marketing strategies. $c_{\tau k}$ denotes the capacity of class $k \in K$ on type $\tau \in \tilde{T}$.

The offering of the airline is assumed to consist of specific and flexible products. Specific products are described by a set $I = \{1, \ldots, |I|\}$ of origin–destination connections and the assigned class with revenue $r_{ik}, i \in I, k \in K$. A set $J = \{1, \ldots, |J|\}$ with revenue f_j, $j \in J$, indicates flexible products where an execution-mode set $M_j \subseteq I$ fixes a possible allocation of $j \in J$ to a subset of appropriate origin–destination connections.

A matrix A (with entries a_{hi} equal to 1, if the origin–destination connection i needs flight leg h, and equal to 0, otherwise, for $h \in H$ and $i \in I$, respectively, a_{hm} equal to 1, if execution-mode m uses flight leg h, and equal to 0, otherwise, for $h \in H$ and $m \in M_j$) describes the assignment between flight legs and products characterized via origin–destination connections.

The class-matrix B (with b_{ik} equal to 1, if the origin–destination connection i offers class k, and equal to 0, otherwise, respectively, b_{mk} equal to 1, if execution-mode m provides class k, and equal to 0, otherwise) indicates the class as important characteristic within the offering of an airline.

The values p^s_{hk} and p^f_{hk} denote expected show-probabilities, i.e., whether customers show up before departure (divided into specific and flexible bookings) of flight leg h and class k. As overbooking will be considered, d_{hk} describes the costs for a denied service concerning flight leg h and class k. A parameter ρ_k indicates an upper bound of the share of flexible customers in class k.

Available capacities must be sold within a given time-horizon which is discretized into T time-points or accompanying periods which are numbered backward from T to 1 (time to departure).

The expected aggregated demand data up to time period t are denoted by $\overline{D}^s_{ikt}$ for the specific product described by the tuple (i, k), respectively, by $\overline{D}^f_{jt}$ in the case of flexible products and may change when t approaches the departure times of corresponding origin–destination connections.

1.2.2 Model Formulation and Selected Features

Now, the formulation of an approach, abbreviated as "class-specific DLP$_t$-flex-over with aircraft-type assignment" (see also the appendix), can be described as follows:

$$max \quad \sum_{i \in I} \sum_{k \in K} r_{ik} \cdot x_{ik} + \sum_{j \in J} f_j \cdot \sum_{m \in M_j} \sum_{k \in K} y_{jmk}$$

$$- \sum_{h \in H} \sum_{k \in K} d_{hk} \cdot u_{hk} - \sum_{h \in H} \sum_{\tau \in \tilde{T}} cost_{h\tau} \cdot l_{h\tau}$$

$$s.t. \sum_{i \in I} a_{hi} \cdot b_{ik} \cdot x_{ik} + \sum_{j \in J} \sum_{m \in M_j} a_{hm} \cdot b_{mk} \cdot y_{jmk} \leq z_{hk} \quad \forall h \in H, k \in K \quad (1.1)$$

$$\sum_{j \in J} \sum_{m \in M_j} a_{hm} \cdot b_{mk} \cdot y_{jmk} \leq \rho_k \cdot z_{hk} \quad \forall h \in H, k \in K \quad (1.2)$$

$$x_{ik} \leq \overline{D}_{ikt}^s \quad \forall i \in I, k \in K \quad (1.3)$$

$$x_{ik} \in \mathbb{Z}_+ \quad \forall i \in I, k \in K \quad (1.4)$$

$$\sum_{m \in M_j} \sum_{k \in K} y_{jmk} \leq \overline{D}_{jt}^f \quad \forall j \in J \quad (1.5)$$

(**class-specific DLP$_t$ -flex-over** $\quad y_{jmk} \in \mathbb{Z}_+ \quad \forall j \in J, k \in K, \quad (1.6)$

with aircraft-type assignment) $\quad\quad\quad\quad\quad m \in M_j$

$$\sum_{\tau \in \tilde{T}} l_{h\tau} = 1 \quad \forall h \in H \quad (1.7)$$

$$l_{h\tau} \in \{0, 1\} \quad \forall h \in H, \tau \in \tilde{T} \quad (1.8)$$

$$u_{hk} \geq z_{hk} \cdot \left(\rho_k \cdot p_{hk}^f + (1 - \rho_k) \, p_{hk}^s \right) - \sum_{\tau \in \tilde{T}} l_{h\tau} \cdot c_{\tau k} \quad \forall h \in H, k \in K \quad (1.9)$$

$$z_{hk} \geq \sum_{\tau \in \tilde{T}} l_{h\tau} \cdot c_{\tau k} \quad \forall h \in H, k \in K \quad (1.10)$$

$$u_{hk} \in \mathbb{Z}_+ \quad \forall h \in H, k \in K \quad (1.11)$$

$$z_{hk} \in \mathbb{Z}_+ \quad \forall h \in H, k \in K \quad (1.12)$$

where $\mathbb{Z}_+$ denotes the non-negative integers.

As most important part within the description of the model formulation (1.1)–(1.12) only the demand data are explicitly indexed by time period t although also other model descriptors (e.g., flight legs, number of aircraft-types, costs, revenues, show-probabilities) can undergo alterations and—of course—the outcomes of the decision variables of the approach are dependent on the time chosen.

For simplicity, however, the index t is omitted except for consumer demand. The decision-variables of the approach are $l_{h\tau}$, u_{hk}, x_{ik}, y_{jmk}, and z_{hk}. The binary variables $l_{h\tau}$ indicate which aircraft-type $\tau \in \tilde{T}$ is assigned to flight leg $h \subset H$. As usual in overbooking situations u_{hk} describes the number of denied boardings on flight leg $h \in H$ in class $k \in K$. x_{ik} (specific) and y_{jmk} (flexible) denote the numbers of passengers (within execution-mode $m \in M_j$ in the case of flexible products) that can be allocated with respect to the offering of specific and flexible products. Although not contained in the objective function, the overbooking-limits z_{hk}, for which ρ_k (share of flexible passengers) and p_{hk}^s, p_{hk}^f (expected show-probabilities divided into specific and flexible passengers) have to be taken into consideration, are also calculated.

The description of the mathematical formulation of the model can be enriched with the help of the following explanations:

The objective function maximizes the revenues from allocating specific and flexible products taking into account the costs of denied boardings and the assignment of aircraft-types to flight legs, described by restrictions of type (1.7) and (1.8). Conditions of type (1.1) secure that the numbers of allocated seats with respect to the offering of specific and flexible products do not exceed the overbooking-limits, which are bounded from below by the capacity of the aircraft-types used (constraints (1.10)) while conditions (1.2) state that only predefined parts of the overbooking-limits can be allocated to flexible customers. The expected aggregated demands up to time period t are additional upper bounds (constraints (1.3) and (1.5)), which allows to calculate solutions of the model for different time-points. Restrictions of type (1.9) provide lower bounds for denied boardings as difference of the customers who show up (i.e., the overbooking-limits weighted with the show-probabilities (divided into specific and flexible demand and multiplied by the shares of flexible and specific customers)) minus the capacity of the aircraft-types used. Restrictions to nonnegative integers (constraints (1.4), (1.6), (1.11), and (1.12)) are straightforward (see the appendix for additional explanations concerning the historical development of DLP and adaptions to overbooking and the incorporation of flexible products).

Beside the notation used and the mathematical model formulation just presented some selected features of the new approach merit additional attention:

- A bunch of airline network revenue management constraints helps to analyze aviation data (in which changing consumer demand plays the most important role) by linear programming.
- Overbooking and the incorporation of flexible products are jointly treated and not solved one after the other.
- The consideration of different booking-classes is new which allows, e.g., to specify class-dependent revenues and denied boarding costs and to check expected show-

probabilities (divided into specific and flexible bookings) as well as shares of flexible customers for different classes at different time-points. As a result, class-dependent overbooking-limits can be calculated which are of importance for the decisions of the management of an airline.

– The recalculation of solutions dependent on a time point/period t at which expected aggregated demand data are updated also allows other model parameters (e.g., the shares of flexible customers or the show-probabilities) to be reconsidered and possibly adapted.

– The assignment of flight legs to available aircraft-types with different seat capacities can be used to better balance overbooking-decisions and avoid empty seats.

1.3 Example

An example, in which an airline network environment with 3 airports P, Q, and R, 6 flight legs, 3 booking-classes, 27 specific products, and 1 flexible product is used will illustrate the application of our approach.

The example is kept small on purpose to demonstrate which kind of data has to be provided, which results can be expected, and how the management of an airline can react to changes of the data that describe the underlying situation.

1.3.1 Data Setting

The network environment of an airline offering different connections between airports P, Q, and R within a 1 day period is depicted in Fig. 1.1.

The $|H| = 6$ flight legs are denoted by $(1), \ldots, (6)$. The continuous arrows describe flight legs, the dotted ones show possible stopovers/transfers in the airport of Q between flight legs. There are $|I| = 9$ possible origin–destination connections: P to

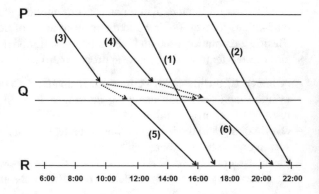

Fig. 1.1 Considered airline network environment

R: (1); (2) (without stopover/transfer in Q), P to Q: (3); (4), Q to R: (5); (6), P via Q to R: (3)→(5); (3)→(6); (4)→(6) (numbered as (7), (8), and (9)).

A traveler from P to R can take one of the direct flight legs (1) or (2) or a connection via one transfer in Q: (3)→(5), (3)→(6) or (4)→(6). With three booking-classes per origin–destination connection (e.g., first/business/economy) there are 27 specific products.

To make the description easy, we define one flexible product consisting of a trip from P to R with assignment to one of the five possible origin–destination connections $i \in M_1 = \{1, 2, 7, 8, 9\}$.

To exclude the possibility to assign flexible customers to first-class seats, $\rho_1 = 0$ is assumed in order to retain the exclusiveness of the high-priced segment.

To allow an as simple as possible description, further data are as follows: Direct flight legs from P to R have revenues of 1200, 600, and 300 monetary units (mu), depending on the three booking-classes (first/business/economy). There is a discount if travelers from P to R accept a transfer (30%). For the shorter distance flight legs ((3), (4), (5), (6)) the revenues are 720, 360, and 180 mu, again, depending on the three booking-classes. The revenue for the flexible product is 175 mu.

$|\tilde{T}| = 3$ different aircraft-types are considered to operate the flight legs with overall capacities of 200 ($\tau = 1$, small), 300 ($\tau = 2$, medium), and 400 ($\tau = 3$, large) seats, for which shares of 5%/10%/85% depending on the booking-classes first/business/economy are assumed.

The assignment of aircraft-type τ to flight leg h induces costs with $cost_{h1} = 18.000$ mu, $cost_{h2} = 28.000$ mu and, $cost_{h3} = 42.000$ mu (assumed independent (for simplicity) of flight legs h).

For the overbooking situation, the compensation-costs (in mu) for denied services concerning the longer distance flight legs ((1), (2)) are $d_{h1} = 1500$, $d_{h2} = 750$, and $d_{h3} = 400$, while for the shorter distance flight legs ((3), (4), (5), (6)) $d_{h1} = 800$, $d_{h2} = 420$, and $d_{h3} = 240$ mu are taken into consideration.

For the show-probabilities of flexible passengers p_{hk}^f we take—for all flight legs h and classes k—a value of 1.0 (because we argue, that, if someone can make a flexible booking, s(he) is such flexible that s(he) can always come to departure). For the show-probabilities of specific customers p_{hk}^s, we deliberately use values smaller than 1.0, as explained in the discussion of the solution of this example, in detail $p_{hk}^s = 0.95/0.99/0.95/0.95/0.9/0.9$ for $h = 1, \ldots, 6$ (assumed independent (for simplicity) of class k). As upper bound for the share of flexible customers we consider $\rho_1 = 0$ (already explained earlier), $\rho_2 = 0.2$, and $\rho_3 = 0.25$ (i.e., we consider up to 25% for the share of flexible customers in class 3).

We use two demand situations to explain how model solutions change:

In case 1, at a certain time point/period t_1 we take values of $\overline{D}_{1kt_1}^s = \overline{D}_{2kt_1}^s = 20/40/326$ for $k = 1, 2, 3$ (direct connections from P to R), $\overline{D}_{3kt_1}^s = 6/13/123$ and $\overline{D}_{4kt_1}^s = 5/12/92$ for $k = 1, 2, 3$ (flight legs from P to Q), $\overline{D}_{5kt_1}^s = 7/15/139$ and $\overline{D}_{6kt_1}^s = 5/10/110$ for $k = 1, 2, 3$ (flight legs from Q to R). For connections from P to R via transfer in Q, we use $\overline{D}^{s}_{7kt_1} = 3/6/52$, $\overline{D}^{s}_{8kt_1} = 1/2/13$, and $\overline{D}^{s}_{9kt_1} = 4/7/49$ for $k = 1, 2, 3$, and, finally, $\overline{D}^{f}_{1t_1} = 95$ as flexible demand.

In case 2, at a time point/period $t_2 \neq t_1$, we assume (for simplicity) that only demand values for the specific products $i = 3$ and $i = 5$ have been altered to $\overline{D}_{3kt_2} = 8/18/172$ and $\overline{D}_{5kt_2} = 10/21/195$ under otherwise unchanged conditions.

1.3.2 Results

Tables 1.1 and 1.2 illustrate the results of our model for the different demand situations.

For both tables in the first column—dependent on single flight legs as well as connections with transfer together with the booking-class—the numbers of allo-

Table 1.1 Results for demand situation 1

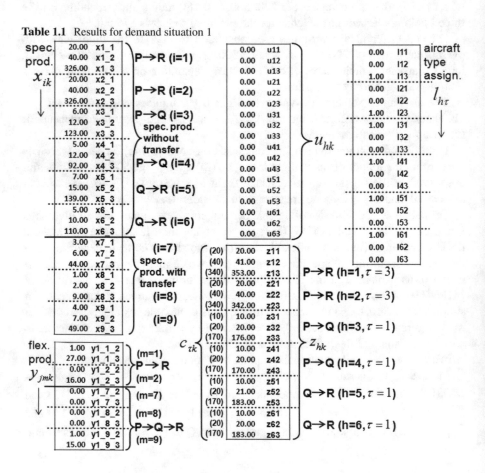

Table 1.2 Results for demand situation 2

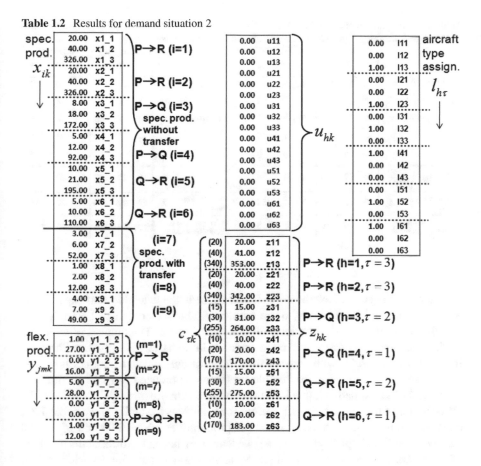

cated specific products x_{ik} (upper part) and flexible products y_{jmk} (lower part with indication of the execution-mode) are depicted (Notice, that flexible demand is not allocated to class 1.). In the third column, $l_{h\tau}$ values assign flight legs to aircraft-types. Finally, in the second column in the middle of the tables the upper part indicates the denied boardings u_{hk} while the lower part provides the overbooking-limits z_{hk} together with the available seat capacities of the aircraft-types used.

The underlying small example and the alteration of consumer demand data for only two origin–destination connections $i = 3$ and $i = 5$ were selected on purpose to allow an easy check how solutions change. The comparison of the results of Tables 1.1 and 1.2 show that while in demand situation 1 flight legs $h = 3, \ldots, 6$ are operated by small aircraft-types (with large types assigned to flight legs $h = 1, 2$), in demand situation 2 medium aircraft-types replace the small ones for flight legs $h = 3$ and $h = 5$ dependent on balancing the costs for operating airplanes of different types with the revenues obtainable by providing larger numbers of seats.

Table 1.3 Comparison of seat allocations of flight leg 3 (for which the aircraft-type assignment is changed) and flight leg 6 (for which the aircraft-type assignment is unchanged)

	k	situation 1			situation 2		
		1	2	3	1	2	3
flight leg (3)	x_{3k}	6	12	123	8	18	172
(7): (3)→(5)	x_{7k}	3	6	44	3	6	52
(8): (3)→(6)	x_{8k}	1	2	9	1	2	12
	y_{17k}	--	0	0	--	5	28
	y_{18k}	--	0	0	--	0	0
$\sum_{l\in\{3,7,8\}} x_{ik} + \sum_{m\in\{7,8\}} y_{1mk}$		10	20	176	12	31	264
overbooking-limits	z_{3k}	10	20	176	15	31	264
seat capacities of aircraft-type used		10	20	170	15	30	255

(a)

	k	situation 1			situation 2		
		1	2	3	1	2	3
flight leg (6)	x_{6k}	5	10	110	5	10	110
(8): (3)→(6)	x_{8k}	1	2	9	1	2	12
(9): (4)→(6)	x_{9k}	4	7	49	4	7	49
	y_{18k}	--	0	0	--	0	0
	y_{19k}	--	1	15	--	1	12
$\sum_{l\in\{6,8,9\}} x_{ik} + \sum_{m\in\{8,9\}} y_{1mk}$		10	20	183	10	20	183
overbooking-limits	z_{6k}	10	20	183	10	20	183
seat capacities of aircraft-type used		10	20	170	10	20	170

(b)

Although the sum of overbooking-limits exceeds the available seat capacity by 49 in situation Table 1.1 and by 61 in situation Table 1.2 no denied boardings are calculated based on the description used for this example. The number of allocated flexible customers increases from 60 in situation Table 1.1 to 90 in situation Table 1.2 in which two aircrafts with larger seat capacities are selected (Notice, however, that the seat allocation y_{193} for flexible customers is reduced from 15 in situation Table 1.1 to 12 in situation Table 1.2 in accordance with the purpose intended by the introduction of flexible offerings.). Except for flight legs (3) and (6) (see Table 1.3a, b and remember that we selected the show-probabilities $p_{3k}^s = 0.95$ and $p_{6k}^s = 0.9$ (independent (for simplicity) of class k)) we leave it to the reader to check the numbers of customers allocated to a single flight leg, to an origin–destination connection with transfer, and/or to inspect the numbers of customers with specific or flexible demand, to add these numbers in those cases where different types of customers are jointly seated in a certain class, and to compare the results obtained with the overbooking-limits and the physical available seat capacities in the classes of the aircraft-types used.

Already this small example demonstrates which kind of model descriptors have to be provided by the airline management and how changing consumer demand influences the data analysis situation.

Additional alterations of the data-setting can be caused because, e.g., the numbers of available aircraft-types change, the offering of booking-classes has to be adapted, new flight legs are added and/or not attractive flight legs are canceled. Normally, show-probabilities and shares of flexible customers are fixed on the basis of information from comparable situations in the past but—of course—actual consumer behavior influences the underlying airline network environment. In many cases, however, changes in consumer demand for certain origin–destination connections are the main reason for recalculations—as explained in the example just described.

1.4   Conclusion

We combined the concepts of overbooking, incorporation of flexible products, class-dependent seat allocation, and aircraft-type assignments to flight legs in an integrated airline network revenue management approach in which the overlapping science directions of data analysis and operations research were used to solve a sophisticated aviation data analysis problem by linear programming. The data part does not use classical data analysis techniques, instead, time-varying data of consumers are analyzed with the help of operations research. The consideration of different booking-classes allows, e.g., to exclude higher valued classes from an assignment to flexible offerings. Another aspect is the integration of aircraft-type assignment constraints as seat allocation is connected with actually available aircraft-type capacity. We formulated an extended DLP (deterministic linear programming)-model in which these aspects are handled simultaneously and used the marking "class-specific DLP-flex-over-model with aircraft-type assignment" (see also the appendix).

To demonstrate how the new DLP-formulation would behave in an application situation we tested our model within a revenue management network and illustrated how model results react to changing consumer demand data.

A   Appendix

In this appendix, we consider different DLP (deterministic linear programming)-model adaptions for airline network revenue management to support the understanding of our new approach that we have called "class-specific DLP_t-flex-over with aircraft-type assignment". The notation corresponds to what has already been explained in Chap. 2. Based on the formulation of the starting DLP-model (A1) in which only the offering of specific products was described (Simpson 1989), the incorporation of overbooking-decisions (A2) has been rewritten (Bertsimas and Popescu 2003) as well as the extension by Gallego et al. (2004) to include a distinction between specific and flexible offerings (A3). We further state a DLP-model with simultaneous treatment of both aspects (A4) which was the starting point for our treatise on aviation data analysis by linear programming.

A1   The Basic DLP-Model

An early formulation of a DLP-model in network revenue management is as follows (see Simpson 1989):

$$max \sum_{i \in I} r_i \cdot x_i$$

$$s.t. : \sum_{i \in I} a_{hi} \cdot x_i \leq c_h \quad \forall h \in H \tag{A1.1}$$

(DLP_t)
$$x_i \leq \overline{D}_{it} \quad \forall i \in I \tag{A1.2}$$
$$x_i \geq 0 \quad \forall i \in I \tag{A1.3}$$

The objective function maximizes the revenue. The constraints (A1.1) secure that the sum of sold products does not exceed the capacities c_h. Constraints (A1.2) and (A1.3) stand for upper bounds $\overline{D}_{it}$ of expected demand data and the nonnegativity of the number of accepted products.

A2 DLP Adaption to Overbooking

An extension of DLP to overbooking is expressed by the following model (see Bertsimas and Popescu 2003):

$$max \sum_{i \in I} r_i \cdot x_i - \sum_{h \in H} d_h \cdot u_h$$

$$s.t. : \sum_{i \in I} a_{hi} \cdot x_i \leq z_h \quad \forall h \in H \tag{A2.1}$$

$$x_i \leq \overline{D}_{it} \quad \forall i \in I \tag{A2.2}$$

$(DLP_t\text{-}over)$
$$x_i \geq 0 \quad \forall i \in I \tag{A2.3}$$
$$u_h \geq p_h \cdot z_h - c_h \quad \forall h \in H \tag{A2.4}$$
$$z_h \geq c_h \quad \forall h \in H \tag{A2.5}$$
$$u_h \geq 0 \quad \forall h \in H \tag{A2.6}$$

In the objective function, additionally, the costs d_h for denied services are subtracted. Constraints (A2.1) are similar as (A1.1) but, now, overbooking-limits z_h for all resources $h \in H$ as right-hand side replace the capacities. While constraints (A2.2) and (A2.3) are unchanged compared to formulation (A1) conditions (A2.4)–(A2.6) delineate the overbooking-extension where p_h denotes show-probabilities.

A3 DLP Adaption to Flexible Offerings

The extension to the case in which specific as well as flexible products can be offered (see Gallego et al. 2004) can be stated as follows:

$$max \sum_{i \in I} r_i \cdot x_i + \sum_{j \in J} f_j \cdot \sum_{m \in M_j} y_{jm}$$

$$s.t.: \sum_{i \in I} a_{hi} \cdot x_i + \sum_{j \in J} \sum_{m \in M_j} a_{hm} \cdot y_{jm} \leq c_h \quad \forall h \in H \qquad (A3.1)$$

$$x_i \leq \overline{D}_{it}^{s} \quad \forall i \in I \qquad (A3.2)$$

(DLP$_t$-flex) $$x_i \geq 0 \quad \forall i \in I \qquad (A3.3)$$

$$\sum_{m \in M_j} y_{jm} \leq \overline{D}_{jt}^{f} \quad \forall j \in J \qquad (A3.4)$$

$$y_{jm} \geq 0 \quad \forall j \in J, m \in M_j \qquad (A3.5)$$

Now, the objective function maximizes the revenue from specific and flexible products with $M_j \subseteq I$ as execution-mode set that describes the allocation of the flexible product j to a subset of specific products. Also, capacity-constraints (A3.1) are extended by a term for flexible products. While (A3.2) and (A3.3) are already known from the (A1) and (A2) model descriptions (with a distinction of the demand data terms ($\overline{D}_{it}^{s}$ for specific products and D_{jt}^{f} for flexible products)) the restrictions for flexible products have now to take into consideration the execution-mode sets M_j in (A3.4) and (A3.5).

A4 DLP-Formulation of Integrated Specific/Flexible Offerings and Overbooking

A DLP-model in which both aspects (overbooking as well as specific and flexible products) are handled simultaneously is now easy to formulate and was the starting point for our treatise of aviation data analysis within airline network revenue management:

$$max \sum_{i \in I} r_i \cdot x_i + \sum_{j \in J} f_j \cdot \sum_{m \in M_j} y_{jm} - \sum_{h \in H} d_h \cdot u_h$$

$$s.t.: \sum_{i \in I} a_{hi} \cdot x_i + \sum_{j \in J} \sum_{m \in M_j} a_{hm} \cdot y_{jm} \leq z_h \quad \forall h \in H \qquad (A4.1)$$

$$x_i \leq \overline{D}_{it}^{s} \quad \forall i \in I \qquad (A4.2)$$

(DLP$_t$-flex-over) $$x_i \geq 0 \quad \forall i \in I \qquad (A4.3)$$

$$\sum_{m \in M_j} y_{jm} \leq \overline{D}_{jt}^{f} \quad \forall j \in J \qquad (A4.4)$$

$$y_{jm} \geq 0 \quad \forall j \in J, m \in M_j \qquad (A4.5)$$

$$u_h \geq p_h \cdot z_h - c_h \quad \forall h \in H \qquad (A4.6)$$

Fig. 1.2 From DLP to class-specific DLP-flex-over with aircraft-type assignment

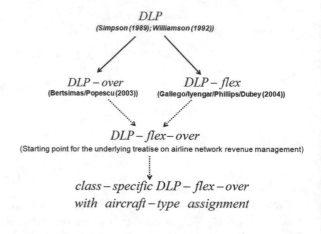

$$z_h \geq c_h \quad \forall h \in H \qquad \text{(A4.7)}$$
$$u_h \geq 0 \quad \forall h \in H \qquad \text{(A4.8)}$$

The objective function maximizes the revenue of both kinds of offered products from which the costs for rejected customers have to be subtracted. Constraints (A4.1) secure that the sum of allocated flexible and specific products does not exceed the overbooking-limits for all resources $h \in H$. Constraints (A4.2) and (A4.4) consider the expected demand data of specific and flexible products. (A4.3) and (A4.5) are nonnegativity restrictions. Conditions (A4.6)–(A4.8) stand for the overbooking-extension already formulated in (A2).

Strictly speaking, the formulation described in (A4) provides already a new approach in which overbooking and the incorporation of flexible products are integrated. Even more challenging is the additional consideration of booking-classes (which, e.g., allows to restrict flexible products to lower valued classes) and aircraft-type assignments (which, e.g., combines demand with physically available seat capacity) in the DLP-model handled in this paper (see Fig. 1.2).

References

Adelman, D. (2007). Dynamic bid-prices in revenue management. *Operations Research, 55*(4), 647–661.

Amaruchkul, K., & Sae-Lim, P. (2011). Airline overbooking models with misspecification. *Journal of Air Transport Management, 17*(2), 143–147.

Barz, C., & Gartner, D. (2016). Air cargo network revenue management. *Transportation Science, 50*(4), 1206–1222.

Beckmann, M. J. (1958). Decision and team problems in airline reservations. *Econometrica, 26*(1), 134–145.

Bertsimas, D., & de Boer, S. V. (2005). Simulation-based booking limits for airline revenue management. *Operations Research, 53*(1), 90–106.

Bertsimas, D., & Popescu, I. (2003). Revenue management in a dynamic network environment. *Transportation Science, 37*(3), 257–277.

El-Haber, S., & El-Taha, M. (2004). Dynamic two-leg airline seat inventory control with overbooking, cancellations and no-shows. *Journal of Revenue and Pricing Management, 3*(2), 143–170.

Erdelyi, A., & Topaloglu, H. (2009). Separable approximations for joint capacity control and overbooking decisions in network revenue management. *Journal of Revenue and Pricing Management, 8*(1), 3–20.

Erdelyi, A., & Topaloglu, H. (2010). A dynamic programming decomposition method for making overbooking decision over an airline network. *INFORMS Journal on Computing, 22*(3), 443–456.

Erdelyi, A., & Topaloglu, H. (2011). Using decomposition methods to solve pricing problems in network revenue management. *Journal of Revenue and Pricing Management, 10*(4), 325–343.

Gallego, G., Iyengar, G., Phillips, R., & Dubey, A. (2004). Managing flexible products on a network. CORC Technical Report TR-2004-01, IEOR Department, University of Columbia.

Gallego, G., & Phillips, R. (2004). Revenue management of flexible products. *Manufacturing & Service Operations Management, 6*(4), 321–337.

Gaul, W. (2007). *Data analysis and operations research.* Studies in classification, data analysis, and knowledge organisation (pp. 357–366). Berlin: Springer.

Goensch, J. (2017). A survey on risk-averse and robust revenue management. *European Journal on Operational Research, 263*(2), 337–348.

Gosavi, A., Ozkaya, E., & Kahraman, A. F. (2007). Simulation optimization for revenue management of airlines with cancellations and overbooking. *OR Spectrum, 29*(1), 21–38.

Karaesmen, I., & van Ryzin, G. J. (2004a). *Coordinating overbooking and capacity control decisions on a network.* Technical Report, Columbia Business School.

Karaesmen, I., & van Ryzin, G. J. (2004b). Overbooking with substitutable inventory classes. *Operations Research, 52*(1), 83–104.

Klein, R. (2007). Network capacity control using self-adjusting bid prices. *OR Spectrum, 29*(1), 39–60.

Koch, S., Goensch, J., & Steinhardt, C. (2017). Dynamic programming decomposition for choice-based revenue management with flexible products. *Transportation Science, 51*(4), 1031–1386.

Kunnumkal, S., Talluri, K., & Topaloglu, H. (2012). A randomized linear programming method for network revenue management with product-specific no-shows. *Transportation Science, 46*(1), 90–108.

Kunnumkal, S., & Topaloglu, H. (2010). Computing time-dependent bid prices in network revenue management problems. *Transportation Science, 44*(1), 38–62.

Lan, Y., Ball, M. O., & Karaesmen, I. (2011). Regret in overbooking and fare-class allocation for single leg. *Manufacturing & Service Operations Management, 13*(2), 194–208.

Lapp, M., & Weatherford, L. (2014). Airline network revenue management: Considerations for implementation. *Journal of Revenue and Pricing Management, 13*(2), 83–112.

Liu, Q., & Van Ryzin, G. J. (2008). On the choice-based linear programming model for network revenue management. *Manufacturing & Service Operations Management, 10*(2), 288–310.

Petrick, A., Goensch, J., Steinhardt, C., & Klein, R. (2010). Dynamic control mechanisms for revenue management with flexible products. *Computers & Operations Management, 37*(11), 2027–2039.

Petrick, A., Steinhardt, C., Goensch, J., & Klein, R. (2012). Using flexible products to cope with demand uncertainty in revenue management. *OR Spectrum, 34*(1), 215–242.

Simpson, R.W. (1989). Using network flow techniques to find shadow prices for market and seat inventory control. MIT, Flight Transportation Laboratory Memorandum m89-1, MIT, Cambridge, Massachusetts.

Subramanian, J., Stidham, S., & Lautenbacher, C. J. (1999). Airline yield management with overbooking, cancellations, and no-shows. *Transportation Science, 33*(2), 147–167.

Talluri, K. T., & van Ryzin, G. J. (1998). An analysis of bid-price controls for network revenue management. *Management Science*, *44*(11), 1577–1593.

Talluri, K. T., & van Ryzin, G. J. (1999). A randomized linear programming method for computing network bid prices. *Transportation Science*, *33*(2), 207–216.

Talluri, K. T., & van Ryzin, G. J. (2004). *The theory and practice of revenue management*. New York: Springer.

Topaloglu, H. (2008). A stochastic approximation method to compute bid prices in network revenue management problems. *INFORMS Journal on Computing*, *20*(4), 596–610.

Topaloglu, H. (2009a). Using Lagrangian relaxation to compute capacity-dependent bid prices in network revenue management. *Operations Research*, *57*(3), 637–649.

Topaloglu, H. (2009b). On the asymptotic optimality of the randomized linear program for network revenue management. *European Journal of Operational Research*, *197*(3), 884–896.

Van Ryzin, G. J., & Vulcano, G. (2008). Simulation-based optimization of virtual nesting controls for network revenue management. *Operations Research*, *56*(4), 865–880.

Vossen, T., & Zhang, D. (2015). Reductions of approximate linear programs for network revenue management. *Operations Research*, *63*(6), 1352–1371.

Wannakrairot, A., & Phumchusri, N. (2016). Two-dimensional air cargo overbooking models under stochastic booking request level, showup rate and booking request density. *Computers & Industrial Engineering*, *100*, 1–12.

Williamson, E. L. (1992). Airline network seat control. Ph.D. thesis MIT, Cambridge, Massachusetts.

Chapter 2
Bayesian Reduced Rank Regression for Classification

Heinz Schmidli

Abstract Many classical multivariate analysis methods are special cases of reduced rank regression, including canonical correlation analysis, redundancy analysis, and R. A. Fisher's linear discriminant analysis. The latter classifies an object based on a few linear combinations of its multivariate measurements. Classical inference for linear discriminant analysis may be based on asymptotic theory or resampling methods. A Bayesian linear discriminant analysis is proposed using Bayesian reduced rank regression as a starting point. The model can be implemented in Bayesian software with Markov chain Monte Carlo approaches and is easily extendable.

2.1 Introduction

Two commonly encountered problems in applied statistics are the description of class differences and the allocation of objects into classes. The data on which description or classification is based consist of a number of objects, which are known to belong to several classes or groups and for which multiple quantitative measurements are available. Classification is widely applied (Weihs and Schmidli 1990, 1991; Weihs et al. 1993, 2007; Aggarwal 2014), and possibly the first example is the classification of different species of Iris flowers based on measurements of their sepal/petal lengths and widths (Fisher 1936). In medicine, examples include the classification of tumors based on gene expression data (Dudoit et al. 2002), the diagnosis of diseases based on metabolomic data (Madsen et al. 2010) and the classification of biological activity of putative drugs based on data related to their chemical structure (Dudek et al. 2006).

Linear discriminant analysis (LDA) has been introduced by Fisher (1936) both for the description of class differences and for the classification of new objects. LDA is also known as canonical variate analysis or canonical discriminant analysis. Three aspects of an LDA are particularly important. First, it provides a *dimension reduction*

H. Schmidli (✉)
Novartis Statistical Methodology, Basel, Switzerland
e-mail: heinz.schmidli@novartis.com

© Springer Nature Switzerland AG 2019   19
N. Bauer et al. (eds.), *Applications in Statistical Computing*,
Studies in Classification, Data Analysis, and Knowledge Organization,
https://doi.org/10.1007/978-3-030-25147-5_2

by constructing a few latent variables, the discriminant scores, that describe the variation between classes. Second, class differences can be visualized by a *graphical analysis* of the low-dimensional space of discriminant scores. And third, a new object is *classified* by first projecting it into the discriminant space and then allocating it to the nearest class. LDA is a tool for exploratory data analysis, although methods for statistical inference have also been proposed (Krzanowski 1989; Weihs 1993).

Reduced rank regression (Anderson 1951; Schmidli 1995) refers to statistical models that encompass LDA (Izenman 1986) and many other classical multivariate analysis methods such as canonical correlation analysis (Tso 1981) and redundancy analysis (Leeden 1990). Hence approaches for statistical inference developed for reduced rank regression may also be applied for LDA. Reduced rank regression consists of a multivariate linear regression model (i.e., with several dependent variables), where the coefficient matrix is not of full rank. The singularity of the coefficient matrix makes statistical inference challenging both from a classical and from a Bayesian perspective.

Within the classical framework, a first attempt at statistical inference was made by Velu et al. (1986). However, Anderson showed 15 years later that these results are incorrect (Anderson 2002), and correct results are in line with those obtained for models assuming normally distributed errors (Ryan et al. 1992; Schmidli 1995). For LDA, confidence and tolerance regions for the discriminant scores based on analytical approximations (Krzanowski 1989) or resampling procedures may be used (Krzanowski and Radley 1989; Weihs and Schmidli 1990; Weihs et al. 1993).

A Bayesian approach was first proposed by Geweke (1996). However, 8 years later, Geweke reported numerous flaws in the code provided with his article (Geweke 2004). Another 13 years later, Karlsson noted that some of the conditional posterior distributions in Gewekes original article are incorrect (Karlsson 2017). An alternative approach to Bayesian inference for reduced rank regression models has been described in Schmidli (1996).

Bayesian LDA is straightforward when no reduced rank constraints are imposed, as LDA is then equivalent to a standard linear multivariate regression. For example, when the population consists of just two classes, then the coefficient matrix of the multivariate regression model is always of full rank one. However, a Bayesian LDA has not been discussed in the literature when there are rank constraints, e.g., if the population consists of ten classes and a one-dimensional discriminant space is desired. This article describes the methodology and software implementation of a Bayesian LDA with rank constraints.

The article is structured as follows. Section 2.2 contains a description of the classification example used for illustrating the methods. The example involves flowers as this seems most appropriate for a *Festschrift*. In Sects. 2.3 and 2.4, classical LDA is shortly reviewed, and the equivalence of LDA and reduced rank regression is explained, respectively. In Sect. 2.5, a Bayesian LDA is described. The article closes with a discussion.

I would like to thank Claus Weihs for helpful discussions of a draft version of this article at a time when he lived in Basel, Switzerland.

Fig. 2.1 Iris flowers near Villa Merian, Basel, Switzerland

2.2 Fisher's Iris Data

Discrimination between different species of Iris flowers requires expert knowledge (Fig. 2.1). To simplify this task, one would like to first take quantitative measurements of a flower of unknown species and then use a numerical rule to classify this flower to a species. For three Iris species (setosa, versicolor, virginica), four measurements (sepal length and width, petal length, and width) are available for 50 flowers of each species. Based on this learning data, one would like to describe differences among species and to obtain a classification rule. This Iris data example motivated R.A. Fisher to develop linear discriminant analysis (Fisher 1936). The example will be used below for illustrating the methodology. The data are available in the *R* software environment (R Core Team 2019) as dataset *iris* or *iris3*.

2.3 Linear Discriminant Analysis

The learning data used for classification consist of q variables measured for n objects, where n_k of these objects fall into the kth of g classes. As described, for example, in Chap. 11 of Mardia et al. (1979), LDA finds a linear combination of the measured variables, such that the between-classes variance is maximal compared with the pooled within-class variance. Then a second linear combination is searched which optimizes the same criterion, and so on. The number of selected linear combinations is called the dimensionality r of the model. The linear combinations are gathered in a $q \times r$ matrix L, the discriminant loadings.

Table 2.1 Discriminant loading vector L obtained by LDA for the Iris data

Sepal. L.	Sepal. W.	Petal. L.	Petal. W.
-0.8293776	-1.5344731	2.2012117	2.8104603

More precisely, let Y be the $n \times q$ data matrix, W the $q \times q$ pooled within-class covariance matrix and B the $q \times q$ between-class covariance matrix. The discriminant loading matrix L maximizes the trace $tr(L'BL)$ subject to the constraint $L'WL = I_r$, where I_r is the identity matrix. The discriminant scores are given by the $n \times r$ matrix $(Y - 1_n m'_Y)L$, with mean $m_Y = 1/n \, Y' 1_n$, where $1_n = (1, 1, \ldots, 1)'$. The scores are projections from the high-dimensional space to a low-dimensional space, in such a way that most information about the group differences is kept. Note that in the discriminant space, the pooled within-class covariance is the unit matrix and the between-class covariance matrix is diagonal.

Graphical methods are used to analyze the low-dimensional discriminant space. The discriminant rule of LDA classifies a new object based on its measurements y as follows: y is projected in the discriminant space by $(y - m_Y)'L$, and then classified to the class whose corresponding projected mean is nearest.

In the R software environment (R Core Team 2019), an LDA analysis is easily obtained for the Iris data (see Appendix). For dimension $r = 1$, the discriminant loading matrix L is a 4-dimensional vector and is shown in Table 2.1.

2.4   Reduced Rank Regression Linked to LDA

A statistical model closely related to LDA is based on reduced rank regression (see e.g., Schmidli 1995). The q-dimensional vector of measured quantities y is assumed to follow a multivariate normal distribution with different means for different classes but the same covariance matrix. Formally one can write for object $i = 1, \ldots, n$ and class $k(i) \in \{1, \ldots, g\}$,

$$p(y_i | \mu_{k(i)}, \Sigma) = N(\mu_{k(i)}, \Sigma), \tag{2.1}$$

where μ_k is an unknown q-dimensional vector of class means and Σ an unknown $q \times q$ covariance matrix. In matrix form, model (2.1) can be written as a multivariate regression model:

$$Y = 1_n \mu' + XM + E, \ E \sim N(0, \Sigma \otimes I_n), \tag{2.2}$$

where $Y = (y_1 \ldots y_n)'$ is an $n \times q$ matrix of measured variables. X is a centered indicator matrix of dimensionality $n \times (g - 1)$, obtained by centering the matrix $X^\star$ with

$X_{ik}^\star = 1$ if object i is in class k, $X_{ik}^\star = 0$ otherwise. $M = ((\mu_1 - \mu_g) \ldots (\mu_{g-1} - \mu_g))'$ is a $(g-1) \times q$ coefficient matrix, and $\mu = \mu_g$.

In many applications, the class means μ_k are not arbitrary vectors in q-dimensional space, but lie approximately in an r-dimensional linear subspace, where $r < \min (q, g-1)$. This is equivalent to the statement that the coefficient matrix M in (2.2) is of reduced rank r. Such a multivariate regression model with rank constraints on the coefficient matrix is called a reduced rank regression model. Although this multivariate model was first proposed by Anderson (1951), the name was introduced by Izenman (1986).

A parametrization of the coefficient matrix of reduced rank is necessary for estimation, i.e. by writing $M = AB'$ with $p \times r$ matrix A and $q \times r$ matrix B, where both A and B have rank r. Since this decomposition is not unique, one needs to constrain A to be in a subset of matrices Ψ such that uniqueness is guaranteed. For example, one may require that $A = (I_r \, \tilde{A})$, where $\tilde{A}$ is a $(q - r) \times r$ matrix; for example, if $r = 1$, A is a vector with first element equal to 1. This type of constraint facilitates implementation in standard software. However, many other choices are possible (Geweke 1996; Schmidli 1996). For example, classical multivariate methods often use the *orthogonality constraint*

$$A'X'XA = I_r. \tag{2.3}$$

This does not fully remove all identifiability problems, as, e.g., for rank $r = 1$ multiplying the vector A by (-1) gives an equivalent solution. For higher dimensions $(r > 1)$, any rotated matrix A still fullfills the orthogonality constraint, and provides an equivalent model.

The reduced rank regression model can now be written as

$$p(Y|A, B, \Sigma, \mu) = N(1_n\mu' + XAB', \Sigma \otimes I_n), \quad A \in \Psi. \tag{2.4}$$

Maximum likelihood (ML) estimates of μ, A, B, and Σ can be easily obtained based on a canonical correlation analysis of X and Y (Tso 1981), as shown in the Appendix. The ML discriminant rule classifies a new object based on the measurements y to the likeliest class, i.e., to the class k for which $(y - \mu - BA'x_k)'\Sigma^{-1}(y - \mu - BA'x_k)$ is maximal, where x_k is the centered indicator vector corresponding to class k. Anderson (1951) noted that reduced rank regression models can be used for discriminant analysis. As essentially shown by Izenman (1986), linear discriminant analysis with r dimensions is equivalent to the reduced rank regression model of rank r, in the sense that the ML discriminant rule is the same as the discriminant rule of LDA. In fact, if the orthogonality constraint (2.3) is used, the discriminant loadings can be obtained directly from the ML estimates of the unknown parameters in (2.4) as $L = \Sigma^{-1}B(B'\Sigma^{-1}B)^{-1/2}$.

2.5 Bayesian LDA

A Bayesian analysis of model (2.4) is intractable analytically. However, the Gibbs sampling procedure (Gelfand and Smith 1990) can be used for the calculation of the posterior distribution of the parameters. To use this Markov chain Monte Carlo method, the full conditional posterior distributions have to be specified (Geweke 1996; Karlsson 2017; Schmidli 1996). Weakly informative normal priors may be used for A, B, μ, and a weakly informative inverse Wishart prior for Σ (see Appendix for details).

As explained in the previous section, a Bayesian reduced rank regression may be used for a Bayesian LDA. Suppose that a large sample $\{A_j, B_j, \Sigma_j, \mu_j; j = 1, \ldots, M\}$ from the posterior distribution $p(A, B, \Sigma, \mu | Y)$ of the reduced rank regression model is available using the Gibbs sampling method. This sample from the posterior distribution can be used for the description of class differences and for the prediction of class membership.

LDA describes class differences in a low-dimensional space characterized by discriminant scores, and discriminant loadings are obtained as $L = \Sigma^{-1}B$ $(B'\Sigma^{-1}B)^{-1/2}$, if an orthogonality constraint (2.3) is used on A. Hence, a sample from the posterior distribution of the loadings is obtained as $L_j = \Sigma_j^{-1}B_j(B_j'\Sigma_j^{-1}B_j)^{-1/2}$.

For the prediction of class membership, a new object described by the q-dimensional vector y is classified to the class for which the predictive density is highest (Aitchison and Dunsmore 1975). More formally, the predictive density for class k is given by

$$\rho_k(y) = \int N(y|\mu' + x_k'AB', \Sigma)p(A, B, \Sigma, \mu | Y)d\omega, \tag{2.5}$$

where the $(g - 1)$-dimensional vector x_k characterizes class k as explained above in (2.2). The predicted class for the new object is then given by $k_0 = argmax_k \, \rho_k(y)$. Based on a sample from $p(A, B, \Sigma, \mu | Y)$, the density $\rho_k(y)$ in (2.5) can be estimated as $1/M \sum N(y|\mu_j' + x_k'A_jB_j', \Sigma_j)$.

For illustration, classical and Bayesian analysis results are shown for the Iris data based on the code provided in the Appendix. The ML analysis can be obtained within the R software environment (R Core Team 2019), while the Bayesian analysis can be obtained through WinBUGS (Lunn et al. 2012). The ML estimates and summaries of the Bayesian posterior distribution (2.5, 50, and 97.5% quantiles) for the key parameters (A, B, L) are shown in Table 2.2. As seen from the results, ML estimates are close to the median of the posterior distribution. This is not unexpected, as weakly informative priors were used, and sample sizes are moderately large in this example. The ML estimate for the discriminant loading vector L is exactly the same as the one obtained using LDA (see Table 2.1).

Table 2.2 Summary of ML and Bayesian LDA analysis for the Iris data

A (%)	A1	A2		
ML	0.1946495	0.05752927		
Bayes 2.5	0.1930086	0.05110868		
Bayes 50	0.1946015	0.05732960		
Bayes 97.5	0.1959980	0.06352294		
B (%)	B1	B2	B3	B4
ML	−7.882773	2.781000	−20.90178	−8.913941
Bayes 2.5	−8.912916	2.060404	−21.75969	−9.355788
Bayes 50	−7.886958	2.784618	−20.89771	−8.910545
Bayes 97.5	−6.855253	3.525173	−20.02679	−8.462303
L (%)	L1	L2	L3	L4
ML	−0.8293776	−1.5344731	2.201212	2.810460
Bayes 2.5	−1.3890620	−2.0679588	1.734913	1.774936
Bayes 50	−0.8639108	−1.4963499	2.283539	2.651137
Bayes 97.5	−0.3318067	−0.9133734	2.843259	3.540898

2.6 Discussion

The Bayesian linear discriminant analysis described in this article has been closely aligned with R. A. Fisher's linear discriminant analysis. However, the methodology may be extended. For example, rather than assume the same covariance matrix for each class, these could be class specific. Also, the assumption of a multivariate normal distribution may be changed, for example, by assuming a more heavy-tailed distribution. The assumption of a known model dimensionality could also be relaxed.

In the past 30 years, Bayesian approaches have been increasingly applied in practice. This trend can, for example, be clearly seen in pharmaceutical industry (Ashby 2006; Grieve 2007). Before the seminal paper by Racine et al. (1986), Bayesian methods were almost nonexistent in drug research and development. In the early days, Bayesian applications focused on preclinical or early clinical phases of drug development and on pharmacokinetics (Racine-Poon 1986; Dempster et al. 1983; Gelfand et al. 1990; Racine-Poon et al. 1991; Seewald 1994; Bennett and Wakefield 1996). In the twenty-first century, Bayesian methods are now used in all phases of drug development, covering evidence synthesis and prediction, adaptive clinical trial designs, decision-making, and many other areas (Grieve and Krams 2005; Schmidli et al. 2007; Neuenschwander et al. 2008, 2010; Pozzi et al. 2013; Gsteiger et al. 2013; Schmidli et al. 2013; Gsponer et al. 2014; Schmidli et al. 2014; Fisch et al. 2015; Lange and Schmidli 2015; Schmidli et al. 2017; Bornkamp et al. 2017; Weber et al. 2018; Mielke et al. 2018; Holzhauer et al. 2018).

In the twentieth century, a major hurdle for the use of Bayesian methods in practice has been the computational difficulties. However, with the introduction of Markov chain Monte Carlo approaches (Gelfand and Smith 1990), the availability of Bayesian software packages (Lunn et al. 2012; Carpenter et al. 2017), and increasing computing power, this hurdle has been overcome. Lindley's prediction of a Bayesian twenty-first century (Lindley 1975) may become a reality!

Appendix

The script below is for use with the *R* software environment (R Core Team 2019). It provides LDA analysis, corresponding ML analysis of the reduced rank regression model, and Bayesian LDA for the Iris data (dimension $r = 1$).

For the Bayesian analysis, code based on WinBUGS (Lunn et al. 2012) is embedded in this script.

```
### R script
dir = getwd()
library(MASS); library("R2WinBUGS")

### LDA
data.iris = data.frame(
    rbind(iris3[,,1], iris3[,,2], iris3[,,3]),
    Sp = rep(c("s","c","v"), rep(50,3)))
fit.lda = lda(Sp ~ ., data.iris, prior = c(1,1,1)/3)
loading.lda = fit.lda$scaling[,1]; loading.lda

### Rearrange IRIS data for Reduced Rank Regression
Y = as.matrix(data.frame(Y1=iris[,1],Y2=iris[,2],
                          Y3=iris[,3],Y4=iris[,4]))
SPECIES = as.character(iris[,5])
X1 = as.numeric(SPECIES=="setosa")
X2 = as.numeric(SPECIES=="versicolor")
X1 = X1-mean(X1); X2 = X2-mean(X2)
X = as.matrix(data.frame(X1=X1,X2=X2))
n = nrow(Y)

### Reduced Rank Regression: ML analysis
fit.cca = cancor(X,Y)
A = fit.cca$xcoef[,1,drop=F]
Z = cbind( rep(1,n) , X %*% A )
muB = solve( t(Z) %*% Z ) %*% t(Z) %*% Y
mu = muB[1,]
B = t(muB[2,,drop=F])
Sigma = cov(Y - Z %*% muB)
```

```
loading.rrr = c(-solve(Sigma) %*% B /
               c(sqrt( t(B) %*% solve(Sigma) %*% B)))

### Reduced Rank Regression: Bayesian analysis
OMEGA = solve(100*Sigma)
data = list("Y","X1","X2","n","OMEGA")
modn = "mod.txt"
cat(" model {
for (i in 1:n) {
  Y[i,1:4] ~ dmnorm( theta[i,] , R[,])
  Z[i] <- (X1[i] + A2*X2[i])
  theta[i,1] <- mu[1] + B[1]*Z[i]
  theta[i,2] <- mu[2] + B[2]*Z[i]
  theta[i,3] <- mu[3] + B[3]*Z[i]
  theta[i,4] <- mu[4] + B[4]*Z[i]
  }
R[1:4,1:4] ~ dwish(OMEGA[,],4)
A2 ~ dnorm(0,1.0E-06)
for (j in 1:4) {
  mu[j] ~ dnorm(0,1.0E-06)
  B[j] ~ dnorm(0,1.0E-06)
  }
} ",file=paste(dir,'/',modn,sep=''))
parms = c("A2","B","mu","R")
inits = function() { list(A2=A[2,1]/A[1,1],
                          B=c(B*A[1,1]), mu=mu) }
set.seed(1)
mod = bugs(data, inits=inits, parameters=parms,
           model.file=paste(dir,'/',modn,sep=''),
           n.chains=5, n.burnin=5000,
           n.iter=25000, n.thin=10,
           bugs.seed=1, DIC=F,debug=F)
print(mod,3)
posterior = mod$sims.matrix
m = nrow(posterior)
posterior.A = matrix(1,nrow=m,ncol=2,
   dimnames=list(NULL,c('A1','A2')))
posterior.B = matrix(NA,nrow=m,ncol=4,
   dimnames=list(NULL,c('B1','B2','B3','B4')))
posterior.mu = matrix(NA,nrow=m,ncol=4,
   dimnames=list(NULL,c('mu1','mu2','mu3','mu4')))
posterior.Sigma = array(NA,dim=c(m,4,4),
   dimnames=list(NULL,c('S1','S2','S3','S4'),
                      c('S1','S2','S3','S4')))
posterior.loading = matrix(NA,nrow=m,ncol=4,
```

```
      dimnames=list(NULL,c('L1','L2','L3','L4')))
 for (k in 1:m) {
   posterior.A[k,2] = posterior[k,1]
   posterior.B[k,] = posterior[k,2:5]
   scaleAB = sqrt(c(t(X %*% posterior.A[k,]) %*%
                    (X %*% posterior.A[k,])))
   posterior.A[k,] = posterior.A[k,]/scaleAB
   posterior.B[k,] = posterior.B[k,]*scaleAB
   posterior.mu[k,] = posterior[k,6:9]
   R = matrix(posterior[k,10:25],nrow=4,ncol=4)
   posterior.Sigma[k,,] = solve(R)
   Bv = t(posterior.B[k,,drop=F])
   posterior.loading[k,] = -c(R %*% Bv /
        c(sqrt( t(Bv) %*% R %*% Bv)))
   }
mu; apply(posterior.mu,2,quantile,
     probs=c(0.025,0.5,0.975))
Sigma; apply(posterior.Sigma,c(2,3),quantile,
     probs=c(0.5))
t(A); apply(posterior.A,2,quantile,
     probs=c(0.025,0.5,0.975))
t(B); apply(posterior.B,2,quantile,
     probs=c(0.025,0.5,0.975))
loading.lda; apply(posterior.loading,2,quantile,
     probs=c(0.025,0.5,0.975))
```

References

Aggarwal, C. C. (Ed.). (2014). *Data classification: Algorithms and applications*. Boca Raton: CRC Press.

Aitchison, J., & Dunsmore, I. (1975). *Statistical prediction analysis*. Cambridge: University Press.

Anderson, T. W. (1951). Estimating linear restrictions on regression coefficients for multivariate normal distributions. *Annals of Mathematical Statistics, 22*, 327–351.

Anderson, T. W. (2002). Canonical correlation analysis and reduced rank regression in autoregressive models. *Annals of Statistics, 30*, 1134–1154.

Ashby, D. (2006). Bayesian statistics in medicine: A 25 year review. *Statistics in Medicine, 25*, 3589–3631.

Bennett, J. E., & Wakefield, J. C. (1996). A comparison of a Bayesian population method with two methods as implemented in commercially available software. *Journal of Pharmacokinetics and Biopharmaceutics, 24*, 403–432.

Bornkamp, B., Ohlssen, D., Magnusson, B. P., & Schmidli, H. (2017). Model averaging for treatment effect estimation in subgroups. *Pharmaceutical Statistics, 16*, 133–142.

Carpenter, B., Gelman, A., Hoffman, M. D., Lee, D., Goodrich, B., Betancourt, M., et al. (2017). Stan: A probabilistic programming language. *Journal of Statistical Software, 76*(1).

Dempster, A., Selwyn, M., & Weeks, B. (1983). Combining historical and randomized controls for assessing trends in proportions. *Journal of the American Statistical Association, 78*, 221–227.

Dudek, A. Z., Arodz, T., & Galvez, J. (2006). Computational methods in developing quantitative structure-activity relationships (QSAR): A review. *Combinatorial Chemistry and High Through-put Screening, 9,* 213–228.

Dudoit, S., Fridlyand, J., & Speed, T. P. (2002). Comparison of discrimination methods for the classification of tumors using gene expression data. *Journal of the American Statistical Association, 97,* 77–87.

Fisch, R., Jones, I., Jones, J., Kerman, J., Rosenkranz, G. K., & Schmidli, H. (2015). Bayesian design of Proof-of-Concept trials. *Therapeutical Innovations and Regulatory Science, 49,* 155–162.

Fisher, R. A. (1936). The use of multiple measurements in taxonomic problems. *Annals of Eugenics, 7,* 179–188.

Gelfand, A. E., & Smith, A. F. M. (1990). Sampling-based approaches to calculating marginal densities. *Journal of the American Statistical Association, 85,* 398–406.

Gelfand, A. E., Hills, S. E., Racine-Poon, A., & Smith, A. F. M. (1990). Illustration of Bayesian inference in normal data models using Gibbs sampling. *Journal of the American Statistical Association, 85,* 972–985.

Geweke, J. (1996). Bayesian reduced rank regression in econometrics. *Journal of Econometrics, 75,* 121–146.

Geweke, J. (2004). Getting it right: Joint distribution tests of posterior simulators. *Journal of the American Statistical Association, 99,* 799–804.

Gsponer, T., Gerber, F., Bornkamp, B., Ohlssen, D., Vandemeulebroecke, M., & Schmidli, H. (2014). A practical guide to Bayesian group sequential designs. *Pharmaceutical Statistics, 13,* 71–80.

Gsteiger, S., Neuenschwander, B., Mercier, F., & Schmidli, H. (2013). Using historical control information for the design and analysis of clinical trials with overdispersed count data. *Statistics in Medicine, 32,* 3609–3622.

Grieve, A. P., & Krams, M. (2005). ASTIN: A Bayesian adaptive dose-response trial in acute stroke. *Clinical Trials, 2,* 340–351.

Grieve, A. P. (2007). 25 years of Bayesian methods in the pharmaceutical industry: A personal, statistical bummel. *Pharmaceutical Statistics, 6,* 261–281.

Holzhauer, B., Wang, C., & Schmidli, H. (2018). Evidence synthesis from aggregate recurrent event data for clinical trial design and analysis. *Statistics in Medicine, 37,* 867–882.

Izenman, A. J. (1986). Reduced rank regression procedures for discriminant analysis. *American Statistical Association Proceedings of the Statistical Computing Section,* 249–253.

Karlsson, S. (2017). Corrigendum to Bayesian reduced rank regression. *Journal of Econometrics, 201,* 170–171.

Krzanowski, W. J. (1989). On confidence regions in canonical variate analysis. *Biometrika, 76,* 107–116.

Krzanowski, W. J., & Radley, D. (1989). Nonparametric confidence and tolerance regions in canonical variate analysis. *Biometrics, 45,* 1163–1173.

Lange, M. R., & Schmidli, H. (2015). Analysis of clinical trials with biologics using dose-time-response models. *Statistics in Medicine, 34,* 3017–3028.

Lindley, D. (1975). The future of statistics—a Bayesian 21st century. *Supplement Advances in Applied Probability, 7,* 106–115.

Lunn, D., Jackson, C., Best, N., Thomas, A., Spiegelhalter, D. (2012). The BUGS book: A practical introduction to Bayesian analysis. Chapman and Hall/CRC

Madsen, R., Lundstedt, T., & Trygg, J. (2010). Chemometrics in metabolomics—a review in human disease diagnosis. *Analytica Chimica Acta, 659,* 23–33.

Mardia, K. V., Kent, J. T., & Bibby, J. M. (1979). *Multivariate Analysis.* London: Academic.

Mielke, J., Schmidli, H., & Jones, B. (2018). Incorporating historical information in biosimilar trials: Challenges and a hybrid Bayesian-frequentist approach. *Biometrical Journal, 60,* 564–582.

Neuenschwander, B., Branson, M., & Gsponer, T. (2008). Critical aspects of the Bayesian approach to phase I cancer trials. *Statistics in Medicine, 27,* 2420–2439.

Neuenschwander, B., Capkun-Niggli, G., Branson, M., & Spiegelhalter, D. J. (2010). Summarizing historical information on controls in clinical trials. *Clinical Trials, 7,* 5–18.

Pozzi, L., Schmidli, H., Gasparini, M., & Racine-Poon, A. (2013). A Bayesian adaptive dose selection procedure with an overdispersed count endpoint. *Statistics in Medicine, 32*, 5008–5027.

Racine, A., Grieve, A. P., Fluehler, H., & Smith, A. F. M. (1986). Bayesian methods in practice: Experiences in the pharmaceutical industry (with Discussion). *Applied Statistics, 35*, 93–150.

Racine-Poon, A. (1986). A Bayesian approach to nonlinear random effects models. *Biometrics, 4*, 1015–1023.

Racine-Poon, A., Weihs, C., & Smith, A. F. M. (1991). Estimation of relative potency with sequential dilution errors in radioimmunoassay. *Biometrics, 47*, 1235–1246.

R Core Team. (2019). R: A language and environment for statistical computing. R Foundation for Statistical Computing. Vienna, Austria. https://www.R-project.org/.

Ryan, D. A. J., Hubert, J. J., Carter, E. M., Sprague, J. B., & Parrott, J. (1992). A reduced rank multivariate regression approach to aquatic joint toxicity experiments. *Biometrics, 48*, 155–162.

Seewald, W. (1994). Time trend in historical controls for tumour incidences in long-term animal studies. *Journal of the Royal Statistical Society Series C (Applied Statistics), 43*, 127–137.

Schmidli, H. (1995). *Reduced rank regression with applications to quantitative structure-activity relationships*. Heidelberg: Springer.

Schmidli, H. (1996). Bayesian analysis of reduced rank regression. *Test, 51*, 159–186.

Schmidli, H., Bretz, F., & Racine-Poon, A. (2007). Bayesian predictive power for interim adaptation in seamless phase II/III trials where the endpoint is survival up to some specified timepoint. *Statistics in Medicine, 26*, 4925–4938.

Schmidli, H., Wandel, S., & Neuenschwander, B. (2013). The network meta-analytic-predictive approach to non-inferiority trials. *Statistical Methods in Medical Research, 22*, 219–240.

Schmidli, H., Gsteiger, S., Roychoudhury, S., O'Hagan, A., Spiegelhalter, D., & Neuenschwander, B. (2014). Robust meta-analytic-predictive priors in clinical trials with historical control information. *Biometrics, 70*, 1023–1032.

Schmidli, H., Neuenschwander, B., & Friede, T. (2017). Meta-analytic-predictive use of historical variance data for the design and analysis of clinical trials. *Computational Statistics and Data Analysis, 113*, 100–110.

Tso, M. K.-S. (1981). Reduced rank regression and canonical analysis. *Journal of the Royal Statistical Society B, 43*, 183–189.

Velu, R. P., Reinsel, G. C., & Wichern, D. W. (1986). Reduced rank models for multiple times series. *Biometrika, 73*, 109–118.

van der Leeden, R. (1990). *Reduced rank regression with structured residuals*. Leiden: DSWO Press.

Weber, S., Gelman, A., Lee, D., Betancourt, M., Vehtari, A., & Racine-Poon, A. (2018). Bayesian aggregation of average data: An application in drug development. *Annals of Applied Statistics, 3*, 1583–1604.

Weihs, C., & Schmidli, H. (1990). OMEGA (Online Multivariate Exploratory Graphical Analysis): Routine searching for structure (with discussion). *Statistical Science, 5*, 175–226.

Weihs, C., & Schmidli, H. (1991). Multivariate exploratory data-analysis in chemical industry. *Mikrochimika Acta, 2*, 467–482.

Weihs, C., Baumeister, W., & Schmidli, H. (1993). Classification methods for multivariate quality parameters. *Journal of Chemometrics, 7*, 131–142.

Weihs, C. (1993). Canonical discriminant analysis: Comparison of resampling methods and convex-hull approximation. In O. Opitz, B. Lausen, & R. Klar (Eds.), *Information and classification*. Dortmund: Springer.

Weihs, C., Ligges, U., Moerchen, F., et al. (2007). Classification in music research. *Advances in Data Analysis and Classification, 1*, 255–291.

Chapter 3
Modelling and Classification of GC/IMS Breath Gas Measurements for Lozenges of Different Flavours

Claudia Wigmann, Laura Lange, Wolfgang Vautz and Katja Ickstadt

Abstract The composition of exhaled breath contains information about diet, oral hygiene and other environmental influences as well as on the state of health and on medications. Therefore, rapid and sensitive breath analysis would be a helpful tool, for example, for medical diagnosis or therapy control. Ion mobility spectrometry coupled with gas-chromatographic pre-separation (GC/IMS) meet those requirements and can be used to distinguish between healthy and diseased persons or to detect drug usage, for example, based on characteristic exhaled metabolites. So far, the detection of peaks in IMS measurements and the assignment of compounds is done manually and an automated procedure is urgently needed. In this article, we analyse breath gas measurements by GC/IMS from a volunteer having consumed lozenges of 12 different flavours. The IMS measurements are modelled along drift time with an additive model of unimodal regressions to describe each peak. The regressions are afterwards combined across all spectra and all datasets to determine typical peak locations and the respective heights of the peaks in each measurement are inferred. The obtained matrix of peak intensities is then used to classify the measurements into the 12 flavour groups using support vector machines. Since the true class labels are known, we can assess the mis-classification rate using cross-validation.

C. Wigmann (✉)
IUF – Leibniz-Institut für umweltmedizinische Forschung gGmbH,
Auf'm Hennekamp 50, 40225 Düsseldorf, Germany
e-mail: claudia.wigmann@iuf-duesseldorf.de

L. Lange
OptiMedis AG, Burchardstraße 17, 20095 Hamburg, Germany
e-mail: l.lange@optimedis.de

W. Vautz
ION-GAS GmbH, Konrad-Adenauer-Allee 11, 44263 Dortmund, Germany

Leibniz-Institut für Analytische Wissenschaften – ISAS – e.V., Bunsen-Kirchhoff-Straße 11,
44139 Dortmund, Germany
e-mail: vautz@isas.de

K. Ickstadt
Faculty of Statistics, TU Dortmund University, 44221 Dortmund, Germany
e-mail: ickstadt@tu-dortmund.de

© Springer Nature Switzerland AG 2019
N. Bauer et al. (eds.), *Applications in Statistical Computing*,
Studies in Classification, Data Analysis, and Knowledge Organization,
https://doi.org/10.1007/978-3-030-25147-5_3

3.1 Introduction

Ion Mobility Spectrometry (IMS) plays an important role in a variety of application areas, where the amount of volatile organic compounds in an analyte is to be measured. These areas comprise, for example, forensics, the detection of explosives, moulds or drugs, and medical purposes such as the diagnosis of diseases (especially of the lung) (Eiceman and Karpas 2005; Handa et al. 2014).

In the latter example, metabolites in exhaled human breath are particularly informative about infections and diseases as well as medication and nutrition (Vautz et al. 2009). IMS breath gas analysis has, therefore, been used for various medical research questions, for example, on how to distinguish lung cancer patients from healthy probands (Handa et al. 2014).

The metabolites in breath gas are represented by intensity peaks in the IMS measurements, which are 3-dimensional matrices of intensities versus reduced ion mobility and retention time. Thus, a necessary step in the analysis of IMS datasets is the detection and quantification of these peaks. While it is still the prevailing procedure, the manual execution of this task by experts using interactive visualization software is very tedious as well as subjective and not necessarily reproducible (Horsch et al. 2015). Therefore, automatic peak detection procedures are required and have already been proposed over the past years by several authors (Bödeker et al. 2008; Bödeker and Baumbach 2009; Bunkowski 2012; Purkhart et al. 2012; Hauschild et al. 2013; Kopczynski and Rahmann 2015). Horsch et al. (2017) give an extensive comparison of automated analysis processes for the purpose of binary classification of IMS measurements. Some of these compared approaches model the peaks using distributions such as the (shifted) inverse Gaussian distribution (see e.g. Bödeker and Baumbach 2009; Kopczynski and Rahmann 2015). Although these distributions allow, for example, for skewness, this representation of a peak shape might be too restrictive.

In this article, we present the results of the Master's thesis (Lange 2015) written by Laura Lange at the Faculty of Statistics of TU Dortmund University. Here, the peaks are instead modelled by more flexible unimodal spline functions. The analysis process is used for classification of IMS breath gas measurements in a multi-class setting with a volunteer having consumed lozenges of 12 different flavours. The following Section gives an introduction to ion mobility spectrometry coupled with gas-chromatographic pre-separation and describes the data acquisition of the measurements used for our analyses. The essential statistical methods are also presented. Afterwards, we describe the steps for modelling and classification of the IMS datasets as well as the obtained results. We conclude with a summary and discussion.

3.2 Materials and Methods

In this Section, we describe the GC/IMS technology as well as the acquired datasets and the basic statistical methods used for the analysis in Sect. 3.3.

3.2.1 GC/IMS Technology

The data analysed in this article has been collected using ion mobility spectrometry coupled with Gas-Chromatographic (GC) pre-separation, which enables to identify and quantify volatile organic compounds in complex mixtures like breath gas samples even down to the ppt (parts per trillion) concentration range.

In the ion mobility spectrometer (see Fig. 3.1), the sample is first led into an ionization chamber, where the so-called reactant ions—protonated water clusters obtained from the continuous ionization of the available drift gas (synthetic air)—transfer their proton to the sample molecules. Due to a weak electric field, the ions move through a periodically opening shutter grid into the drift region and towards a detector (Faraday-plate). Their speed is determined by the electric field strength and by collisions with molecules of the so-called drift gas, which flows in the opposite direction, thus resulting in a characteristic drift velocity. Therefore, the ions arrive at the detector after a drift time specific for their size, shape and charge. The drift time also depends on temperature and pressure of the drift gas, but both parameters are used to normalize the measured drift times to the so-called reduced ion mobility K_0 given in $cm^2 V^{-1} s^{-1}$ (which is anti-proportional to the drift time). To avoid additional measurements of temperature and pressure in the spectrometer, this normalization is achieved using the drift time of the reactant ions with their known ion mobility, which cause the so-called Reactant Ion Peak (RIP) in the measurement. The RIP is characteristic for the drift gas and present in all measurements (Vautz et al. 2009).

In very complex mixtures, molecules may have similar reduced ion mobilities which makes their identification difficult or even impossible. Furthermore, the simultaneous presence of many compounds leads to the formation of undefined cluster ions which also hinders their identification. Therefore, additional gas-chromatographic pre-separation is applied prior to IMS (Vautz et al. 2018a). The time it takes the sample molecules to pass the coated column, the so-called retention time, depends on their polarity. With this experimental setup, the different molecules enter the IMS after different retention times in the ideal case totally separated. The retention time is also influenced by the type of the column, the carrier gas and the temperature but gives another characteristic value to identify the related compounds together with the ion mobility. The use of Multi-Capillary Columns (MCC) consisting of about 1000 glass capillaries makes the gas-chromatographic pre-separation very rapid and enables a high sample flow. In addition, it is advantageous when analysing humid samples like breath gas (Vautz and Baumbach 2008).

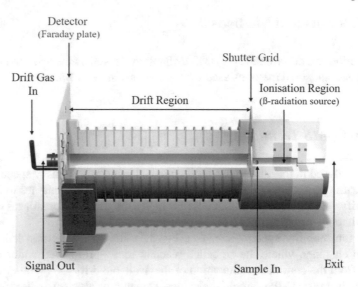

Fig. 3.1 Design of the ion mobility spetrometer

For each combination of retention time and reduced ion mobility—which enable the identification of the related molecule ion—the ion current (or intensity) measured by the detector represents the concentration of the respective molecule in the sample using a calibration carried out earlier (Bergen et al. 2018). Thus, each sample leads to a data matrix containing the measured intensities for a series of reduced ion mobilities (columns) and retention times (rows). The 3-dimensional plot of the ion current versus inverse reduced ion mobility and retention time is called IMS-chromatogram. For a fixed retention time, the intensities along the inverse reduced ion mobility (see also Fig. 3.4) are called a spectrum (Vautz and Baumbach 2008). More detailed descriptions of the MCC-IMS technology can be found elsewhere (Eiceman and Karpas 2005; Vautz and Baumbach 2008; Vautz et al. 2009; Vautz et al. 2018a, b).

The experimental parameters applied for the acquisition of the here analysed data are summarized in Table 3.1. The GC/IMS measurements are represented by 600 × 2500 matrices of intensity values for 600 retention time and 2500 ion mobility values. See Fig. 3.3 for exemplary heatmaps of such intensity matrices. Due to technical reasons, the first 34 reduced ion mobilities (leftmost columns) of each measurement always exhibit high intensities, which do not represent sample molecules. These columns are discarded for the analyses. In addition, the last 200 spectra (topmost rows) in each of our measurements show no peaks except for the reactant ion peak. Thus, to reduce the computational burden, these 200 spectra as well as the spectra corresponding to every other retention time were discarded for the analyses. Nevertheless, the heatmaps show the full intensity matrices.

Table 3.1 Experimental parameters for data acquisition

Parameter	Setting
Electric field	4380 V
Drift length	12 cm
Drift and carrier gas	Synthetic air
Drift gas flow	100 ml/min
Carrier gas flow	150 ml/min
Shutter grid opening time	300 µs
MCC type	Multichrom, Novosibirsk, Russland, Typ OV-5
MCC column length	20 cm
MCC temperature	40 °C

3.2.2 Data Acquisition and Resulting Data Sets

Data acquisition was conducted at the ISAS—Institute for Analytical Sciences between September and November 2011. The data material comprises breath gas measurements from a volunteer having consumed lozenges of 12 different flavours (see Table 3.2). The measurements were carried out 20 min after consumption to avoid overloading of the GC/IMS. Furthermore, measurements of room air were carried out in combination with the breath measurements to enable exclusion of compounds inhaled from the pattern of compounds characteristic for the lozenges. After each particular measurement of room air or breath, another GC/IMS analysis of clean humid air was carried out to control the instrument for contaminations. These intermediate measurements are not used in our analyses. Each of the 12 flavours were analysed 10 times. One measurement of lozenge 4 was recorded with a finer resolution of the reduced ion mobility than the other measurements and was discarded. One missing value and one wrongly encoded numerical value in a measurement of lozenge 6 were replaced by adequate values. In total, $n = 119$ GC/IMS measurements are used in the following analyses. These measurements were all recorded with the same IMS device, the same kind of MCC (and thus the same flow velocity) and at the same temperature, so that we expect the measurements to be very appropriate for a comparison of the different lozenge flavours.

The volunteer gave written informed consent in the frame of a positive ethics vote (IfADo, Dortmund, No. 66-01.07.2013). The data was recorded, stored, and evaluated anonymously.

3.2.3 Statistical Methods

In the following, we describe the basic statistical concepts needed for the here presented analysis of the GC/IMS datasets. An overview of the analysis steps is given in

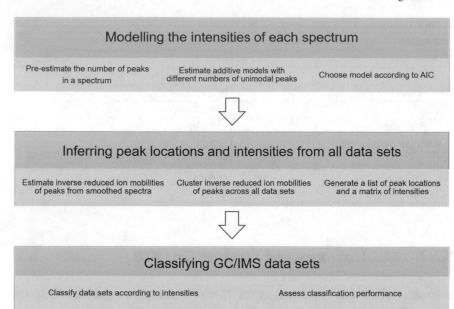

Fig. 3.2 Steps for modelling and classification of the breath gas measurements corresponding to Sects. 3.3.1–3.3.3

Fig. 3.2. The concepts comprise unimodal spline regression for modelling a peak in the measured intensities and an additive model to describe several maybe overlapping peaks of one IMS spectrum. Clustering, in explicit the k-means algorithm, is used to cluster the peaks of all spectra and all datasets according to their inverse reduced ion mobility. From the clusters and all retention times, a list of peak locations is generated. The intensities at these locations are used to classify each data set to one of the 12 flavours. The classification is performed using support vector machines.

Let the intensity matrix of a GC/IMS measurement be denoted by

$$\mathcal{Y} = \left(y_{i,j}\right)_{i=1,\dots,n_r,\ j=1,\dots,n_m},$$

where n_r is the number of retention times and n_m the number of reduced ion mobilities. The row vectors of this matrix (spectra) are denoted by $y_{i\cdot} = (y_{i,1}, \dots, y_{i,n_m})^T$ and the column vectors by $y_{\cdot j} = (y_{1,j}, \dots, y_{n_r,j})^T$. In addition, let $x = (x_1, \dots, x_{n_r})^T$ be the vector of retention times and $z = (z_1, \dots, z_{n_m})^T$ the vector of inverse reduced ion mobilities. There is one intensity matrix $\mathcal{Y}_s$ for each of the $s = 1, \dots, 119$ IMS measurements. We drop the index s for simplicity since the following methods only work on one intensity matrix at a time.

Unimodal spline regression. Suppose for now that a GC/IMS spectrum $y_{i\cdot}$ only has one peak. This means that it can be described by a function which is first increasing and then decreasing, i.e. which is unimodal. If no specific parametric unimodal form of such peaks is known, a semi-parametric approach like unimodal spline regression is appropriate as already demonstrated in Köllmann et al. (2014). Without a shape constraint like unimodality, spline regression reduces to a simple linear regression model

$$y_{i.} = \mathscr{B}_i \boldsymbol{\beta}_i + \boldsymbol{\varepsilon}_i,$$

where $\boldsymbol{\beta}_i \in \mathbb{R}^d$ is the vector of spline coefficients, $\boldsymbol{\varepsilon}_i \in \mathbb{R}^{n_m}$ is the vector of measurement errors assumed to stem from a $\mathscr{N}(\mathbf{0}, \sigma^2 I_{n_m})$-distribution and $\mathscr{B}_i \in \mathbb{R}^{n_m \times d}$ is the matrix of d B-spline basis functions evaluated at the regressor values $z_1, \ldots, z_{n_m}$ (here: inverse reduced ion mobilities). We assume that the same B-spline basis functions are used for all spectra i, that is, $\mathscr{B}_i = \mathscr{B} \ \forall i \in \{1, \ldots, n_r\}$. The assumption of independent errors might not hold for possibly auto-correlated spectra as in our application. But since we only use the spline regression model for smoothing, the respective least squares approach still yields reasonable results.

It was shown in Köllmann et al. (2014) that the spline function is unimodal, if the B-spline coefficients $\boldsymbol{\beta}_i$ form a unimodal sequence with mode $g \in \{1, \ldots, d\}$, that is,

$$\beta_{i,1} \leq \cdots \leq \beta_{i,g-1} \leq \beta_{i,g} \geq \beta_{i,g+1} \geq \cdots \geq \beta_{i,d}.$$

This constraint can be written linearly as $\mathscr{A}_g \boldsymbol{\beta}_i \geq \mathbf{0}$, where $\mathscr{A}_g \in \mathbb{R}^{(d-1) \times d}$ is a constraint matrix of the form

$$\mathscr{A}_g = \begin{pmatrix} -1 & 1 & & & & \\ & \ddots & \ddots & & \mathbf{0} & \\ & & -1 & 1 & & \\ & & & 1 & -1 & \\ & \mathbf{0} & & & \ddots & \ddots \\ & & & & 1 & -1 \end{pmatrix} \quad \leftarrow g\text{th row.}$$

Instead of minimizing the residual sum of squares over all possible coefficients vectors $\boldsymbol{\beta}_i \in \mathbb{R}^d$ as in linear regression, the minimization is done subject to the linear constraint, which can be achieved using quadratic programming algorithms. Since it is unknown which $g \in \{1, \ldots, d\}$ is the best, this step is carried out for all possible values and the model with smallest residual sum of squares is chosen (cp. Köllmann et al. 2014).

The flexibility of a spline function depends on the number d of spline coefficients (which is in turn dependent on the degree of the spline and the number of knots) and the positions of the knots. Köllmann et al. (2014) follow Eilers and Marx (1996) in proposing cubic (degree 3) splines with a large number of equally spaced knots and a difference penalty on the B-spline coefficients to introduce smoothness. The choice of the tuning parameter for the penalty term is not straightforward when using a shape constraint and the authors show how to use restricted maximum likelihood theory to find a compromise between goodness of fit and smoothness. To find this compromise, an estimate of the residual standard error σ is needed prior to model fitting or has to be calculated iteratively during the fitting process. We refer to Köllmann et al. (2014) for a more detailed description of unimodal penalized spline regression. The approach is implemented in the R package `unireg` (Köllmann 2016b).

Additive model with unimodal components. In reality the spectra of a GC/IMS measurement will mostly contain more than one intensity peak and the peaks of two or more metabolites might also be so close to each other that the corresponding intensities accumulate. We propose to model this accumulation with an additive regression model with unimodal spline components for spectrum i:

$$\boldsymbol{y}_{i\cdot} = \alpha_i \cdot \mathbf{1}_{n_m} + \sum_{v=1}^{V_i} \mathscr{B}^{(v)} \boldsymbol{\beta}_i^{(v)} + \boldsymbol{\varepsilon}_i,$$

where α_i is the overall mean of spectrum i, $\boldsymbol{\varepsilon}_i \in \mathbb{R}^{n_m}$ is again the vector of measurement errors and V_i is the number of the unimodal spline components that are given by $\mathscr{B}^{(v)} \boldsymbol{\beta}_i^{(v)}$, $v = 1, \ldots, V_i$. Again, we assume that the same B-spline basis functions are used for each spline component, so that $\mathscr{B}^{(v)} = \mathscr{B} \ \forall v$. Additive models can be estimated using the so-called backfitting algorithm (see, e.g. Hastie et al. 2009). In the following, we describe the algorithm for the context at hand:

1. Set $\hat{\alpha}_i := \dfrac{1}{n} \sum_{j=1}^{n_m} y_{i,j}$ and $\boldsymbol{\beta}_i^{(v)} := \mathbf{0} \ \forall v = 1, \ldots, V_i$.

2. For $v = 1, \ldots, V_i$: estimate $\boldsymbol{\beta}_i^{(v)}$ from data $(\boldsymbol{z}, \tilde{\boldsymbol{y}}_{i\cdot})$ using penalized unimodal spline regression, where $\tilde{\boldsymbol{y}}_{i\cdot} = \boldsymbol{y}_{i\cdot} - \hat{\alpha} \cdot \mathbf{1}_{n_m} - \sum_{w \neq v} \mathscr{B} \hat{\boldsymbol{\beta}}_i^{(w)}$.

3. Centre the spline functions (or respectively their coefficients) around zero:
$$\hat{\boldsymbol{\beta}}_i^{(v)} := \hat{\boldsymbol{\beta}}_i^{(v)} - \sum_{v=1}^{V_i} \mathscr{B} \hat{\boldsymbol{\beta}}_i^{(v)}.$$

4. Repeat steps 2 and 3 until convergence.

k-means clustering. Cluster analysis is used to find groups (or 'clusters') of objects so that objects from different clusters are as dissimilar to each other as possible and so that objects from the same cluster are as similar as possible according to one or several characteristics measured on these objects. In our analysis, clustering is used to partition intensity peaks (objects) across all spectra and datasets according to their inverse reduced ion mobilities (characteristic). Thus, we describe the k-means approach only for the one-dimensional case of n_c objects with measurements $z_1, \ldots, z_{n_c}$ of one characteristic.

k-means is a so-called partitioning clustering method, where the number of clusters k is fixed in advance. It seeks to minimise the within-cluster sum of squares

$$\sum_{\ell=1}^{k} \sum_{z_j \in C_\ell} (z_j - \bar{z}_\ell)^2,$$

where C_ℓ ($\ell = 1, \ldots, k$) are the clusters and $\bar{z}_\ell = \dfrac{1}{|C_\ell|} \sum_{z_j \in C_\ell} z_j$ are the clusters' centres. There exist several heuristic algorithms that approximate the optimal solution, though typically converging to a local optimum. After (randomly) choosing a set of start clusters and their centres, the algorithms iterate between two steps:

1. Assign each object z_j to the cluster C_ℓ for which the squared Euclidean distance $(z_j - \bar{z}_\ell)^2$, $\ell = 1, \ldots, k$, is minimal.
2. Recalculate the cluster centres $\bar{z}_\ell = \frac{1}{|C_\ell|} \sum_{z_j \in C_\ell} z_j$.

The iteration stops, when the objects are no longer re-assigned to another cluster (James et al. 2013). k-means clustering can, for example, be performed with the help of the R function `kmeans` (package `stats`, R Core Team 2018).

If the number of clusters cannot be inferred from any prior knowledge, the clustering is performed for a range of cluster numbers and a scree plot is drawn with the number of clusters on the x-axis and the within-cluster sum of squares on the y-axis (in analogy to a scree plot for principal component analysis, cp. James et al. 2013). At some point, the within-cluster sum of squares will no longer be reduced by additional clusters, which is represented by an 'elbow' in the graph. This elbow criterion gives the final number of clusters.

Support Vector Machines. The $n = 119$ GC/IMS measurements will be classified to the 12 different flavours using Support Vector Machines (SVMs) on the $(n \times p)$-matrix $\mathcal{M}$ holding in its rows $m_i \in \mathbb{R}^p$ the measured intensities at p peak locations (features) for each measurement $s = 1, \ldots, n$.

In general, SVMs can be used to find a hyperplane for a group of n (training) points $m_1, \ldots, m_n$ with class labels $\omega_1, \ldots, \omega_n \in \{-1, 1\}$, which best separates the points of the two classes. The best-separating hyperplane is the one with the widest margin, that is, the one with maximal distance to the nearest data points of each class. The data points falling exactly onto the boundary of this margin are called support vectors (cp. Cortes and Vapnik 1995).

In the following, we will describe the principle of SVMs for classification problems with two classes. For multi-class problems, one SVM can be applied for each pair of classes ('one-versus-one' approach) and the data point is classified to the class which wins the most comparisons (James et al. 2013).

The set of labelled training points is linearly separable if there exists a vector w and a scalar b, for which

$$\langle w, m_s \rangle + b \geq 1, \text{ if } \omega_s = 1 \text{ and } \langle w, m_s \rangle + b \leq -1, \text{ if } \omega_s = -1$$

or equivalently $\omega_s(\langle w, m_s \rangle + b) \geq 1$, for all $s = 1, \ldots, n$, where $\langle \cdot, \cdot \rangle$ denotes the inner product. It can be shown that the optimal (maximal margin) hyperplane with equation $\langle w_o, m \rangle + b_0 = 0$ is the one which minimizes $||w||_2^2$ under the above constraints (cp. Cortes and Vapnik 1995).

If the given set of data points is not linearly separable, one can allow for separation errors in the form of $\omega_s(\langle w, m_s \rangle + b) \geq 1 - \xi_s \; \forall \, s$ with scalar error variables $\xi_s \geq 0 \; \forall \, s$. The optimal hyperplane is the one, which minimizes the separation error while at the same time maximizing the margin between the correctly separated data points (that is, minimizing the length of w):

$$\min\left\{\frac{1}{2}||w|| + C \cdot \sum_{s=1}^{n} \xi_s\right\}$$

subject to the above linear inequality constraints. The tuning parameter C controls the trade-off between separation errors and margin width. This constrained minimization is a convex optimization problem. Using Lagrange multipliers, SVM algorithms usually optimize the objective function of the corresponding dual problem

$$\max_{\gamma}\left\{-\frac{1}{2}\sum_{s=1}^{n}\sum_{\iota=1}^{n}\gamma_s\gamma_\iota\omega_s\omega_\iota\langle m_s, m_\iota\rangle + \sum_{s=1}^{n}\gamma_s\right\},$$

subject to constraints $0 \leq \gamma_s \leq C$ and $\sum_{s=1}^{n}\gamma_s\omega_s = 0 \,\forall\, s$, which is a simpler convex quadratic programming problem (cp. Hastie et al. 2009).

Given solutions w_0 and b_0 the label of a new data point m is given by

$$\omega = \text{sign}(\langle w_0, m\rangle + b_0).$$

To achieve a more flexible separation of data points than linear, there is the possibility to transform the points m_s with a function $\phi : \mathbb{R}^n \mapsto \mathbb{R}^N$ ($N > n$) into a higher dimensional space and perform the linear separation using SVM in this space. Since the dual optimization problem depends on the transformed points only through inner products $\langle\phi(m_s), \phi(m_\iota)\rangle$, it is sufficient to specify a kernel function $\mathfrak{K}(m_s, m_\iota) = \langle\phi(m_s), \phi(m_\iota)\rangle$ instead of specifying ϕ. One example is the radial basis kernel function $\mathfrak{K}(m_s, m_\iota) = \exp(-\delta||m_s - m_\iota||_2^2)$ (cp. Hastie et al. 2009).

To asses the classification performance of an SVM rule, leave-one-out cross-validation (LOOCV) can be used, for instance. Analogous to LOOCV for other prediction tasks (cp. Hastie et al. 2009), the separation rule is learned on all data points but m_s and the class label ω_s of the sth point is predicted by the SVM. This is done for all s and the percentage of discrepancies between predicted and observed label is an estimate of the mis-classification rate of the classification procedure.

SVMs in conjunction with LOOCV can be applied with the help of the R package `mlr` (Bischl et al. 2016), which reverts for multi-class problems to function `ksvm` of package `kernlab` (Karatzoglou et al. 2004). The default settings of `ksvm` (also used in our analysis) are $C = 1$ and the radial basis kernel function with automatic heuristic estimation of the kernel width δ.

3.3 Modelling and Classification of GC/IMS Measurements

The 119 datasets described in Sect. 3.2.2 are modelled and classified according to the steps depicted in Fig. 3.2, making use of the statistical concepts just described. As pointed out already, the data matrix used for analysis contains only the intensities for $n_m = 2500 - 34 = 2466$ inverse reduced ion mobilities $z = (z_1, \ldots, z_{2466})$ and $n_r = 200$ retention times $x = (x_1, \ldots, x_{200})$. All analyses are performed using R (R Core Team 2018).

Roughly, the idea is to model the GC/IMS measurements in a way that enables inferring typical locations of intensity peaks (representing certain molecules) from all measurements. For a specific data set, the intensities measured at these locations—which can in parts also be zero or small, if the respective molecule in not present in the sample—can then be used to classify the measurement to one of the lozenge flavours. Looking at the distinguished and differing peak patterns of two measurements with lozenges of two different flavours in Fig. 3.3, this approach seems promising.

3.3.1 Modelling the Intensities of Each Spectrum

To infer typical locations of intensity peaks throughout all datasets $s = 1, \ldots, 199$, the spectra of each measurement are first modelled individually using additive models with unimodal components as described in Sect. 3.2.3. Since the number of unimodal components has to be fixed beforehand, this number is pre-estimated heuristically from the raw data. Afterwards, three additive models are fitted: one with this number of components, one with one less and one with an additional component. The most appropriate model is chosen with the help of Akaike's Information Criterion (AIC). These steps are carried out for each measurement s and within one measurement for each spectrum $y_{i.}$. The index s is again dropped for simplicity.

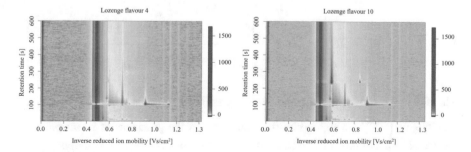

Fig. 3.3 Heatmaps for comparison of the intensities and peak patterns in the measurements with lozenges of two different flavours: lozenge 4 (left) and lozenge 10 (right). The warmer the colours the higher the intensity at the respective combination of inverse reduced ion mobility and retention time. The palette is logarithmically scaled relative to the median intensity of the data set

Pre-estimate the number of peaks in a spectrum. For each spectrum $y_{i.}$, $i = 1, \ldots, n_r$, the number V_i of peaks in this spectrum is pre-estimated using a heuristic of five criteria which thoroughly study the intensity values in the neighbourhoods of local maxima and have to be fulfilled for an inverse reduced ion mobility value z_j to be called a peak location. Details on the criteria can be found in Lange (2015).

Estimate additive models with different numbers of unimodal peaks. For each spectrum $y_{i.}$, $i = 1, \ldots, n_r$, additive models with $V_i - 1$, V_i and $V_i + 1$ unimodal components are fitted along the lines described in Sect. 3.2.3. The estimate of σ_i^2 needed for each spectrum to fit the model is calculated from the first 680 intensities:

$$\hat{\sigma}_i^2 = \frac{1}{679} \sum_{j=1}^{680} \left(y_{i,j} - \frac{1}{680} \sum_{j=1}^{680} y_{i,j} \right)^2 .$$

These intensities correspond to drift times shorter than those of the reactant ions and thus, no molecule is expected to show up in this region. In Fig. 3.3, the intensities in the respective range before the RIP (smaller than an inverse reduced ion mobility of 0.41 VS/cm^2) indeed do not exhibit peaks and can be seen as proxy for the noise of the whole measurement.

The additive model is then fitted only to the remaining 1786 intensity values of the spectrum $(y_{i,681}, \ldots, y_{i,2466})$, using the mean of these values to estimate the overall mean parameter α_i:

$$\hat{\alpha}_i = \frac{1}{1786} \sum_{j=681}^{2466} y_{i,j}.$$

Each component of the additive model is fitted using a cubic unimodal spline regression with a difference penalty of order 0 (i.e. ridge penalty) and 200 inner knots (resulting in $d = 204$ B-spline coefficients), so that there is one knot for approximately every tenth observation. The backfitting algorithm is run with a maximum of 20 iterations. It is stopped early if $S < 10$ with the discrepancy measure

$$S := \sum_{j=681}^{2466} \left(\hat{y}_{i,j}^{(new)} - \hat{y}_{i,j}^{(old)} \right)^2 \cdot \max \left\{ \frac{\sum_j \left(y_{i,j} - \hat{y}_{i,j}^{(old)} \right)^2}{\sum_j \left(y_{i,j} - \hat{y}_{i,j}^{(new)} \right)^2}, \frac{\sum_j \left(y_{i,j} - \hat{y}_{i,j}^{(new)} \right)^2}{\sum_j \left(y_{i,j} - \hat{y}_{i,j}^{(old)} \right)^2} \right\},$$

where S is the sum of the squared differences between the actual $\left(\hat{y}_{i,j}^{(new)} \right)$ and the previous fit $\left(\hat{y}_{i,j}^{(old)} \right)$ weighted by the maximum of the goodness of fit ratio and its inverse. The usefulness of this convergence criterion was checked visually.

Choose model according to AIC. Among these three models the final model for spectrum i is chosen according the model selection criterion AIC, that is, we choose the one with the smallest AIC value. Here, the AIC with V_i unimodal components is given by

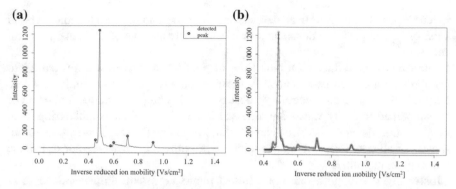

Fig. 3.4 Spectrum at retention time 96s of a measurement after consumption of lozenge 8. (**a**) depicts the whole spectrum (including the first 680 intensities) and the peaks found by the heuristic peak detection (red circles). (**b**) shows the part of the spectrum used for model fitting (excluding the first 680 intensities) together with the fitted values of the additive model (blue line) and the fitted five unimodal spline components (black lines)

$$n_m \cdot \log\left(n_m ||y_i. - \hat{y}_i.||^2\right) + 2 \sum_{t=1}^{V_i} \rho_t,$$

where the (effective) number of parameters of the additive model is given by the sum the effective dimensions of the unimodal spline components ρ_t, $t = 1, \ldots, V_i$. The calculation of the effective dimension of a penalized unimodal spline regression model is described in Lemma 4 of Köllmann (2016a).

Figure 3.4 shows the results of the above three steps for modelling the intensities of a spectrum exemplary for retention time 96 s of a measurement after consumption of lozenge 8. The pre-estimate of the number of peaks in the spectrum is six (see the 6 red circles in Fig. 3.4a), so that additive models with five, six and seven unimodal components are fitted to the spectrum. According to AIC the choice falls on the model with only 5 components. A very small bump next to the third peak is thus not modelled as a peak by our approach. The five unimodal spline functions of the chosen model are depicted in Fig. 3.4b (in black) and they accumulate to the fitted values (in blue). The overall fit is suitable and smoother than the raw data. In addition, the unimodal spline functions describe the peak locations appropriately.

3.3.2 Inferring Peak Locations and Intensities from All Data Sets

The smoothed spectra from the last step are used to infer typical locations of intensity peaks throughout all datasets. Since the peak locations of the same molecule may vary slightly between adjacent retention times and between the measurements, the spectra's peak locations from all datasets are identified and grouped together to build a list of typical peak locations and a matrix of the respective peak intensities in all data sets.

Estimate inverse reduced ion mobilities of peaks from smoothed spectra. First, we apply another heuristic to identify those inverse reduced ion mobilities, where a peak in the smoothed spectrum $\hat{y}_{i\cdot}$ occurs. For this purpose, the local maxima of the smoothed intensity values are assessed in relation to their neighbouring local maxima. Details on the heuristic can be found in Lange (2015). This is done for all spectra of all data sets, resulting in a long—possibly repetitive—vector of inverse reduced ion mobilities.

Cluster inverse reduced ion mobilities of peaks across all datasets. Second, k-means clustering (using R function kmeans with algorithm = "Lloyd") is applied to the vector of inverse reduced ion mobilities to find groups of peaks probably belonging to the same molecule. The clusters' centres (intensity-weighted cluster means) can then be used to replace the peaks' inverse reduced ion mobilities which fall into the respective cluster. Since it is unknown how many clusters/typical peak locations exist, the k-means algorithm is applied for all numbers of clusters between 1 and 40. For all numbers of clusters the partitioning is repeated with different random start clusters and the solution with minimal within-cluster sum of squares is chosen (as suggested in James et al. 2013). The number of repetitions is chosen in dependence of the final number of clusters: # start cluster $= 5 + 5 \cdot$ round (# final clusters/10). Furthermore, the algorithm is run with a maximum of 20 iterations.

On the basis of the scree plot in Fig. 3.5 we choose 28 clusters. One could argue that there is an 'elbow' in the within-cluster sum of squares already for five clusters. However, in a breath sample the number of molecules and thus the number of distinct drift times will certainly be higher than five. We included the lower numbers of clusters anyway to show the overall trend in the within-cluster sum of squares. When zooming in on the lower within-cluster sum of squares it becomes clear that the reduction in the within-cluster sum of squares slows down between 20 and 40 clusters. We then choose (more or less arbitrarily) the solution with 28 clusters, which yields for the start clusters used the lowest sum of squares of the clusterings with up to 30 clusters. Further reductions for higher numbers of clusters seem negligible.

Generate a list of peak locations and a matrix of intensities. As already mentioned in Sect. 3.2.1, the gas-chromatographic pre-separation enables to further distinguish molecules with similar drift times. Thus, the retention times of the intensity peaks should definitely be taken into account when generating a list of typical peak locations. As a starting point, the list consist of all 5600 combinations of the 28 cluster centres and the 200 analysed retention times. This preliminary list is matched with the peaks found by the heuristic in the smoothed spectra. A list entry is removed, if there exists no peak corresponding to it in any of the datasets. For the remaining 3553 peak locations in the list, the peak intensity is recorded for all $s = 1, \ldots, 119$ datasets. The intensity is taken to be zero for a certain dataset, if there is no peak at this location in the respective dataset. This procedure results in a 119×3553 matrix of intensities for all datasets and all peak locations (cluster–retention time

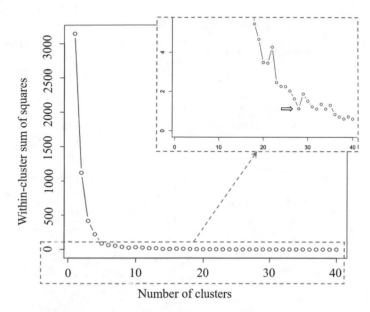

Fig. 3.5 Scree plot for the k-means clusterings of the inverse reduced ion mobilities with 1–40 clusters and close-up for within-cluster sum of squares below 6. The arrow in the close-up marks the chosen solution with 28 clusters

combinations), which can be used for classification of the measurements in the next step.

3.3.3 Classifying GC/IMS Data Sets

In the last step of our analysis of the GC/IMS measurements, the just created matrix of peak intensities is used to classify each of the $s = 1, \ldots, 119$ measurements to one of the 12 lozenge flavours. The performance of the classifier is assessed as well. **Classify datasets according to intensities**. The 119 measurements are classified by applying SVM to the peak intensity matrix. The features in this particular case are the peak locations (3553 combinations of cluster centres and retention times) with their corresponding intensity values.

Assess classification performance. The performance of the classification is assessed via leave-one-out cross-validation yielding a mis-classification rate of 11 out of 119 measurements or 9.244%, respectively. When taking a closer look at the mis-classified measurements (see Table 3.2), one notices that one measurement of lozenge 1 'Liquorice' is mistaken as lozenge 10 'Aniseed' and two measurements of lozenge 10 the other way around. This suggests that these measurements exhibit similar patterns and thus, that lozenges 1 and 10 are similar in terms of the molecules found in the breath after consumption. In addition, lozenges 4 'Cool Cinnamon without

Table 3.2 Classification results for the 119 lozenges of 12 different flavours

Lozenge no.	Flavour	Mis-classification rate	Assigned lozenge no.
1	Liquorice	1/10	10
2	Extra fresh lemon w/o sugar	1/10	11
3	Salmiak w/o sugar	0/10	–
4	Cool cinnamon w/o sugar	2/9	12
5	Extra strong (eucalyptus)	1/10	8
6	Mint	0/10	–
7	Mint w/o sugar	0/10	–
8	Extra strong (eucalyptus) w/o sugar	2/10	4, 12
9	Cherry w/o sugar	2/10	3, 8
10	Aniseed	2/10	1
11	Blackcurrant	0/10	–
12	Citrus w/o sugar	0/10	–
Σ		11/119	

sugar', 8 'Extra strong (eucalyptus) without sugar' and 12 'Citrus without sugar' seems to be easily confused with each other: two measurements of lozenge 4 and one of lozenge 8 are classified to lozenge 12 and one measurement of lozenge 8 is classified to lozenge 4. Though these three flavours are not the only ones without sugar, this might be the reason for their confusion.

3.4 Summary and Discussion

In this article, we proposed to employ an additive model with unimodal spline components in conjunction with clustering to detect typical metabolite-peaks in GC/IMS measurements of breath gas, which can afterwards be used to classify the breath gas samples to different groups, for example, healthy or diseased people. To demonstrate the performance of our approach, we applied it to breath gas samples analysed after consumption of lozenges of 12 different flavours. In the modelling step, we saw that the intensity peaks in the spectra of the measurements, which represent different metabolites in the breath, are well described by the unimodal spline components. By using k-means clustering of the peaks' inverse reduced ion mobilities across the smoothed spectra in all measurements and combining the cluster centres with the retention times, we were able to locate typical intensity peaks in the measurements. The matrix of intensities at those typical peak locations was used for classification of the measurements by SVM. The resulting leave-one-out mis-classification rate of only 9.244% is very promising for applying our approach also in other classification tasks based on GC/IMS data. Other authors have reported mis-classification rates between 27 and 32%, when applying different peak detection algorithms and using SVM for binary classification (Hauschild et al. 2013). The low mis-classification rate of our approach could result from the high number of peak locations used as features, which might not be nicely interpretable but beneficial for classification.

Our efforts were focussed on the step of modelling the spectra and generating the list of typical peak locations as well as the respective matrix of intensities. Surely, there are numerous different classification methods as well as possibilities to choose the corresponding tuning parameters, some of which might improve the classification performance even further. However, an optimisation of the classification method was beyond the scope of this work and we chose SVM for their straightforward applicability. See also Horsch et al. (2017) for a thorough comparison of analysis procedures combining different peak detection, peak clustering and classification methods.

A drawback of the presented analysis approach is the fact that for the additive model the number of unimodal components and the locations of the peaks in the smoothed spectra have been estimated using a heuristic. A model which is able to simultaneously estimate the number, locations and the shapes of the peaks is presented in Köllmann (2016a) and it would be interesting to see, if the final classification performance could be improved by modelling the spectra in this way.

Acknowledgements The financial support of the Bundesministerium für Bildung und Forschung and the Ministerium für Innovation, Wissenschaft und Forschung des Landes Nordrhein-Westfalen is gratefully acknowledged. This work has also been supported by Deutsche Forschungsgemeinschaft (DFG) within the Collaborative Research Center SFB 876 'Providing Information by Resource-Constrained Analysis', Project C4.

References

Bergen, I., Liedtke, S., Güssgen, S., Kayser, O., Hariharan, C., Drees, C., et al. (2018). Calibration of complex mixtures in one sweep. *International Journal for Ion Mobility Spectrometry, 21*(3), 55–64.

Bischl, B., Lang, M., Kotthoff, L., Schiffner, J., Richter, J., Studerus, E., et al. (2016). mlr: Machine learning in R. *Journal of Machine Learning Research, 17*(170), 1–5.

Bödeker, B., & Baumbach, J. I. (2009). Analytical description of IMS-signals. *International Journal for Ion Mobility Spectrometry, 12*(3), 103–108.

Bödeker, B., Vautz, W., & Baumbach, J. I. (2008). Peak finding and referencing in MCC/IMS-data. *International Journal for Ion Mobility Spectrometry, 11*(1–4), 83–87.

Bunkowski, A. (2012). MCC-IMS data analysis using automated spectra processing and explorative visualisation methods. Ph.D. thesis, Bielefeld University.

Cortes, C., & Vapnik, V. (1995). Support-vector networks. *Machine Learning, 20*(3), 273–297.

Eiceman, G., & Karpas, Z. (2005). *Ion mobility spectrometry* (2nd ed.). London: CRC Press.

Eilers, P. H. C., & Marx, B. D. (1996). Flexible smoothing with B-splines and penalties. *Statistical Science, 11*(2), 89–121.

Handa, H., Usuba, A., Maddula, S., Baumbach, J. I., Mineshita, M., & Miyazawa, T. (2014). Exhaled breath analysis for lung cancer detection using ion mobility spectrometry. *PLoS ONE, 9*(12), e114555.

Hastie, T., Tibshirani, R., & Friedman, J. (2009). *The elements of statistical learning: Data mining, inference, and prediction* (2nd ed.)., Springer series in statistics. Berlin: Springer.

Hauschild, A. C., Kopczynski, D., D'Addario, M., Baumbach, J. I., Rahmann, S., & Baumbach, J. (2013). Peak detection method evaluation for ion mobility spectrometry by using machine learning approaches. *Metabolites, 3*, 277–293.

Horsch, S., Kopczynski, D., Baumbach, J.I., Rahnenführer, J., & Rahmann, S. (2015). From raw ion mobility measurements to disease classification: A comparison of analysis processes. PeerJ PrePrints *3*, e1294v1.

Horsch, S., Kopczynski, D., Kuthe, E., Baumbach, J. I., Rahmann, S., & Rahnenführer, J. (2017). A detailed comparison of analysis processes for mcc-ims data in disease classification - automated methods can replace manual peak annotations. *PLOS ONE, 12*(9), 1–16.

James, G., Witten, D., Hastie, T., & Tibshirani, R. (2013). *An introduction to statistical learning— with applications in R* (1st ed.). New York: Springer.

Karatzoglou, A., Smola, A., Hornik, K., & Zeileis, A. (2004). kernlab—an S4 package for kernel methods in R. *Journal of Statistical Software, 11*(9), 1–20.

Köllmann, C. (2016a). Unimodal spline regression and its use in various applications with single or multiple modes. Ph.D. thesis, TU Dortmund. https://doi.org/10.17877/DE290R-17270.

Köllmann, C. (2016b). uniReg: Unimodal penalized spline regression using B-splines. http://cran. R-project.org/package=uniReg. R package version 1.1

Köllmann, C., Bornkamp, B., & Ickstadt, K. (2014). Unimodal regression using Bernstein-Schoenberg-splines and penalties. *Biometrics, 70*, 783–793.

Kopczynski, D., & Rahmann, S. (2015). An online peak extraction algorithm for ion mobility spectrometry data. *Algorithms for Molecular Biology, 10*, 17.

Lange, L. (2015). Analyse von GC/IMS-Atemluftmessungen unter Berücksichtigung verschiedener Atemerfrischer. Master's thesis, Faculty of Statistics, TU Dortmund University.

Purkhart, R., Hillmann, A., Graupner, R., & Becher, G. (2012). Detection of characteristic clusters in IMS-spectrograms of exhaled air polluted with environmental contaminants. *International Journal for Ion Mobility Spectrometry, 15*(2), 1–6.

R Core Team. (2018). R: A language and environment for statistical computing. R foundation for statistical computing, Vienna, Austria. https://www.R-project.org/

Vautz, W., & Baumbach, J. I. (2008). Exemplar application of multi-capillary column ion mobility spectrometry for biological and medical purpose. *International Journal for Ion Mobility Spectrometry, 11*, 35–41.

Vautz, W., Baumbach, J. I., & Westhoff, M. (2009). An implementable approach to obtain reproducible reduced ion mobility. *International Journal for Ion Mobility Spectrometry, 12*, 47–57.

Vautz, W., Nolte, J., Fobbe, R., & Baumbach, J. I. (2009). Breath analysis-performance and potential of ion mobility spectrometry. *Journal of Breath Research, 3*(3), 036004.

Vautz, W., Franzke, J., Zampolli, S., Elmi, I., & Liedtke, S. (2018a). On the potential of ion mobility spectrometry coupled to GC pre-separation—a tutorial. *Analytica Chimica Acta, 1024*, 52–64.

Vautz, W., Liedtke, S., Martin, B., & Drees, C. (2018b). Data interpretation in GC-ion mobility spectrometry: Pecunia non olet? *International Journal for Ion Mobility Spectrometry, 21*(3), 97–103.

Chapter 4
The Cosine Depth Distribution Classifier for Directional Data

Houyem Demni, Amor Messaoud and Giovanni C. Porzio

Abstract Directions, rotations, axes, clock, or calendar measurements can be represented as angles or equivalently as unit vectors. As points lying on the boundary of circles, spheres, or hyper-spheres, they are also referred as directional data, and they require dedicated methods to be analyzed. In the framework of supervised classification, this work introduces a directional data classifier based on a data depth function. Depth functions provide an inner–outer ordering of the data in a reference space according to some centrality measure, and have appeared as a powerful tool in many fields of multivariate statistics. The recently introduced distance-based depth functions for directional data are considered here. More specifically, this work introduces a cosine depth based distribution method which aims at assigning directional data to classes, given that a training set with class labels is already available. A simulation study evaluating the performance of the proposed method is provided.

4.1 Introduction

Directional information arises whenever observations are recorded as directions in two or three dimensions, or as points lying on the surface of d-dimensional hyper-spheres. They can be recorded using either angles or unit vectors in the d-dimensional

H. Demni (✉)
Institut Supérieur de Gestion de Tunis, Université de Tunis, Tunis, Tunisia
e-mail: houyem.demni@tunis-business-school.tn

University of Cassino and Southern Lazio, Cassino, Italy

A. Messaoud
Tunis Business School, Université de Tunis, Tunis, Tunisia
e-mail: amor.messaoud@tunis-business-school.tn

G. C. Porzio
Department of Economics and Law, University of Cassino and Southern Lazio, Cassino, Italy
e-mail: porzio@unicas.it

© Springer Nature Switzerland AG 2019
N. Bauer et al. (eds.), *Applications in Statistical Computing*,
Studies in Classification, Data Analysis, and Knowledge Organization,
https://doi.org/10.1007/978-3-030-25147-5_4

49

Euclidean space. A one-to-one mapping between these two representations exists, so that the Cartesian coordinates of the ending point of a unit vector can be transformed in spherical coordinates, and vice versa.

A two-dimensional observation (circular data) can be represented by an angle or as a point on the circumference of the unit circle centered at the origin, or as a unit vector connecting the origin to this point. Three-dimensional directional data (spherical data) can be depicted by two angles or as a unit vector in three dimensions, or as a point on the unit sphere, and so on.

Directional data can be found in many scientific areas in the context of measuring directions or cycles. For instance, they are used to estimate the relative rotations of tectonic plates in Earth sciences (Chang 1993), to measure the electrical cardiological activity during a heartbeat in Medicine (Downs and Liebman 1969), to study the animal navigation patterns in Biology (Batschelet 1981), and to analyze wind and ocean directions in Metrology (Bowers et al. 2000). For a deeper introduction to directional data, the reader is referred to the book (Mardia and Jupp 2009).

In directional data analysis, a special role is played by data depth functions. They characterize the centrality of a point with respect to a distribution or a sample so that a center-outward ordering of the points can be obtained. Depth functions are available for linear, functional, and directional data. For a review of directional depth functions, see the recent work by Pandolfo et al. (2018).

Depth functions have found interesting applications in supervised classification, where the aim is to assign new observations to labeled classes. Among the many proposals that exploit data depth functions, the most popular are probably the max depth classifier (widely investigated in Ghosh and Chaudhuri 2005), and the depth versus depth (denoted by DD) classifier (Li et al. 2012). Unlike the many parametric and semi-parametric classification methods, they neither assume any particular type of probability distribution, nor consider any specified parametric form for the separating surface.

In the directional data domain, both these popular depth-based classifiers have been developed, at least to a certain extent. The performance of directional max depth classifiers when different depth functions are adopted has been studied in Pandolfo et al. (2018), while the DD classifier for circular data has been introduced in Pandolfo et al. (2018).

An additional depth-based classifier for linear data has been recently introduced by Makinde and Fasoranbaku (2018): the depth distribution classifier. The proposal seems promising. On the one hand, it is optimal for a wider class of distributions if compared with the max depth classifier. On the other hand, unlike the DD classifier, it is naturally able to deal with multiclass classification issues.

For these reasons, the aim of this work is twofold. First, it provides a review of the max depth, the DD classifier, and the depth distribution techniques. Then, it introduces a new supervised classification tool for directional data: the cosine depth distribution classifier. A simulation study to evaluate its performance with respect to the max depth classifier is offered to the reader.

The work is organized as follows. Section 4.2 provides background on statistical depth functions and on depth-based classifiers. In Sect. 4.3, the proposed directional depth distribution classifier is introduced, while its performance is assessed through a simulation study in Sect. 4.4. Finally, some final remarks are offered in Sect. 4.5.

4.2 Depth-Based Classifiers

Generally speaking, the depth of a point measures to what extent that point is inner with respect to a given distribution or to a multivariate sample. The most internal point is called the deepest, and it is considered a measure of centrality. More precisely, by definition of a depth function (Liu et al. 1999), if a given distribution has a symmetry point, this point should be the deepest above all of that distribution, and a depth will decrease whenever the distance from the symmetry point increases.

Formally, we have that a depth function $D(.) : \mathbb{R}^q \to \mathbb{R}$ with respect to a distribution F is a mapping of a vector $x \in \mathbb{R}^q$ to a real-valued number $D(x; F) \in \mathbb{R}$. Within the literature, several depth functions have been introduced. Among the many, Tukey's half-space depth (Tukey 1975), the simplicial depth (Liu 1990), and the zonoid depth (Koshevoy and Mosler 1997) have reached some popularity.

The data depth concept provides center-outward ordering of points in any dimension and it allows some nonparametric multivariate statistical analysis to be performed, in which no distributional assumptions are needed. That is, the distribution F in the expression $D(x; F)$ is typically substituted by its empirical counterpart $\hat{F}$, with no needs to assume a parametric form for it. Applications of data depth arise in statistical inference (location and scatter estimation) (Romanazzi 2009), statistical quality control (Messaoud et al. 2008), outlier detection and data visualization (Rousseeuw et al. 1999; Buttarazzi et al. 2018).

Depth functions have been also employed in supervised learning, where a classification rule is constructed from labeled training data to assign an arbitrary new data point to one of the labels. The main underlying idea is that the centrality of a new point with respect to the labeled classes (i.e., its depth) is a measure of the degree of closeness to each label.

Dissimilar to parametric and semi-parametric classification methods, the depth-based classifiers neither assume any particular type of probability distribution nor consider any specified parametric form for the separating surface. Generally speaking, any depth function can be adopted to define a depth-based classifier.

In this section, the two main depth-based classifiers are reviewed. Namely, the max depth classifier and the DD classifier. Other depth-based classifiers have been proposed for linear data, such as the $DD - \alpha$ classification approach (Lange et al. 2014), and the class of depth-based functions associated with the kNN classification rule introduced by Paindaveine and Van Bever (2015). A recent comprehensive overview of depth-based classifiers can be found in Vencálek (2017). More recently, a depth distribution classifier was introduced by Makinde and Fasoranbaku (2018). Given the focus of this work, it will be also briefly described in this section.

4.2.1   The Max Depth Classifier

The max depth classifier assigns the new data point to the class with respect to which it attains the highest depth value. This is because higher depth values correspond to more central areas with respect to the class. After (Liu 1990), the concept of max depth in classification has bee developed by Ghosh and Chaudhuri (2005).

Let y be the new point to be assigned, and let $D(y, \hat{F})$ be the depth of y with respect to the empirical distribution $\hat{F}$. For the sake of simplicity, let us consider the case of two groups (i.e., we have available $\hat{F}_1$ and $\hat{F}_2$ from a training data set). Then, the max depth classification rule is given by

$$\begin{cases} D(y, \hat{F}_1) > D(y, \hat{F}_2) \implies \text{assign } y \text{ to } \text{population 1} \\ D(y, \hat{F}_1) < D(y, \hat{F}_2) \implies \text{assign } y \text{ to } \text{population 2} \end{cases}$$

If $D(y, \hat{F}_1) = D(y, \hat{F}_2)$, the classification rule will randomly assign the observation to one of the two groups with equal probability.

The max depth classifier is equivalent to the optimal Bayes classifier with equal prior probabilities in case all the populations are elliptically distributed with density function strictly decreasing when moving away from the ellipsoid center, and with populations differing only in location parameters (Ghosh and Chaudhuri 2005). For this condition to hold, the adopted depth functions must be continuous, positive over the entire d-dimensional space and decreasing too.

When the populations differ not only in location, a modified version of the classification approach based on a function of the half-space depth has been proposed (Ghosh and Chaudhuri 2005), where the empirical half-space depth of a point y with respect to a multivariate sample is given by the minimum number of points that lie in any closed half-space containing y (Tukey 1975). A modified version of the max depth classifier based on the projection depth function has been introduced as well (Cui et al. 2008). It outperforms the method by Ghosh and Chaudhuri (2005) only in normal settings.

Finally, the use of the max depth classifier for directional data has also been studied (Pandolfo et al. 2018). By means of a simulation study, it has been shown that the distance-based depth classifiers outperform classifiers based on the angular Tukey's (Liu and Singh 1992) and on the angular simplicial depth (Liu and Singh 1992) if data are drawn from a von Mises–Fisher distribution (Mardia and Jupp 2009), either with equal or different concentration levels.

4.2.2   The DD Classifier

The depth versus depth classifier (or DD classifier) is a nonparametric two-class classification method introduced by Li et al. (2012). It is based on the depth versus depth (or DD) plot, which is a graphical tool allowing the comparison of two multivariate distributions or samples through their corresponding depth values.

Briefly, the DD plot is a scatterplot where each plotted point has coordinates given by the depths of the corresponding point in the original multivariate space with respect to the two examined groups. In this way, it is possible to transform two multivariate samples to a simple two-dimensional scatter plot regardless of the dimensions of the original sample space.

The main idea behind the DD classifier is to find the best polynomial separating function in a DD plot. Consequently, the generic form of the DD classifier is given as follows. Let $r(.)$ be some real increasing function, and $\hat{F}_1$ and $\hat{F}_2$ be the empirical CDF's of two multivariate samples (the two samples are the training set, where each of the two sample has its own class label). Then, the classification rule is defined by

$$\begin{cases} D(y, \hat{F}_1) > r(D(y, \hat{F}_2)) \implies \text{assign} \quad y \quad \text{to} \quad \text{population 1} \\ D(y, \hat{F}_1) < r(D(y, \hat{F}_2)) \implies \text{assign} \quad y \quad \text{to} \quad \text{population 2} \end{cases}$$

In case of equality, y will be randomly classified to group 1 or 2 with equal probability.

If $r(.)$ is set equal to the 45 degree line, and apart from the case of equality, the DD classification rule would assign y to group 1 if $D(y, \hat{F}_1) > D(y, \hat{F}_2)$ and assign y to group 2 otherwise. This will reduce the DD classifier to the max depth classifier described above. If F_1 and F_2 differ and they both admit a density from the elliptically contoured family, then the DD classifier will be optimal in the Bayes sense whenever the used depths are strictly increasing functions of the densities themselves.

The performance of the DD classifier associated with different depth functions (Mahalanobis depth Mahalanobis 1936, projection depth Zuo 2003, half-space depth Tukey 1975, and simplicial depth Liu 1990) has been compared by Li et al. (2012) with the performance of other classifiers such as the Linear Discriminant Analysis (James et al. 2013), the Quadratic Discriminant Analysis (James et al. 2013), the Support Vector Machine (Vapnik 1998), and the max depth (Ghosh and Chaudhuri 2005) classifier. It seems the DD classifier shows a better performance in many of the cases, or a similar performance otherwise.

For this reason, a recent interest eventually arised on the use of the DD classifier for directional data. It has been investigated for the case of circular data by Pandolfo et al. (2018).

4.2.3   The Depth Distribution Classifier

Based on the cumulative distribution function (CDF) of depths, a new depth-based classifier has been very recently introduced (Makinde and Fasoranbaku 2018). That is, the depth distribution classifier.

Let $F_D^G(x)$ be the cumulative distribution function of a depth function $D(X, G)$ evaluated in x:

$$F_D^G(x) := P(D(X, G) \leq D(x, G))$$

where X is a random variable, and G is a generic distribution with respect to which the depth is evaluated. Both X and G are defined on the original sample space.

The value of $F_D^G(x)$ provides information on how central is x with respect to G. If x is a central observation, then $D(x, G)$ will be large, and hence $F_D^G(x)$ will be large too. At the extreme, if x_0 is a deepest point of G, we will have $F_D^G(x_0) = 1$. On the other hand, if x is far from the center of G (from its deepest point), then $F_D^G(x)$ will be small.

Accordingly, a depth distribution classifier can be defined (Makinde and Fasoranbaku 2018). Let y be the new point to be assigned. And, for the sake of simplicity, let us consider the case of two groups G_1 and G_2. Then, the depth distribution classification rule is given by

$$\begin{cases} F_D^{\hat{G}_1}(y) > F_D^{\hat{G}_2}(y) \implies \text{assign} \quad y \quad \text{to} \quad \text{population 1} \\ F_D^{\hat{G}_1}(y) < F_D^{\hat{G}_2}(y) \implies \text{assign} \quad y \quad \text{to} \quad \text{population 2} \end{cases}$$

If $F_D^{\hat{G}_1}(y) = F_D^{\hat{G}_2}(y)$, the classification rule will randomly assign the observation to one of the two groups with equal probability.

4.3   The Cosine Depth Distribution Classifier

The cosine depth distribution classifier is proposed here as a tool to classify points lying on the surface of hyper-spheres, in analogy with the work by Makinde and Fasoranbaku (2018).

Directions in q dimensions can be represented as unit vectors z on the sphere $S^{(q-1)} = \{z : z^T z = 1\}$ with unit radius and center at the origin. Let $H_1, ..., H_J$ be a set of directional distributions, and let $F_D^H(z)$ be the cumulative distribution function of a depth function defined on hyper-spheres $D(Z, H)$ evaluated in z:

$$F_D^H(z) := P(D(Z, H) \le D(z, H))$$

Let w be the new point to be assigned, and, again for the sake of simplicity, let us consider the case of two groups (i.e., G_1 and G_2). Then, the directional depth distribution classification rule will be given by

$$\begin{cases} F_D^{\hat{G}_1}(w) > F_D^{\hat{G}_2}(w) \implies \text{assign} \quad w \quad \text{to} \quad \text{population 1} \\ F_D^{\hat{G}_1}(w) < F_D^{\hat{G}_2}(w) \implies \text{assign} \quad w \quad \text{to} \quad \text{population 2} \end{cases}$$

If $F_D^{\hat{G}_1}(w) = F_D^{\hat{G}_2}(w)$, the classification rule will randomly assign the observation to one of the two groups with equal probability.

The performance of the just introduced directional depth distribution classifier depends on the choice of the depth function. Many depths for directional data were introduced and are available in the literature (Pandolfo et al. 2018). Here, distance-based directional depths will be considered. They are briefly reviewed below.

Let $d()$ be a bounded and nonnegative directional distance function with $d^{\sup}$ its supremum over $S^{(q-1)}$. By definition, a directional distance-based depth of a point $z \in S^{(q-1)}$ with respect to a distribution H on $S^{(q-1)}$ is given by

$$D(z, H) := d^{\sup} - E_H[d(z, W)],$$

where $E[.]$ is the expected value, and W is a random variable from a distribution H.

To obtain a directional depth function enjoying nice properties, suitable distances should be adopted. Particularly, they must be rotation invariant. As a consequence, they will be of the form $\delta(z's)$ for some function $\delta = [-1, 1] \rightarrow \mathbb{R}^+$, where s is also a point on $S^{(q-1)}$. For instance, three of these distances that will yield rotational-invariant directional depths are discussed in Pandolfo et al. (2018), and briefly reported below.

- **Cosine depth**: Adopting the cosine distance, i.e., $\delta(t) = 1 - t$, the cosine depth is obtained as $D_{cos} := 2 - E_H[(1 - z'W)]$.
- **Arc distance depth**: Adopting the arc distance, i.e., $\delta(t) = \arccos(t)$, the arc distance depth is obtained as $D_{arc} := \pi - E_H[\arccos(z'W)]$
- **Chord depth**: Adopting the chord distance, i.e., $d_{chord} = ||z - t|| = \sqrt{2(1 - z't)}$, the chord depth is obtained as $D_{chord} := 2 - E_H[\sqrt{2(1 - z'W)}]$

Finally, among these three directional depth functions, the cosine depth is preferred here. This is because of two reasons. First, it can be easily computed. Then, it provided good performances when associated to the max depth classifier on hyperspheres (Pandolfo et al. 2018). This yields the cosine depth distribution classifier.

4.4 Simulation Study

The performance of the cosine depth distribution classifier is evaluated by means of a simulation study. A comparison with the max depth classifier for directional data based on the same depth function is also offered to the reader.

4.4.1 Study Design

This study is based on the assumption of equal prior probabilities and considering a two-class classification problem.

Let G_1 and G_2 be two von Mises–Fisher distributions (vMF). That is, G_1 and G_2 have their probability density function given by

$$h(z; \mu, c) = \left(\frac{c}{n}\right)^{p/2-1} \frac{1}{\Gamma^{(p/2)} I_{p/2-1}(c)} \exp\{c\mu^T z\},$$

where $c \geq 0$, $\|\mu\| = 1$, and I_v denote the modified Bessel function (Mardia and Jupp 2009) of the first kind and order v. The parameters μ and c are the mean direction and the concentration parameter, respectively.

For the sake of comparison, the simulation scheme was designed in analogy with the simulations in Pandolfo et al. (2018), where the performance of the directional max depth classifier was investigated with respect to the choice of the depth function.

Within the first two setups, the two groups are both unimodal with different locations parameters and same/different concentration levels (setup 1 and 2, respectively). On the contrary, the third setup will investigate the case of bimodality for one of the two groups. The other will be unimodal with its mode lying between the modes of the first group.

- In Setup 1, we study the case of difference in location, same concentration parameter for both G_1 and G_2, data on the sphere ($q = 3$) and on a hypersphere in dimension $q = 10$. The location parameters are set $\mu_1 = (0, 0, 1)$ and $\mu_2 = (1, 0, 0)$ for dimension $q = 3$, and $\mu_1 = (0, 0, 0, 0, 0, 0, 0, 0, 0, 1)$ and $\mu_2 = (1, 0, 0, 0, 0, 0, 0, 0, 0, 0)$ for dimension $q = 10$, respectively. The performance is then investigated for different concentration levels. We set $c \in \{2, 5\}$.
- In Setup 2, the two distributions differ also in concentration. With the same location parameters of Setup 1, the concentration c is set equal to 2 for group 1, and 5 for group 2, considering again dimensions $q \in \{3, 10\}$.
- In Setup 3, we consider discrimination between a vMF distribution with $\mu = \mu_1$, and a bimodal density obtained as an equal weight mixture of two von Mises–Fisher, with means μ_{21} and μ_{22}. For dimension $d = 3$, we set $\mu_1 = (0, 0, 1)$, $\mu_{21} = (1, 0, 0)$, $\mu_{22} = (1, 0, 1)$. For dimension $d = 10$, we set $\mu_1 = (0, 0, 0, 0, 0, 0, 0, 0, \cos 7\pi/4, \sin 7\pi/4)$, $\mu_{21} = (0, 0, 0, 0, 0, 0, 0, 0, 1, 0)$ and $\mu_{22} = (0, 0, 0, 0, 0, 0, 0, 0, 0, 1)$. All the vMFs have the same concentration levels $c = 4$.

As in Pandolfo et al. (2018), we set the training set size equal to 200 with 100 observations generated from G_1 and 100 observations generated from G_2 and the testing set size equal to 100 with 50 observations generated from G_1 and 50 observations generated from G_2. For each simulation condition, the experiment is replicated 250 times.

4.4.2 Results

The detailed result of our simulation studies are reported in this Section. The performance of the classifiers is evaluated by means of the misclassification rate. That is, the number of observations misclassified over the sample size in each replicated sample.

Table 4.1 Average misclassification rate (AMR) and standard deviations of the depth distribution (DistD) and the max depth (MaxD) classifiers in different simulation setups. Best achieved results are highlighted in bold

Setup			AMR (*standard deviation*)	
			DistD	MaxD
Setup 1	$q = 3$	$c = 2$	**0.236** *(0.037)*	0.257 *(0.045)*
		$c = 5$	**0.066** *(0.028)*	0.074 *(0.022)*
	$q = 10$	$c = 2$	**0.375** *(0.052)*	0.380 *(0.052)*
		$c = 5$	**0.150** *(0.030)*	0.167 *(0.037)*
Setup 2	$q = 3$		**0.104** *(0.022)*	0.170 *(0.030)*
	$q = 10$		**0.238** *(0.045)*	0.274 *(0.037)*
Setup 3	$q = 3$		0.496 *(0.022)*	**0.440** *(0.030)*
	$q = 10$		**0.170** *(0.037)*	0.185 *(0.037)*

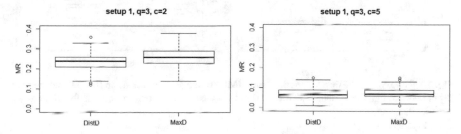

Fig. 4.1 Boxplots of the misclassification rates (MR) of the depth distribution (DistD) and the max depth (MaxD) classifiers obtained from 250 independent replications in Setup 1 with concentration parameters $c = 2$ and $c = 5$ in dimension $q = 3$

For each simulation setup, the distribution of the misclassification rates obtained by the cosine depth distribution classifier (DistD) and by the max depth classifier (MaxD) are summarized through boxplots (Figs. 4.1, 4.2, 4.3 and 4.4). The corresponding average misclassification rates (AMR) and standard deviations are instead reported in Table 4.1, where the best achieved results are highlighted in bold.

Considering Setup 1 in dimension 3 (results in Fig. 4.1), the cosine depth distribution classifier achieved a slightly better performance than the max depth classifier for both scenarios of concentration parameters, i.e., $c = 2$ and $c = 5$. On the other hand, the misclassification rate is lower when the concentration parameter is higher, and this is because data are more separated and less sparse on the sphere, and hence they can be better discriminated.

For $q = 10$, results from Setup 1 indicate that the cosine depth distribution classifier performs better than the max depth classifier, in case of equal concentration, also in higher dimensions (Fig. 4.2). However, the overall performance deteriorates, especially for quite sparse data ($c = 2$).

In the case of different concentration levels across the two groups (Setup 2), the cosine depth distribution classifier shows highly satisfactory performance, with a

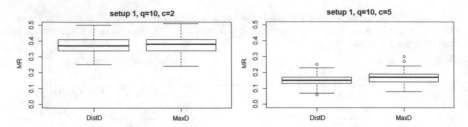

Fig. 4.2 Boxplots of the misclassification rates (MR) of the depth distribution (DistD) and the max depth (MaxD) classifiers obtained from 250 independent replications in Setup 1 with concentration parameters 2 or 5 for dimension $q = 10$

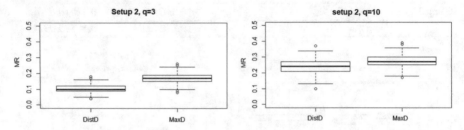

Fig. 4.3 Boxplots of the misclassification rates (MR) of the depth distribution (DistD) and the max depth (MaxD) classifiers obtained from 250 independent replications in Setup 2 for dimensions $q \in \{3, 10\}$

much lower misclassification rate compared to the max depth classifier, especially in lower dimension (Fig. 4.3).

In the third setup, the case of the bimodal group, if both the dimension and the concentration level are low ($q = 3$, $c = 4$), the two classifiers essentially fail (Fig. 4.4, left). Although the max depth classifier slightly outperform the new introduced method, both their average misclassification rates approximate the 50% rate, which can be attained by just assigning randomly each new observations to one of the two groups. This happens because data from the two groups are largely mixed up on the sphere. When the dimension increases ($q = 10$), data are not largely mixed up any more, the two classifiers perform better, with the cosine distribution depth slightly beating the max depth classifier (Fig. 4.4, right).

4.5 Final Remarks

This work first reviews supervised classification methods based on data depth, and then it introduces a procedure to classify directional data. Directional data are a special class of quantitative measures which requires dedicated methods to be analyzed properly. They refer to point lying on the surface of hyper-spheres.

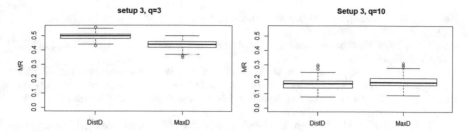

Fig. 4.4 Boxplots of the misclassification rates (MR) of the depth distribution (DistD) and the max depth (MaxD) classifiers obtained from 250 independent replications in Setup 3 for dimension $q \in \{3, 10\}$

The proposed classifier is based on the distribution function of the cosine depth, and for this reason it is called cosine depth distribution classifier. The performance of the proposed classification method is investigated in lower and higher dimension settings with a comparison to the max depth classifier. The simulation results suggest the cosine depth distribution classifier might improve over the max depth classifier in many scenarios.

The use of the cosine depth classifier for directional data seems thus promising. As a further study, it would be of interest to investigate to what extent, and in which cases, such a classifier would provide better performances if different depth functions are adopted in place of the cosine depth. Some real data applications would also be of interest. The authors intend to report on that elsewhere.

Acknowledgements The authors wish to thank the two anonymous referees for their precious comments on a first version of this work. Thanks are also due to Giuseppe Pandolfo and Ondrej Vencalek for their valuable support and suggestions.

References

Batschelet, E. (1981). *Circular statistics in biology*. London: Academic.

Bowers, J. A., Morton, I. D., & Mould, G. I. (2000). Directional statistics of the wind and waves. *Applied Ocean Research*, 22(1), 1330.

Buttarazzi, D., Pandolfo, G., & Porzio, G. C. (2018). A boxplot for circular data. *Biometrics*, 74(4), 14921501.

Chang, T. (1993). Spherical regression and the statistics of tectonic plate reconstructions. *International Statistical Review/Revue Internationale de Statistique*, 61(2), 299316.

Cui, X., Lin, L., & Yang, G. (2008). An extended projection data depth and its applications to discrimination. *Communications in Statistics, Theory and Methods*, 37(14), 22762290.

Downs, T. D., & Liebman, J. (1969). Statistical methods for vectorcardiographic directions. *IEEE Transactions on Biomedical Engineering*, 16(1), 8794.

Ghosh, A. K., & Chaudhuri, P. (2005). On maximum depth and related classifiers. *Scandinavian Journal of Statistics*, 32(2), 327350.

James, G., Witten, D., Hastie, T., & Tibshirani, R. (2013). *An introduction to statistical learning*. New York: Springer.

Koshevoy, G., & Mosler, K. (1997). Zonoid trimming for multivariate distributions. *The Annals of Statistics, 25*(5), 19982017.

Lange, T., Mosler, K., & Mozharovskyi, P. (2014). Fast nonparametric classification based on data depth. *Statistical Papers, 55*(1), 4969.

Li, J., Cuesta-Albertos, J. A., & Liu, R. Y. (2012). DD-classifier: Nonparametric classification procedure based on DD-plot. *Journal of the American Statistical Association, 107*(498), 737753.

Liu, R. Y. (1990). On a notion of data depth based on random simplices. *The Annals of Statistics, 18*(1), 405414.

Liu, R. Y., Parelius, J. M., & Singh, K. (1999). Multivariate analysis by data depth: Descriptive statistics, graphics and inference, (with discussion and a rejoinder by Liu and Singh). *The Annals of Statistics, 27*(3), 783–858.

Liu, R. Y., & Singh, K. (1992). Ordering directional data: concepts of data depth on circles and spheres. *The Annals of Statistics, 20*(3), 14681484.

Mahalanobis, P. (1936). On the generalized distance in statistics. *Proceedings of the National Institute of Science of India, 12*, 4955.

Makinde, O. S., & Fasoranbaku, O. A. (2018). On maximum depth classifiers: depth distribution approach. *Journal of Applied Statistics, 45*(6), 11061117.

Mardia, K. V., & Jupp, P. E. (2009). *Directional statistics*. New York: Wiley.

Messaoud, A., Weihs, C., & Hering, F. (2008). Detection of chatter vibration in a drilling process using multivariate control charts. *Computational Statistics and Data Analysis, 52*(6), 32083219.

Paindaveine, D., & Van Bever, G. (2015). Nonparametrically consistent depth-based classifiers. *Bernoulli, 21*(1), 6282.

Pandolfo, G., D'Ambrosio, A., & Porzio, G. C. (2018). A note on depth-based classification of circular data. *Electronic Journal of Applied Statistical Analysis, 11*(2), 447462.

Pandolfo, G., Paindaveine, D., & Porzio, G. C. (2018). Distance-based depths for directional data. *The Canadian Journal of Statistics, Canadian Journal of Statistics, 46*(4), 593609.

Romanazzi, M. (2009). Data depth, random simplices and multivariate dispersion. *Statistics and Probability Letters, 79*(12), 14731479.

Rousseeuw, P. J., Ruts, I., & Tukey, J. W. (1999). The bagplot: A bivariate boxplot. *The American Statistician, 53*(4), 382387.

Tukey, J. W. (1975). Mathematics and the picturing of data. *Proceedings of the International Congress of Mathematicians, Vancouver, 1975*(2), 523531.

Vapnik, V. (1998). *Statistical learning theory*. New York: Wiley.

Vencálek, O. (2017). Depth-based classification for multivariate data. *Austrian Journal of Statistics, 46*(34), 117128.

Zuo, Y. (2003). Projection-based depth functions and associated medians. *The Annals of Statistics, 31*(5), 14601490.

Chapter 5
A Nonconformity Ratio Based Desirability Function for Capability Assessment

Ramzi Talmoudi

Abstract The objective of this paper is to provide a process capability index structure, which respects "the higher the better" rule even when the underlying distribution of the quality characteristic not the normal distribution. An estimator of the univariate capability index is proposed and its statistical properties are studied using Taylor series expansion to monitor an exponential and a lognormal distribution. The used approximation shows that the estimator is unbiased and convergent when the underlying distribution is the exponential one. However, like classical indices, the estimator is biased for the lognormal distribution. A comparative study is carried out with some process capability indices from the literature designed to deal with non- normality. The proposed index performs better than existing indices considering "the higher the better" rule as a benchmark. Finally, a bootstrap confidence interval is implemented for capability judgement and a multivariate extension of the index is introduced.

5.1 Introduction

Process capability indices (PCIs) are an important tool in statistical process control. Minimum values of process capability indices are usually specified before buying any process equipment. Customers often specify quality characteristics for which the suppliers have to maintain a minimum value of process capability indices. For measurable quality characteristics, customers specify specification limits and only items having quality characteristic measurements inside the specification limits are accepted. Moreover, process capability indices are used for continuous quality improvement.

Tang and Than (1999) noticed that under the normality assumption the classical indices C_p and C_{pk} jointly determine the proportion of nonconforming items. As

R. Talmoudi (✉)
University of Carthage, Faculté des Sciences Economiques et de Gestion de Nabeul,
Laboratoire Environnement Economique et Institutionnel de l'Entreprise, Avenue Merazga,
8000 Nabeul, Tunisia
e-mail: ramzitelmoudi@yahoo.com

© Springer Nature Switzerland AG 2019
N. Bauer et al. (eds.), *Applications in Statistical Computing*,
Studies in Classification, Data Analysis, and Knowledge Organization,
https://doi.org/10.1007/978-3-030-25147-5_5

C_p is also called PCR (Process Capability Ratio), it was commonly admitted that p stands for "process" in C_p notation. It is also known that $C_{pk} = (1 - k)C_p$, with $k = \frac{|M-\mu|}{\frac{1}{2}(USL-LSL)}$, where USL and LSL are the specification limits, μ is the process mean, and M is the specification limits midpoint given by $M = \frac{USL+LSL}{2}$. Kotz and Lovelace (1998) noticed that C_{pk} was created in Japan and that k was one of the original Japanese indices, this is why there is a common understanding that k stands for the Japanese word "katayori" which means deviation. The C_p index value is closely connected to the potential proportion of nonconforming items. It means that it is connected to the minimum of the proportion of nonconforming items obtained by adjusting the process mean to the center of the specification limits. However, the C_{pk} index value is connected to the actual proportion of nonconforming items.

If the process distribution is non-normal, this relation is no longer valid. Some methods and indices were presented in the literature to deal with PCIs for non-normal distributions. A skewness correction factor was incorporated to the index C_{pmk} by Wright (1995) in order to create the new index C_s. The capability index C_{pmk} was presented first by Pearn et al. (1992) as $C_{pmk} = (1 - k)C_{pm}$, where C_{pm} is an index based on the Taguchi loss function and presented independently by Hsiang and Taguchi (1985) and by Chan et al. (1988) as noticed Kotz and Lovelace (1998). It is still not clear what m stands for in C_{pm} notation as it was not formally defined, however, Chan et al. (1988) stated that "If the process drifts from its target value (i.e., μ moves away from T)...", it was then understood that m stands for "moves away". Choi and Bai (1996) proposed a weighted variance method which adjusts the PCI value by considering the deviations above and below the process mean. Several authors proposed a new generation of PCIs. Jessenberger (1999) proposed to use a new generation of indices based on the desirability function. She proposed to use the index EDU, which is the expected value of the Derringer and Suich (1980) desirability function as a metric for capability assessment. The extension of the index to the bivariate case is also available. EDU is understood to be the abbreviation for Expected Desirability in the Univariate case.

Multivariate PCIs are unavoidable especially when several quality characteristics determine the quality of a product. This is why multivariate process capability indices are discussed in the literature. Wang et al. (2000) compared different multivariate PCIs and noticed that in the multivariate case there is no common agreement about PCI definition, even when the assumption of multivariate normality holds.

In what follows, a nonconformity ratio based desirability in the univariate case NCDU is used as a capability index and its extension to the multivariate case *NCDM* is discussed. *NCDU* is based on the Derringer and Suich (1980) desirability function. Furthermore, the performance of NCDU for non-normal distributions is compared with the performance of indices from the literature.

5.2 A Nonconformity Ratio Based Desirability Function

Derringer and Suich (1980) desirability function evolves transformation of each response variable Y_i into a desirability value d_i between 0 and 1. The desirability of the response increases as it becomes closer to its target value T_i. It reaches the maximum value of 1 only if the response value is equal to the target T. The overall desirability is then given by the desirability index, which is the geometric mean of the individual desirabilities. It is noticed that the definition of the desirability function does not depend on any distribution assumption. In what follows, the nonconformity ratio is considered as a response variable and the property that the desirability value increases when the response becomes closer to its target is used to make the "higher the better" rule hold for any type of distribution and for any type of specification limit.

The nonconformity-based desirability function associated with the quality characteristic Y_1 is

$$
NCDU_1 = \begin{cases} 0 & \text{if } r_1 \geq USL', \\ \frac{USL' - r_1}{USL' - r_1^{min}} & \text{if } r_1^{min} \leq r_1 < USL', \end{cases} \tag{5.1}
$$

where r_1 is the current nonconformity ratio associated with the quality characteristic Y_1. If $Y_1 \sim N(\mu_1, \sigma_1^2)$, then $r_1 = \Phi(\frac{LSL_1 - \mu_1}{\sigma_1}) + 1 - \Phi(\frac{USL_1 - \mu_1}{\sigma_1})$, where LSL_1 and USL_1 are the specification limits for Y_1. r_1^{min} is the minimum of the nonconformity ratio, it is obtained when $\mu_1 = \frac{USL_1 + LSL_1}{2}$, hence, $r_1^{min} = \Phi(\frac{LSL_1 - USL_1}{2\sigma_1}) + 1 - \Phi(\frac{USL_1 - LSL_1}{2\sigma_1})$. USL' is the upper limit for the nonconformity ratio beyond it the process is not capable. It is common to set $USL' = 64$ p.p.m (parts per million) as it corresponds to 4σ level in six sigma theory. Hence, when the proportion of nonconforming items is less than 64 p.p.m, the NCDU is positive and the process is considered capable. However, if the process is not capable, then the index value is equal to 0. The interpretation of the index depends on the choice of the threshold value. The choice of USL' value is set up by the practitioner and can be reached through the six sigma strategy. r^{min} is considered as the nonconformity ratio that the process most likely will produce after some process location adjustment. Indeed for several processes, location adjustment is much easier than other quality improvement alternatives. If r^{min} value obtained in this way is acceptable, then it is identified as a target value. However, this does not mean that quality improvement efforts should stop at this stage.

It is important to note that using 0 instead of r_1^{min} can lead to misleading interpretations of the index, especially in the case when the index is used to compare between the capability of several processes. Indeed, assume that a capability comparison is carried out between the two processes 1 and 2. If 0 is used instead of r_1^{min} and r_2^{min}, the comparison will be between r_1 and r_2, if $r_2 < r_1$ then process 2 is considered as more capable than process 1. However, including r_1^{min} and r_2^{min} with $r_1^{min} < r_2^{min}$ in the index computation gives the additional information that with some process adjustments process 1 is more capable than process 2. When $r_1^{min} < r_2^{min}$ we say that the potential capability of process 1 is higher than the potential capability of process 2. Hence, the index NCDU is not only used for the comparison of the actual capability, but also to compare the potential capability. This is an analogy to the classical indices where the use of the index C_{pk} which assesses the actual capability is associated with the use of the index C_p which assesses the potential capability. Then, why not comparing individual nonconformity ratios and the individual minimum of the nonconformity ratios separately? This question is equivalent to the question why we are using capability indices. In fact, the capability index is used to characterize in one value the ability of the process to meet the customer requirements. Hence, capability indices are easy to be communicated inside each organization. The use of NCDU avoids the use of two indices for the actual capability and for the potential capability separately as the computation of NCDU is based already on the comparison between the actual capability and the potential capability. The NCDU value gives an idea about how far away the actual nonconformity ratio from the maximum ability of the process r^{min}. NCDU is also easy to interpret as it is sufficient to notice that the index value is positive in order to judge that the process is capable. NCDU numerator is interpreted as the actual capability margin, which is compared to the NCDU denominator considered as the potential capability margin. Hence, a value of 0.1 of NCDU, for example, means that first the process is capable (when $USL' = 64 p.p.m.$) but also only 10% of the potential capability margin is used.

Moreover, assume that two processes are considered. The quality of process 1 and process 2 is expressed in terms of the quality characteristics Y_1 and Y_2, respectively. r_1 and r_2 are the nonconformity ratios associated to Y_1 and Y_2, respectively. It is noticed that when r_1 and r_2 simultaneously reach their minimum, the assigned desirability is 1 for both processes, although both processes do not have the same capability. To overcome this shortcoming, it is proposed to consider $min(r_1^{min}, r_2^{min})$ in the computation of the desirability function. Hence, if $r_1^{min} < r_2^{min}$ the nonconformity-based desirability function associated with Y_2 is

$$
NCDU_2 = \begin{cases} 0 & \text{if } r_2 \geq USL', \\ \frac{USL' - r_2}{USL' - r_1^{min}} & \text{if } r_2^{min} < r_2 < USL', \\ \frac{USL' - r_2^{min}}{USL' - r_1^{min}} & \text{if } r_2 \leq r_2^{min}. \end{cases} \tag{5.2}
$$

Notice that in order to allow comparability between processes when $r_1^{min} < r_2^{min}$ only $NCDU_1$ can reach the maximum value of 1. This is due to the fact that the potential capability of process 1 is higher than the potential capability of process 2. However, $NCDU_2$ value will not exceed $\frac{USL' - r_2^{min}}{USL' - r_1^{min}}$. When more than two quality characteristics are considered, r^{min} is determined for each quality characteristic and the minimum among all r^{min} is used in the computation of each $NCDU$ as explained for (2). In the case, where several quality characteristics express the quality of a single product, the natural extension of $NCDU$ is given by the desirability index $D(r_1, \ldots, r_p) = [\Pi_{k=1}^{p} NCDU_k]^{\frac{1}{p}}$, where $NCDU_k$ are defined in (1) and (2) and p is the number of quality characteristics. The geometric mean assigns an overall desirability of 0 if there exists at least one quality characteristic for which the nonconformity ratio exceeds the USL' value. NCDU allows for capability judgment, it compares the actual capability of a process to its potential capability and it allows the comparison between the capability of several processes for any type of distribution and any type of specification limits. It will be interesting to introduce an estimator for $NCDU$ and to compare its performance with other indices from the literature in a case study.

5.3 NCDU Estimator and Its Statistical Properties

NCDU estimation depends on the nonconformity ratio estimation. Hence as explained in this section, NCDU estimator properties depend on the quality characteristic distribution. In what follows NCDU estimator properties, especially the bias and the variance, are derived for the exponential and the lognormal distributions. These properties are obtained using Taylor series expansion. It is important to remind that if A is a random variable with expected value α and variance σ^2 then the expected value of a function $g(A)$ is approximated in terms of the moments of A.

$$g(A) \approx g(\alpha) + (A - \alpha)g'(\alpha) + \frac{1}{2}(A - \alpha)^2 g''(\alpha) \tag{5.3}$$

$$E[g(A)] \approx g(\alpha) + \frac{1}{2}\sigma^2 g''(\alpha) \tag{5.4}$$

$$Var[g(A)] \approx \{g'(\alpha)\}^2 \sigma^2 \tag{5.5}$$

5.3.1 Properties of NCDU Estimator for the Exponential Distribution

If a quality characteristic X follows an exponential distribution with parameter $\frac{1}{\theta}$ then the probability density function is given by

$$f(x, \theta) = \frac{1}{\theta}e^{-\frac{x}{\theta}}; x \geq 0$$

and the distribution function is $F(x, \theta) = 1 - e^{-\frac{x}{\theta}}$. The nonconformity ratio r is computed as follows: $r = 1 - F(USL) + F(LSL) = 1 + e^{-\frac{USL}{\theta}} - e^{-\frac{LSL}{\theta}}$. An estimator of r is $\hat{r} = 1 + e^{-\frac{USL}{\bar{X}}} - e^{-\frac{LSL}{\bar{X}}}$ as $\bar{X}$ is the maximum likelihood estimator of θ with $E(\bar{X}) = \theta$ and $Var(\bar{X}) = \frac{\theta^2}{n}$. It is noticed that $\hat{r}$ is a function of $\bar{X}$ then it becomes possible to approximate its properties through (5.4) and (5.5). Hence, $\lim\limits_{n \to \infty} E(\hat{r}) = r$

and $\lim\limits_{n\to\infty} Var(\hat{r}) = 0$. This means that $\hat{r}$ is asymptotically unbiased and convergent estimator of r. Before discussing the effect on NCDU estimation, r_{min} should be defined first. Applying the first-order optimality condition to r leads to finding the value of the parameter θ which minimizes r given by

$$\theta_{min} = \frac{USL - LSL}{ln(\frac{USL}{LSL})}$$

It becomes obvious that for this value of θ_{min}, the second order optimality condition is satisfied and that r_{min} is not a random variable. Therefore,

$$\widehat{NCDU} = \frac{USL' - \hat{r}}{USL' - r_{min}}$$

and in this case $\widehat{NCDU}$ is an asymptotically unbiased and convergent estimator for $NCDU$.

5.3.2 Properties of NCDU Estimator for the Lognormal Distribution

If a quality characteristic X follows a lognormal distribution with parameters μ and σ then the probability density function is given by

$$f(x, \mu, \sigma) = \frac{1}{x\sigma\sqrt{2\pi}}exp\{-\frac{(ln(x) - \mu)^2}{2\sigma^2}\}; x \geq 0$$

and the distribution function is $F(x, \mu, \sigma) = \frac{1}{2} + \frac{1}{2}erf[\frac{ln(X)-\mu}{\sqrt{2}\sigma}]$. The nonconformity ratio r is computed as follows: $r = 1 - F(USL) + F(LSL)$, where erf is the error function. An estimator of r called $\hat{r}$ is a function of two random variables: $\hat{r} = f(\bar{X}, S)$, where $\bar{X} = \frac{1}{n}\sum\limits_{i=1}^{n} ln(X_i)$ and $S^2 = \frac{1}{n-1}\sum\limits_{i=1}^{n}(ln(X_i) - \bar{X})^2$. An approximation of $\hat{r}$ is provided using Taylor series as follows:

$$f(\bar{X}, S) \approx f(E(\bar{X}), E(S)) + (\bar{X} - E(\bar{X}))f'_{\bar{X}} + (S - E(S))f'_S +$$

$$\frac{1}{2}[(\bar{X} - E(\bar{X}))^2 f''_{\bar{X}} + 2(\bar{X} - E(\bar{X}))(S - E(S))f''_{\bar{X},S} + (S - E(S))^2 f''_S]$$

with $E(\bar{X}) = \mu$, $Var(\bar{X}) = \sigma_{\bar{X}}^2 = \frac{\sigma^2}{n}$, where n is the sample size. In the same way, $E(S^2) = \sigma^2$, $Var(S^2) = (\beta_2 - \frac{n-3}{n-1})\sigma^4$ even if the underlying distribution is non-normal. Using Taylor series $E(S) \approx \sigma$ and $Var(S) = \sigma_S^2 \approx \frac{\sigma^2}{4}(\beta_2 - \frac{n-3}{n-1})$, where β_2 is the distribution kurtosis.

In what follows, it is supposed that $\bar{X}$ and S are independent. Under this condition,

$$E(\hat{r}) \approx r + \frac{1}{2}[\sigma_{\bar{X}}^2 f_{\bar{X}}''(E(\bar{X}), E(S)) + \sigma_S^2 f_S''(E(\bar{X}), E(S))]$$

$$Var(\hat{r}) \approx (f_{\bar{X}}')^2 \sigma_{\bar{X}}^2 + (f_S')^2 \sigma_S^2$$

It becomes obvious that $\hat{r}$ is a biased estimator.

Under the assumption that $ln(X)$ is normally distributed then r_{min} depends on $\mu_{min} = \frac{ln(LSL)+ln(USL)}{2}$. If σ known then r_{min} is not a random variable. Otherwise,

$$\widehat{r_{min}} = 1 + \frac{1}{2}(erf[\frac{ln(LSL) - \mu_{min}}{\sqrt{2}S}] - erf[\frac{ln(USL) - \mu_{min}}{\sqrt{2}S}]).$$

In this case $\widehat{NCDU}$ is the ratio of two random variables. Using the delta method as explained by Kotz and Johnson (1993), it becomes possible to express the properties of $\widehat{NCDU}$ assuming that $\hat{r}$ and $\widehat{r_{min}}$ are independent.

5.4   Case Study

5.4.1   Process Definition

The object of the case study is a company which is specialized in the production of switches. It evolves several assembly chains that produce several switch versions. The investigated product is an equipment switch. This product will be assembled by the customer in an electrical drill.

The switch is formed by a superposition of two plastic plinthes. Before closing them, some electrical conductors and two screw supports are assembled into one of the plinthes, the other plinth is used to cover the assembled parts. At the end of the assembly process, the switches go through a machine which has to screw two screws, one on each support. Screwing will be made in such a way that a space will be kept between the inferior boundary of the screw and the inferior boundary of the screw support.

The customer will assemble this switch in an electrical drill by the penetration of two conductor cables in the cited spaces.

For lack of measuring directly this space, experiments are evaluated through the screw height. The purpose of the study is to assess the process capability when the specification limits for the screw heights are set to (LSL, T, USL) = (20.15, 20.85, 21.35 mm).

After manually assembling some electrical conductors, two main stages for process control are identified. The first stage is the superimposition of two plastic plinthes. The top plinth will be under pressure to make it go down. When it will reach the level of the inferior plinth the switch is then closed. Four machines which do the same operation are considered at this stage. At the second stage, only one machine is used. It controls some characteristics of the switches and it screws.

5.4.2   Process Capability Indices Computation

Notice that only the capability corresponding to one of the screws is assessed. In order to compute the process capability indices, a sample of $n = 300$ is obtained at the end of each stream. It is known that the classical process capability indices computation is based on the normal distribution assumption. It is interesting to check the normality assumption using the Shapiro–Wilk test. Table 5.1 gives the p-values for each stream. It is obvious that the underlying distribution is not a normal distribution. This is expected as no negative values are possible.

It is interesting to check the process distribution using a goodness of fit test. The chi-squared goodness of fit test is used. Following Moore (1986), the number of bins is given by $2n^{(2/5)}$. The test statistic follows a χ^2_{M-1} distribution, where M is the number of bins. When the tested distribution has p unknown parameters (Moore 1986) assumes that the correct critical points for the test fall between those of $\chi^2_{(M-p-1)}$ and those of $\chi^2_{(M-1)}$. When the value of the test statistic exceeds the critical point value the null hypothesis that the screw height follows the Lognormal distribution is rejected. In our case, 19 bins are used for the goodness of fit test. Because of the problem of rounding to the upper or lower digits some non-equiprobable bins are considered. Indeed, rounding can affect the goodness of fit test, for example, for the stream number 4 we had no observations in one of the bins although it is far from the distribution tails. This can increase the test statistic considerably. Table 5.2 gives the test statistic values and the number of non-equiprobable bins.

Table 5.1 Shapiro–Wilk test

Stream	1	2	3	4	5
p-value	8×10^{-6}	10^{-9}	2×10^{-9}	6×10^{-8}	10^{-10}

Table 5.2 χ^2 goodness of fit test

Stream	1	2	3	4	5
χ^2	24.9	21.73	27.06	26.16	15.88
# non-equiprobable bins	1	4	2	0	5

In order to give an argument to the number of degrees of freedom for the test it is interesting to present the estimation method of the lognormal distribution parameters. The considered distribution is a three parameter lognormal distribution. The parameters are θ, ξ, and σ, where $Z = log(\theta - X) \sim N(\xi, \sigma^2)$. Furthermore, the estimation of θ leads to the estimation of the other parameters of the distribution using maximum likelihood estimation. As explained in Johnson and Kotz (1994a, b), θ is estimated using the quantile method. This estimation is carried out after outliers detection and removal. Hence, the degrees of freedom of the χ^2 distribution is 17. With this degrees of freedom the lognormal distribution is not rejected at the significance level of 5%. In Table 5.3, the parameter θ, the expected value μ and the variance of the lognormal distribution are estimated.

With the parameters of the distribution already estimated, it becomes possible to compute the process capability indices. The considered indices are C_{pmk} introduced by Pearn et al. (1992), C_s introduced by Wright (1995), the weighted variance indices C_{pw} and C_{pkw} introduced by Choi and Bai (1996) and the index $NCDU$. The considered estimators are the following:

$$\hat{C}_{pmk} = \frac{d - |\hat{\mu} - M|}{3\sqrt{S^2 + (\hat{\mu} - T)^2}}$$

$$\hat{C}_s = \frac{min(USL - \hat{\mu}, \hat{\mu} - LSL)}{3\sqrt{\frac{1}{n}\sum_{i=1}^{n}(X_i - T)^2 + |\frac{n^2\hat{\mu}_3}{(n-1)(n-2)} / \sqrt{\frac{n}{n-1}\frac{\hat{\mu}_2}{c_4^2}}|}}$$

$$\hat{C}_{pw} = \frac{USL - LSL}{6S}min(\frac{1}{\sqrt{2\hat{P}_x}}, \frac{1}{\sqrt{2(1 - \hat{P}_x)}})$$

$$\hat{C}_{pkw} = min\{\frac{USL - \hat{\mu}}{3S\sqrt{2\hat{P}_x}}, \frac{\hat{\mu} - LSL}{3S\sqrt{2(1 - \hat{P}_x)}}\}$$

where $d = \frac{USL-LSL}{2}$, $\hat{\mu} = \bar{X}$, $\hat{\mu}_2 = \frac{\sum_{i=1}^{n}(X_i-\bar{X})^2}{n}$, $\hat{\mu}_3 = \frac{\sum_{i=1}^{n}(X_i-\bar{X})^3}{n}$, $c_4 \approx \frac{4(n-1)}{4n-3}$ and $\hat{P}_x$ is the probability that the process variable X is less or equal to $\hat{\mu}$. One of the most important features of the process capability indices is that their values increase when the

Table 5.3 Moments of the lognormal distribution

Stream	1	2	3	4	5
$\hat{\theta}$	21.34143	21.24752	21.17424	21.26072	21.19468
$\hat{\mu}$	0.412464	0.335991	0.241492	0.352547	0.273706
Variance	0.066584	0.010860	0.008529	0.048190	0.029864

Table 5.4 Process capability indices computation

Stream	1	2	3	4	5
$\hat{r}$(p.p.m.)	5.7470	24.820	21.060	0.1519	0.5622
$\hat{C}_{pmk}$	0.2180	0.2170	0.1820	0.2150	0.1910
$\hat{C}_s$	0.1770	0.2150	0.1810	0.1940	0.1850
$\hat{C}_{pkw}$	0.5720	1.4950	1.6300	2.2250	1.8660
$\hat{C}_{pw}$	0.7390	1.8130	2.0220	0.8720	1.0970
$\hat{r}^{min}$(p.p.m.)	1.0300	3.0510	2.2210	0.0047	0.0203
$\widehat{NCDU}$	0.9100	0.6120	0.6700	0.9970	0.9910

proportion of nonconforming items decreases. It will be interesting to check whether the considered indices have this feature in the presence of non-normal distributions. Table 5.4 gives the index values, the actual proportion of nonconforming items for each stream $\hat{r}$ and the minimum proportion of nonconforming items $\hat{r}_{min}$.

From Table 5.4, it becomes obvious that the index $NCDU$ fulfilled "the higher the better" rule requirement. Moreover, its computation takes into account the minimum nonconformity ratio and it has a clear threshold for capability judgment even when the underlying distributions are non-normal. Indeed, when the nonconformity ratio has a positive desirability value then the process is capable, otherwise, NCDU assigns a value of 0 to poor quality processes. This NCDU property does not depend on the quality characteristic distribution. However, it is important to notice that capability judgement in this case is based on a point estimation of the index, it would be more appropriate to base such judgement on a lower limit of a confidence interval.

5.5 Bootstrap Confidence Interval

Bootstrap confidence intervals are constructed for $NCDU$. For that purpose, the observations are gathered in a (300×5) matrix. New samples are obtained by choosing matrix lines with replacement, the number of new samples is equal to 7500. Then, the parameters of the distributions corresponding to each stream have to be estimated at each replication. In this case, observations greater than $\hat{\theta}$ are outliers and they are removed as for three- parameter negatively skewed lognormal distributions the probability to observe values greater than θ is equal to 0. It is expected that $\hat{\theta} > max(x_1, \ldots, x_{300})$ for all replicated samples. After removing the outliers, the distribution parameters are estimated using quantile method and maximum likelihood estimators. The estimation of the distribution parameters allows the nonconformity ratios estimation. Moreover, the location parameter which allows the minimum of the nonconformity ratio determination is calculated. A comparison between all minima of the nonconformity ratios is carried out and the minimum among all minima is used in the $\widehat{NCDU}$ computation for each stream. In order to construct the bootstrap confidence interval $\widehat{NCDU}$ is computed at each replication. In this case, 7500 replications are considered. The lower and upper confidence limits are the $\frac{\alpha}{2}$ and $1 - \frac{\alpha}{2}$ bootstrap distribution quantiles, respectively, in a way that the confidence interval has a $(1 - \alpha)\%$ coverage. In this case $\alpha = 5\%$. The estimated nonconformity ratio, the estimated minimum of the nonconformity ratio, and the estimated capability index show that all streams seem to be capable, however, lower limits of the confidence intervals indicate that no stream is capable. Screw support material should be inspected as other alloys could improve the capability.

5.6 The Multivariate Extension

When p quality characteristics are considered, the NCDU index is defined for each quality characteristic as follows:

$$
NCDU_i = \begin{cases} 0 & \text{if } r_i \geq USL', \\ \frac{USL' - r_i}{USL' - min_{j=1,\ldots,p}(r_j^{min})} & \text{if } r_i^{min} < r_i < USL', \\ \frac{USL' - r_i^{min}}{USL' - min_{j=1,\ldots,p}(r_j^{min})} & \text{if } r_i \leq r_i^{min}. \end{cases} \tag{5.6}
$$

where r_i is the actual nonconformity ratio for the quality characteristic i. r_i^{min} is the minimum of the nonconformity ratio for the quality characteristic i and $min_{j=1,\ldots,p}$ (r_j^{min}) is the minimum among all the minimums of the nonconformity ratios. In the multivariate case the actual nonconformity ratio for a quality characteristic is computed based on of the marginal probability density function.

A natural extension of the $NCDU_i$ to the multivariate case is given by the desirability index. The desirability index is a function of the univariate $NCDU_i$. It will be considered as a multivariate capability index. However, several types of the desirability index were proposed in the literature. It will be interesting to check which type is more appropriate for capability assessment.

Harrington (1965) proposed the geometric mean of the individual desirabilities as a desirability index. It is defined as

$$
D = [\Pi_{i=1}^p d_i]^{\frac{1}{p}}.
$$

In this way if one quality characteristic has a desirability equal to 0 than the overall desirability would be 0. Derringer (1994) proposed a weighted composite desirability which is given by

$$
D = [\Pi_{i=1}^p d_i^{w_i}]^{\frac{1}{\Sigma_{i=1}^p w_i}},
$$

where w_i corresponds to the importance of the quality characteristics i. The weights are determined by individual or group judgement. Kim (2000) proposed the minimum of the desirability values as an assessment for the overall desirability. It is given by

$$D = min_{i=1,\ldots,p}(d_i).$$

In what follows the geometric mean of $NCDU_i$ is considered as a multivariate process capability index. Indeed, this choice is motivated by the fact that the geometric mean of $NCDU_i$ could be written as a function of the joint nonconformity ratio of uncorrelated quality characteristics. When p uncorrelated quality characteristics are considered the joint nonconformity ratio is expressed as

$$R_p = 1 - [(1 - r_1)(1 - r_2)\ldots(1 - r_{p-1})(1 - r_p)].$$

Moreover, it is important to notice that if $r_i < USL'$ for $i = 1, 2, \ldots, p$ does not imply that $R < USL'$, where R is the joint nonconformity ratio computed using a joint probability function f. Hence, univariate capability does not imply multivariate capability. Then it is important to have a link between a multivariate capability index and the joint nonconformity ratio. For that purpose let $NCDM$ the geometric mean of univariate indices, the general expression of $NCDM$ is given by

$$NCDM^p = \frac{USL'^p + (-1)^p \Pi_{i=1}^p r_i}{(USL' - C)^p}$$

$$+ \frac{\sum_{i=1}^{p-1}(USL')^{(p-i)}(-1)^i[\sum_{j=1}^{p-i+1}\sum_{k=2}^{p-i+2}\cdots\sum_{m=l}^{p-i+l}\cdots\sum_{\substack{q=p-1 \\ j<k<\ldots<m<\ldots<q}}^{p} r_j^{1_{i,[1]}} r_k^{1_{i,[2]}}\ldots r_m^{1_{i,[l]}}\ldots r_q^{1_{i,[p-1]}}]}{(USL' - C)^p}.$$

p is the number of quality characteristics, r_i is the actual nonconformity ratio for the quality characteristic i, USL' is an upper limit for the actual nonconformity ratio, r_i^{min} is the minimum of the nonconformity ratio for the quality characteristic i and $C = min_{i=1,\ldots,p}(r_i^{min})$. 1_u is a $[(p-1) \times 1]$ vector and its elements are 0 and 1. Only the first u^{th} elements are 1. $1_{u,[l]}$ is the lth element of the vector and

$$1_{u,[l]} = \begin{cases} 0 & \text{if} \quad l > u, \\ 1 & \text{if} \quad l \leq u. \end{cases} \tag{5.7}$$

In the same way, the general expression of the joint nonconformity ratio for uncorrelated quality characteristics is given by

$$R_p = -(\sum_{i=1}^{p-1}(-1)^i[\sum_{j=1}^{p-i+1}\sum_{k=2}^{p-i+2}\cdots\sum_{m=l}^{p-i+l}\cdots\sum_{\substack{q=p-1 \\ j<k<\ldots<m<\ldots<q}}^{p} r_j^{1_{i,[1]}} r_k^{1_{i,[2]}}\ldots r_m^{1_{i,[l]}}$$

$$\ldots r_q^{1_{i,[p-1]}}] + (-1)^p \Pi_{i=1}^p r_i).$$

Hence, $NCDM$ is expressed as a function of the joint nonconformity ratio of uncorrelated quality characteristics as follows:

$$(USL' - C)^p NCDM^p = USL'^p + (-1)^p \Pi_{i=1}^p r_i$$

$$+ \sum_{i=1}^{p-1} USL'^{(p-i)} (-(\sum_{\substack{u=1 \\ u \neq i}}^{p-1} (-1)^u [\sum_{j=1}^{p-u+1} \sum_{k=2}^{p-u+2} \cdots \sum_{m=l}^{p-u+l} \cdots \sum_{\substack{q=p-1 \\ j<k<\ldots<m<\ldots<q}}^{p}$$

$$r_j^{\mathbb{1}_{u,[1]}} r_k^{\mathbb{1}_{u,[2]}} \cdots r_m^{\mathbb{1}_{u,[l]}} \cdots r_q^{\mathbb{1}_{u,[p-1]}}]) - (-1)^p \Pi_{i=1}^p r_i - R_p).$$

5.7 Conclusion

A capability index is presented. The index value follows "the higher the better" rule even for non-normal distributions. Although the index estimation depends upon the parameters estimation of the underlying distribution, it seems that the index distribution and its statistical properties are still tractable even for non-normal distributions. The case study confirmed that the index estimator has the same advantage as the index structure, this is not the case for classical indices. However, much care should be taken when using bootstrap lower limit confidence interval for capability judgement as the index estimator seems to be biased. An additional advantage of this index is that its multivariate extension is straightforward, indeed an overall capability assessment could be done using the desirability index.

References

Chan, L.K., Cheng, S.W., & Spiring, F.A. (1988). A new measure of process capability: C_{pm}, Journal of Quality Technology, 20(3), 162–173.

Choi, I.S., & Bai, D.S. (1996). Process capability indices for skewed populations. In Proceedings of 20th International Conference on Computer and Industrial Engineering (pp. 1211–1214).

Derringer, G.C. (1994). A balancing act: Optimizing a product's properties. Quality Progress, 51–58.

Derringer, G. C., & Suich, D. (1980). Simultaneous optimization of several response variables. Journal of Quality Technology, 12(4), 214–219.

Harrington, E. (1965). The desirability function. Industrial Quality Control, 494–498.

Hsiang, T. C., & Taguchi, G. (1985). A tutorial on quality control and assurance- The Taguchi methods, ASA annual meeting Nevada: Las Vegas.

Jessenberger, J. (1999). Prozesshäufigkeitsindizes in der Qualitätssicherung, Libri Books on Demand.

Johnson, N.L., Kotz, S., Balakrishnan, N. (1994a). Continuous Univariate Distribution, Wiley Series In Probability And Mathematical Statistics: New York.

Johnson, N. L., Kotz, S., & Pearn, W. L. (1994b). Flexible process capability indices. Pakistan Journal of Statistics, 10(1), 23–31.

Kim, K. J., & Lin, D. K. J. (2000). Simultaneous optimization of mechanical properties of steel by maximizing exponential desirability functions. Applied Statistics, 49(3), 311–325.

Kotz, S., Johnson, N.L. (1993). Process capability indices, Chapman and Hall: London.

Kotz, S., Lovelace, C.R. (1998). Process capability indices in theory and practice, Arnold: London.

Moore, D.S. (1986). Tests of chi-squared type. In R.B. D'Agostino, M.A. Stephens (Eds.), Goodness of fit techniques, Marcel Dekker, Inc: New York.

Pearn, W. L., Kotz, S., & Johnson, N. L. (1992). Distributional and inferential properties of process capability indices. *Journal of Quality Technology, 24*(4), 216–231.

Wang, F. K., Hubele, N. F., Lawrence, F. P., Miskulin, J. D., & Shahriari, H. (2000). Comparison of three multivariate process capability indices. *Journal of Quality Technology, 32*(3).

Wright, P. A. (1995). A process capability index sensitive to skewness. *Journal of Statistical Computation and Simulation, 52*, 195–203.

Part II
Computational Statistics

Chapter 6
Heteroscedastic Discriminant Analysis Using R

Gero Szepannek and Uwe Ligges

Abstract For purposes of dimensionality reduction in classification Linear Discriminant Analysis (LDA) is probably the most common approach. In fact, LDA is a linear dimension reduction technique that also returns a classification rule. In the case of heteroscedasticity of the classes, Quadratic Discriminant Analysis (QDA) can be used to determine an appropriate classification rule, but QDA does not serve for dimensionality reduction. Sliced Average Variance Estimation (SAVE) has been shown to be adequate in such situations as implemented in R in the package dr. This paper presents an alternative approach for linear dimensionality reduction for situations of heteroscedastic intraclass covariances, namely Heteroscedastic Discriminant Analysis (HDA) as well as its R implementation. Furthermore, tests are suggested in order to determine the dimension for the discriminative data subspace and a generalization of HDA by regularization of the covariance matrix estimates is proposed. Examples for application of HDA in R are demonstrated as well as a small simulation study turning out that HDA is preferable to SAVE in a situation where the classes differ in both means and covariances.

6.1 Dimensionality Reduction in Classification

For the analysis of high dimensional data, it is often of interest to represent it in some lower dimensional subspace keeping as much *relevant* information

G. Szepannek (✉)
Stralsund University of Applied Sciences, Stralsund, Germany
e-mail: gero.szepannek@hochschule-stralsund.de

U. Ligges
Department of Statistics, TU Dortmund University, Dortmund, Germany
e-mail: ligges@statistik.tu-dortmund.de

© Springer Nature Switzerland AG 2019   77
N. Bauer et al. (eds.), *Applications in Statistical Computing*,
Studies in Classification, Data Analysis, and Knowledge Organization,
https://doi.org/10.1007/978-3-030-25147-5_6

```
library(mvtnorm)
set.seed(42)
n <- 100   # number of observations per class
# define expectation and covariance of both classes
meana <- meanb <- c(0, 0, 0, 0, 0)
cova <- diag(5)
cova[1, 1] <- 0.2
for (i in 3:4) {
    for (j in (i + 1):5) {
            cova[i, j] <- cova[j, i] <- 0.75^(j - i)
    }}
covb <- cova
diag(covb)[1:2] <- c(1, 0.2)

# simulate observations for both classes
xa <- mvnorm(n, meana, cova)
xb <- mvnorm(n, meanb, covb)
x <- rbind(xa, xb)
classes <- as.factor(c(rep(1, n), rep(2, n)))
# class memberships
# rotate data by eigenvalues of a random matrix
symmat <- matrix(runif(5^2), 5)
symmat <- symmat + t(symmat) # ensures matrix to be symmetric
even <- eigen(symmat)$vectors
rotatedspace <- x %*% even
```

Listing 1 Code for data generation.

as possible. This allows for an easier visualization of the data or simplyfies application of some subsequent statistical method. It is common practice to apply a preliminary variable selection method (for an overview see e.g. Guyon and Elysseeff 2003). A wrapper algorithm for variable subset selection that is compatible to several classification algorithms (`stepclass`) can be found in the R package `klaR` (Weihs et al. 2005; Roever et al. 2018). The resulting subspace after variable selection consists of a subset of the original variables and is thus easy to interpret. The selected variables are not uncorrelated in general. If it is desired to represent the data in a subspace of *uncorrelated* features, linear transforms of the type $\mathbf{y} = \mathbf{A}'\mathbf{x}$ can be used. Probably the most popular linear dimension reduction technique is Principal Component Analysis (PCA, see e.g. Mardia et al. 1979). Applying PCA results in uncorrelated components w.r.t. all data (i.e. some common mean $\bar{x}$). But if the data consists of several classes (i.e., each observation x_n corresponds to some class $c_n \in \{1, \ldots, K\}$) where the conditional distributions of the data given the classes are different, the resulting subspaces will not necessarily be uncorrelated within the classes as soon as the class means are different. Linear Discriminant Analysis (LDA, Fisher 1936) maximizes the average distance of the class means in the transformed space assuming equal covariances in the classes. An estimate of the intraclass covariance matrix is the *pooled covariance matrix*:

$$\mathbf{W} = \frac{1}{N} \sum_{n=1}^{N} (\mathbf{x_n} - \bar{\mathbf{x}}_{\mathbf{c_n}})(\mathbf{x_n} - \bar{\mathbf{x}}_{\mathbf{c_n}})', \tag{6.1}$$

where $\bar{\mathbf{x}}_k$ is the mean of class k and $N_k : \sum_k N_k = N$ are the class frequencies. The resulting subspace is given by the projection $\mathbf{X}'\mathbf{A}$ where $\mathbf{A} \sim \mathbf{p} \times \mathbf{q}$ contains the first $q \leq (K-1)$ eigenvectors with nonzero eigenvalues of the matrix $\mathbf{W}^{-1}\mathbf{B}$, restricted to e.g., $||\mathbf{A}|| = \mathbf{1}$.

$$\mathbf{B} = \frac{1}{N} \sum_{k=1}^{K} N_k (\bar{\mathbf{x}}_k - \bar{\mathbf{x}})(\bar{\mathbf{x}}_k - \bar{\mathbf{x}})' \tag{6.2}$$

is called the *between class* scatter matrix (cf. e.g. Mardia et al. 1979). Consider an example of five dimensional data where the means of two classes are the same but their covariance matrices are different: Data are simulated where the distributions of both classes have different covariances in the first two (uncorrelated) latent variables but are equal in three remaining (correlated) variables. In a second step, these latent variables are randomly rotated in space. The data are generated using R code given in Listing 1. Figure 6.1 shows the simulated example data.

For this example, applying `stepclass` variable selection (using a quadratic discriminant analysis classifier) leads to the results shown in Fig. 6.2 (top left). For the purpose of dimensionality reduction we focus on the resulting first two selected variables ('components'). The selected features are neither uncorrelated nor do they provide a satisfying discrimination of the classes.

```
# Wrapper Variable Selection
library(klaR)
stepcl.res <- stepclass(rotatedspace, classes, method = "qda",
                        improvement = 0, direction = "forward")
# Principal Component Analysis
pca.res <- prcomp(rotatedspace, retx = TRUE)

# Linear Discriminant Analysis
library(MASS)
lda.res <- lda(rotatedspace, classes)

# Sliced Average Variance Estimation
library(dr)
dr.res <- dr(classes ~ ., data.frame(rotatedspace, classes),
            method = "save", numdir = 2)
```
Listing 2 Dimensionality reduction using different models.

The first two principal components neither discriminate the two classes as the discriminative information does not necessarily load on the first principal components (see Fig. 6.2, top right). In case of two classes Linear Discriminant Analysis projects the data onto one single discriminant axis. By the construction of the example (equal means of both classes), the linear discriminant axis does not separate the classes (see Fig. 6.2, bottom left).

The problem in case of heteroscedasticity for dimensionality reduction in classification tasks has been investigated by several authors, (e.g. Young et al. 1987; Schott 1993). Sliced Averaged Variance Estimation (SAVE, Cook and Weisberg 1991),

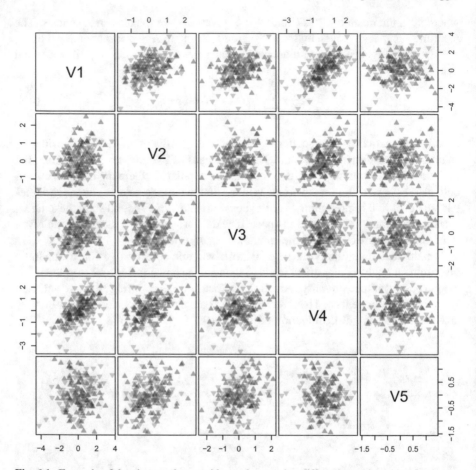

Fig. 6.1 Example of data in two classes with equal means but different covariance matrices

based on the idea of Sliced Inverse Regression (SIR, Li 1991), has been shown to be an appropriate method for dimensionality reduction if the data consists of (normally distributed) classes with different covariance structures (Pardoe et al. 2006). Briefly, the SAVE transformation is obtained after scaling and centering of the data by eigenvalue decomposition of the matrix

$$M = \sum_{k=1}^{K} \frac{N_k}{N} (I - W_k)^2,$$ (6.3)

with class-specific covariance estimations

$$W_k = \frac{1}{N_k} \sum_{n=1}^{N} I_{\{k\}}(c_n)(x_n - \bar{x}_{c_n})(x_n - \bar{x}_{c_n})'.$$ (6.4)

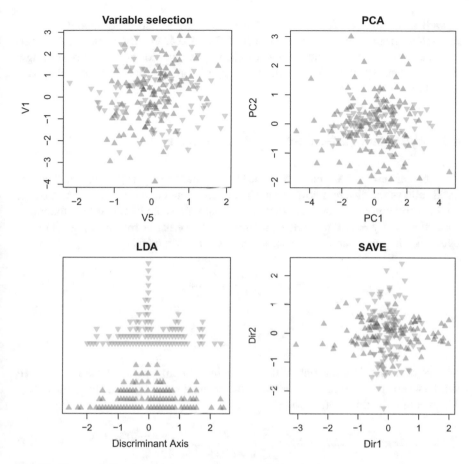

Fig. 6.2 First two components using `stepclass` variable selection (top, left), PCA (top, right), LDA (bottom, left) and SAVE (bottom right) dimensionality reduction on the example data

The variates of the reduced subspace are given by rotation of the standardized data matrix with the eigenvectors corresponding to the largest eigenvalues of **M**. A software implementation of SAVE is provided by the R package `dr` (Weisberg 2002). Figure 6.2 (bottom, right) shows the resulting transformation using SAVE for the example. Discrimination of the two classes is obtained regarding their different variances in the two-dimensional subspace.

Note that the R package `fpc` (Hennig 2018) provides an *Asymmetric Linear Dimension Reduction for Classification* (Hennig 2004), which also allows for different covariances for two classes but it is rather designed to identify a subspace where one of the classes appears to be strongly homogeneous.

In the rest of the paper, an alternative approach for linear dimension reduction is presented that is also suitable for the formerly presented example: In Sect. 6.2 the method of Heteroscedastic Discriminant Analysis (HDA, Kumar and Andreou 1998)

as well as its implementation in R are presented. Further, aspects of model selection and regularization are addressed in Sects. 6.3 and 6.4. Finally, in Sect. 6.5 a real-world example is presented together with a small simulation study demonstrating a data situation, where HDA turns out to be even more appropriate than both SAVE and LDA.

6.2 Heteroscedastic Discriminant Analysis

As an extension of LDA, Heteroscedastic Discriminant Analysis (Kumar and Andreou 1998) allows for different covariance structures of the class distributions: In HDA, a transformation matrix $\mathbf{A}$ is computed that maximizes the likelihood (in the transformed space) $\mathbf{y} = \mathbf{A}'\mathbf{x}$ under the assumption of normality for all classes. For the class-specific means and covariances, it is further assumed that

$$\mu_{\mathbf{k}} = \begin{bmatrix} \mu_{\mathbf{k}}^{\mathbf{q}} \\ \mu_{\mathbf{0}}^{\mathbf{p-q}} \end{bmatrix} \text{ and}$$

$$\Sigma_{\mathbf{k}} = \begin{pmatrix} \Sigma_{\mathbf{k}}^{\mathbf{q}} & 0 \\ 0 & \Sigma_{\mathbf{0}}^{\mathbf{p-q}} \end{pmatrix},$$

i.e., the class distributions only differ in the first q components of the transformed space (where $p > q$ is the dimension of the original space and k denotes the class). The resulting likelihood function is

$$L(\mu_1, \ldots, \mu_K, \Sigma_1, \ldots, \Sigma_K, \mathbf{A}, \mathbf{X}) = \prod_{n=1}^{N} \frac{|\mathbf{A}|}{\sqrt{(2\pi)^p |\Sigma_{\mathbf{c_n}}|}} e^{\left(-\frac{1}{2}(\mathbf{A}'\mathbf{x_n} - \mu_{\mathbf{c_n}})' \Sigma_{\mathbf{c_n}}^{-1}(\mathbf{A}'\mathbf{x_n} - \mu_{\mathbf{c_n}})\right)}. \quad (6.5)$$

For reasons of interpretability and easier application of subsequent statistical methods, it is common practice to require diagonality in the transformed space, i.e., $\Sigma_{\mathbf{k}} = \mathrm{diag}((\sigma_k^1)^2, \ldots, (\sigma_k^q)^2, (\sigma_0^{q+1})^2, \ldots, (\sigma_0^p)^2)$. The log-likelihood becomes

$$l(\mu_1, \ldots, \mu_K, \Sigma_1, \ldots, \Sigma_K, \mathbf{A}, \mathbf{X}) = l(\mu_1, \ldots, \mu_K, \sigma_1, \ldots, \sigma_K, \mathbf{A}, \mathbf{X})$$

$$= \frac{-Np}{2} \log 2\pi + N \log |A|$$

$$- \frac{N}{2} \sum_{d=q+1}^{p} \log \sigma_0^d - \sum_{k=1}^{K} \frac{N_k}{2} \sum_{d=1}^{q} \log \sigma_k^d$$

$$- \frac{1}{2} \sum_{k=1}^{K} \sum_{c_n=k} \sum_{d=1}^{q} \frac{(\mathbf{a_d'}\mathbf{x_n} - \mu_{\mathbf{k}}^{\mathbf{d}})^2}{\sigma_k^d}$$

$$- \frac{1}{2} \sum_{k=1}^{K} \sum_{c_n=k} \sum_{d=q+1}^{p} \frac{(\mathbf{a_d'}\mathbf{x_n} - \mu_{\mathbf{0}}^{\mathbf{d}})^2}{\sigma_0^d}. \quad (6.6)$$

Setting the partial derivatives equal to 0 results in the following maximum likelihood estimates for the distribution parameters as a function of the transformation matrix **A** (see Kumar and Andreou 1998):

$$\hat{\mu}_k^q = (\mathbf{A^q})' \bar{\mathbf{x}}_k, \tag{6.7}$$

$$\hat{\mu}_0^{p-q} = (\mathbf{A^{p-q}})'(\bar{\mathbf{x}}), \tag{6.8}$$

$$\hat{\Sigma}_k^q = \text{diag}((\mathbf{A^q})'\mathbf{W_k}(\mathbf{A^q})), \tag{6.9}$$

$$\hat{\Sigma}_0^{p-q} = \text{diag}((\mathbf{A^{p-q}})'(\mathbf{T})(\mathbf{A^{p-q}})), \tag{6.10}$$

where $\mathbf{A} = (\mathbf{A^q A^{p-q}})$ and $\mathbf{T} = \mathbf{W} + \mathbf{B}$. Note that for $\hat{\mu}_0^{p-q}$ and $\hat{\Sigma}_0^{p-q}$ the class-specific parameter estimates $\bar{\mathbf{x}}_k$ and $\mathbf{W_k}$ in the original data space are not used.

The ML parameter estimates can be used to build plug-in estimates

$$l(\mathbf{A}, \mathbf{x}) = -\frac{N}{2} \log |\text{diag}((\mathbf{A^{p-q}})'(\mathbf{T})(\mathbf{A^{p-q}}))| - \sum_{k=1}^{K} \frac{N_k}{2} \log |\text{diag}((\mathbf{A^q})'\mathbf{W_k}(\mathbf{A^q}))|$$

$$- \frac{Np}{2}(1 + \log(2\pi)) + N \log |\mathbf{A}| \tag{6.11}$$

as a function of **A**. Thus, computing the Heteroscedastic Discriminant Analysis transformation matrix finally ends up in

$$\mathbf{A_{HDA}} = \arg\max_{\mathbf{A}} \left(-\frac{N}{2} \log \left(|\text{diag}\left((\mathbf{A^{p-q}})'(\mathbf{W} + \mathbf{B})(\mathbf{A^{p-q}}) \right)| \right) \right.$$

$$\left. - \sum_{k=1}^{K} \frac{N_k}{2} \log \left(|\text{diag}\left((\mathbf{A_k^q})'(\mathbf{W_k})(\mathbf{A_k^q}) \right)| \right) + N \log |\mathbf{A}| \right). \tag{6.12}$$

Nonsingular transformations of $\mathbf{A^q}$ or $\mathbf{A^{p-q}}$ or linear scaling does not affect the likelihood. Thus, there is no unique solution that maximizes Eq. (6.11). An iterative optimization algorithm is presented in Burget (2004). It can be further shown that the LDA solution is one solution to Eq. (6.12) if in addition the class-specific covariance matrices Σ_k are assumed to be equal (Kumar 1997). For this reason, HDA can be considered as a generalization of Linear Discriminant Analysis. Note that the idea of HDA only implicitly consists in discrimination of the classes but primarily in identification of a subspace that *optimally fits the observed data* assuming that the class distributions are normal and do differ in a subspace of only $q < p$ dimensions of the original data.

If the aim is to build a classification model after dimensionality reduction a Naive Bayes classifier (see e.g., Hastie et al. 2009) is implicitly given by the distribution parameters (μ_k, σ_k^2) in the transformed subspace and corresponds to the appropriate classification rule if the assumptions are met.

The functionality of Heteroscedastic Discriminant Analysis for practical use is provided by the R package hda (Szepannek 2018). Optimization of the likelihood is implemented as it is proposed in Burget (2004). There are two different hda()

methods, the default method and a formula one. Applying hda() to the example from Sect. 6.1 could be done as follows:

```
library(hda)
hda.res <- hda(x = rotatedspace, grouping = classes, newdim = 2,
               initial.loadings = NULL)
```

Listing 3 HDA dimensionality reduction (for $q = 2$) on the example data.

Here x is a matrix or data frame containing the (numeric) explanatory variables. grouping is a factor specifying the class for each observation. Alternatively, formula notation using formula and data can be used (just as in lm()). As an important parameter, newdim specifies the dimension q of the discriminative subspace. By initial.loadings, an initial guess of the matrix of loadings **A** different from the identity (used as default) can be defined.

The output of the call is an object of class hda. Its element $hda.loadings contains the transformation matrix **A** to be post multiplied to the original data. $hda.scores are the transformed input data. The reduced discriminative space is built by its first $reduced.dimension variables. The elements $grouping and class.dist store the original classes of the input data and the estimated class distribution parameters (means and covariance matrices) in the transformed space. As opposed to PCA and LDA, there is no implicit order among the q components. The element $comp.acc contains a matrix of accuracies (i.e., probability of correct a classification of a class according to the estimated distribution) for each component. Furthermore, in order to quantify the impact of single variables $vlift provides the variable importance in terms of ratios to the original accuracies if the loading of a variable would be 0.

A plot method allows to visualize the data in the transformed space: by a second argument one can specify a vector of variable indices to be displayed. The result is shown in Fig. 6.3.

```
plot(hda.res, 1:5, col = classes, pch = (24:25)[classes])
```

Listing 4 First two components using HDA.

It can be seen that HDA transform identifies a subspace of dimension $q = 2$ where both classes show different distributions with respect to their covariance matrices just as in construction of the examples. The difference in the distributions of both classes can be visualized within the first two components of the transform. Of course, permutation of both components would not change the likelihood, i.e., there is no order of the components returned by hda transform.

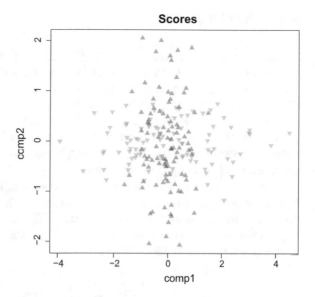

Fig. 6.3 HDA components of example of data

A corresponding predict function can be easily applied to transform new data. Note that apart from dimensionality reduction also classification may be desired. As outlined earlier the class distribution parameters in the new space correspond to Naive Bayes classification models. If specified by the parameter `crule = TRUE` of the `hda()` call a model of class `naiveBayes` Meyer et al. (2019) is returned directly in the `hda` output object for further use (on transformed data). A subsequent Naive Bayes classification based on the identified subspace can be performed easily using the option `task = "c"` (instead of `task = "dr"` for dimensionality reduction). Internally, the `naiveBayes` predict function is called to classify new observations, as shown in the following example:

```
# call hda that also builds a Naive Bayes model
library(e1071)
hda.res <- hda(rotatedspace, classes, 2, crule = TRUE)

# transform new (here original input-)data
newdata <- predict(hda.res, newdata = rotatedspace)

# predict classes using Naive Bayes
class.predictions <- predict(hda.res$naivebayes, newdata = newdata,
                    task = "c")
```

Listing 5 Naive Bayes classification based on a HDA model.

6.3   Model Selection

The result of SAVE dimensionality reduction does not depend on any further param-
eterization (cf. Sect. 6.1). In heteroscedastic discriminant analysis, the result depends
on the choice of q (as prespecified by the parameter newdim)—an assumption on the
dimension of the discriminative subspace: the number of latent components where the
class distributions are different. For the remaining $(p - q)$- dimensional subspace,
the classes' distributions are assumed to be equal. As also normality is assumed for
optimizing the likelihood this is equivalent to equal expectations $\mu_{\mathbf{k}}^{\mathbf{p-q}} = \mu_{\mathbf{0}}^{\mathbf{p-q}}$ and
covariances $\Sigma_{\mathbf{k}}^{\mathbf{p-q}} = \Sigma_{\mathbf{0}}^{\mathbf{p-q}}$ for all classes $k = 1, \ldots, K$.

In order to get an idea of an appropriate choice of q, one can use tests on the
hypotheses of equal means and covariances (see e.g., Mardia et al. 1979). A test on
equal class means (assuming equal covariances $\Sigma_{\mathbf{0}}^{\mathbf{p-q}}$, $\forall \mathbf{k}$) corresponds to performing Multivariate Analysis of Variance (MANOVA). Under H_0 of equal expectations,
Wilk's Λ test statistic is distributed as

$$\Lambda = \frac{|\hat{\Sigma}_0^{p-q}|}{|\hat{\Sigma}_T^{p-q} - \hat{\Sigma}_0^{p-q}|} \sim \Lambda_{p-q, N-K, K-1} \tag{6.13}$$

where $\hat{\Sigma}_0^{p-q}$ and $\hat{\Sigma}_T^{p-q}$ are the pooled covariance estimate and the total covariance
in the nondiscriminative $p - q$ dimensions of the transformed space. In case of
$q = p - 1$, the test reduces to standard analysis of variance. Remind that for this test
equal covariances of the classes are assumed.

A test on equal covariance matrices is given by the test statistic

$$T = \sum_{k=1}^{K} \log |(\hat{\Sigma}_0^{p-q})^{-1} \hat{\Sigma}_k^{p-q}| \sim_{approx.} \chi^2_{(n-q)(n-q+1)(K-1)/2} \tag{6.14}$$

where $\hat{\Sigma}_k^{p-q}$ is the class-specific covariance estimate in the redundant components of
the transformed space as opposed to its pooled estimate. If the equality assumption
holds, the product of both matrices should be similar to identity and thus the logarithm
of its determinant close to 0.

```
hda.res <- hda(rotatedspace, classes, verbose = FALSE)
hda.res$trace.dimensions

                    1            2            3            4
eqmean.tests 9.294659e-01 0.9148176237 0.9958078997 0.9454175
homog.tests  4.241052e-14 0.0001085235 0.0001336032 0.8588074
```

Listing 6 Testing for the dimension of the non-discriminative subspace.

The elements $eqmean.test and $homog.test of an hda output object show the results of both tests. For an appropriate choice of the subspace dimension, q starting with some small value of q the dimension can be iteratively augmented until both tests do not refuse H_0 anymore at some prespecified α level. If no input parameter newdim is specified, the output element $trace.dimensions returns the p values of both tests as described above. As no vector newdim is specified here, starting with a subspace dimension of 1 the call of hda() is repeated while iteratively augmenting newdim by 1. If no unique dimension of the reduced discriminative space is specified, alternatively by the argument sig.levs = c(0.05, 0.05) of the hda call a vector of significance levels for both tests can be specified. But note that this procedure should not replace manual investigation of the appropriate choice of q. Further note that during model selection both tests are computed simultaneously and several times. Therefore, their results should not be interpreted strictly in terms of statistical inference in the context of the subspace dimension selection. In case of any doubt about a good choice of newdim, the best way is to manually compute the results for several different values and compare the results.

6.4 Regularization

For HDA, as a by-product of the extended model flexibility the number of free model parameters increases in contrast to LDA due to the class-specific covariance estimates in the original data space. As a consequence, finding the transformation becomes less stable. For this reason, in Burget (2004) stabilization is proposed by smoothing the covariance estimate (in the original space):

$$\mathbf{W_k^{smoothed}}(\lambda) = \lambda \mathbf{W_k} + (1 - \lambda)\mathbf{W}$$

where $\mathbf{W}$ is the common pooled equal covariance estimate for all classes and $\lambda \in [0, 1]$. For classification tasks an alternative regularization

$$\mathbf{W_k^{Fried}}(\lambda) = \frac{(\lambda)N_k\mathbf{W_k} + (1 - \lambda)N\,\mathbf{W}}{(\lambda)N_k + (1 - \lambda)N}$$

has been introduced that takes into account that covariance estimates of smaller class sizes N_k are less stable (Friedman 1989). Furthermore, there has been proposed an additional shrinkage of $\mathbf{W_k^{Fried}}$ towards diagonality:

$$\mathbf{W_k^{RDA}}(\lambda, \gamma) = \gamma \mathbf{W_k^{Fried}}(\lambda) + (1 - \gamma)\frac{tr(\mathbf{W_k^{Fried}}(\lambda))\mathbf{I}}{p}$$

where $\mathbf{I}$ is the identity and $(\lambda, \gamma) \in [0, 1]^2$. This idea goes back to Di Pillo (1976). A *Regularized Heteroscedastic Discriminant Analysis* can be obtained by computing HDA with covariance estimates $\mathbf{W_k^{RDA}}(\lambda, \gamma)$ in the original space. Note that

$(\lambda, \gamma) = (1, 1)$ represents standard HDA. Regularization of HDA has shown to be beneficial for dimensionality reduction in a speech recognition task (where traditionally strongly unbalanced class sizes do occur) with post processing using Hidden Markov models with diagonal covariance matrices (Szepannek et al. 2009). By the arguments reg.lamb and reg.gamm, HDA-input parameters $\lambda, \gamma \in [0, 1]$ for regularization of the covariance matrix estimates in the original space can be specified (default is no shrinkage).

6.5 Examples

Example 1: As a second example let us consider the Sonar dataset of the UCI benchmark repository (Dheeru and Karra Taniskidou 2017) as provided by the R package mlbench (Leisch and Dimitriadou 2012). The data consists of 214 observations (from 2 classes: metals and rocks) and 60 numeric variables. Each variable represents energy of a sonar signal within a specific frequency band.

```
library(mlbench)
data(Sonar)
lda.res2 <- lda(Class ~ ., data = Sonar)
hda.res2 <- hda(Class ~ ., data = Sonar, newdim = 2, verbose = FALSE)
```

Listing 7 Applying LDA and HDA to sonar data.

Figure 6.4 shows the result of both LDA (left) and HDA (here for exemplary reasons applied for a subspace dimension of $q = 2$, right) on the data. Apparently, LDA's discriminant axe separates the two classes quite well. Nevertheless, allowing for heteroscedasticity using HDA identifies several components to be strongly concentrated, i.e., with very similar realizations for all metals. For reasons of interpretation, one may be interested in having a closer look at the HDA loadings. This can be done using the function showloadings(): Based on the resulting Fig. 6.5 (left) one can get further insights into which variables (i.e. here: which frequency bands) show a characteristic energy for metal cylinders.

Example 2: Pardoe et al. (2006) mention a tendency of SAVE dimensionality reduction to overemphasize intraclass covariance differences in the first components. On the other hand, location differences between the classes are sometimes downplayed. Let us finally consider an example where both are present: differences in means as well as covariances of the classes. Misclassification rates of classifiers built on subspaces are compared in order to evaluate the discriminative power of the projections resulting from LDA, SAVE, and HDA. Again, the example from Sect. 6.1 is used but now location differences are added: the expectation μ_i of class $i \in \{1, 2\}$ is the origin except from its ith coordinate which is shifted to be $1/3$. Training data are generated with 200 observations in each class to train dimensionality reduction and

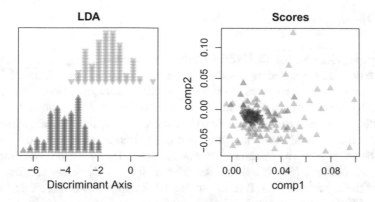

Fig. 6.4 Linear discriminant (left) and HDA components (for $q = 2$, right) of sonar data

subsequent classification models on the first two components rendered by each of the three methods. Separate test data are generated using the same rotation matrix as for the training data. The classification and dimension reduction models built on the training data are used for predicting the test data and the resulting prediction errors are computed. The simulation is repeated 50 times. LDA classification rules are obtained using lda(). For HDA, Naive Bayes classifiers are built on two- dimensional subspaces as described in Sect. 6.2. For SAVE, the appropriate classifiers are constructed in the two-dimensional subspace according to Pardoe et al. (2006) using minimal Mahalanobis distance rule in the transformed space. Naive Bayes classifiers have also been computed on the first two SAVE components leading to similar results.

Figure 6.5 (right) shows the results: Comparing all three methods illustrates the benefits of allowing for different intraclass covariance matrices as SAVE and HDA subspaces show better discrimination than LDA, here. It can be further seen that HDA most efficiently uses both types of information, location differences as well as scale differences of the classes in this case.

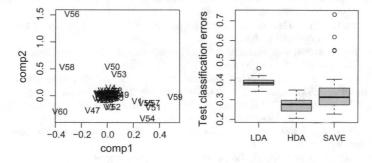

Fig. 6.5 HDA loadings on Sonar data (left) and simulation results (right)

6.6 Summary

In this paper, Heteroscedastic Discriminant Analysis and its R implementation in the package hda are presented as an alternative approach to traditional Linear Discriminant Analysis and sliced average variance estimation for dimensionality reduction for data that consist of several classes. Tests are introduced to help choosing an appropriate dimension for the discriminative data subspace. In addition, an extension of HDA in terms of regularization is presented taking into account that the number of free parameters grows at the same time when allowing for heteroscedastic intraclass covariance matrices. Finally, several examples are presented in order to illustrate the usage of the R package hda. It turns out that HDA appears to be preferable to both LDA and SAVE in an example where both: inter- class location shifts as well as class-specific intra-class covariance matrices are present in the data.

References

Burget, L. (2004). Combination of speech features using smoothed heteroscedastic linear discriminant analysis. In *Proceedings of Interspeech 2004, Jeju/Korea* (pp. 2549–2552).

Cook, R. D., & Weisberg, S. (1991). Comment on Li. *Journal of the American Statistical Association*, *86*, 328–332.

Dheeru, D., & Karra Taniskidou, E. (2017). UCI machine learning repository. http://archive.ics.uci.edu/ml.

Di Pillo, P. (1976). The application of bias to discriminant analysis. *Communications in Statistics - Theory and Methods*, *5*(9), 843–854.

Fisher, R. A. (1936). The use of multiple measures in taxonomic problems. *Annals of Eugenics*, *7*, 179–188.

Friedman, J. (1989). Regularized discriminant analysis. *Journal of the American Statistical Association*, *84*, 165–175.

Guyon, I., & Elysseeff, A. (2003). An introduction to variable selection and feature selection. *Journal of Machine Learning Research*, *3*, 1157–1182.

Hastie, T., Tibshirani, R., & Friedman, J. (2009). *The elements of statistical learning*. New York: Springer.

Hennig, C. (2004). Asymmetric linear dimension reduction for classification. *Journal of Computational and Graphical Statistics*, *13*, 930–945.

Hennig, C. (2018). fpc: Flexible procedures for clustering. R package version 2.1-11.1. http://CRAN.R-project.org/package=fpc.

Kumar, N. (1997). Investigation of silicon auditory models and generalization of linear discriminant analysis for improved speech recognition.

Kumar, N., & Andreou, A. (1998). Heteroscedastic discriminant analysis and reduced rank hmms for improved speech recognition. *Speech Communication*, *25*(4), 283–297.

Leisch, F., & Dimitriadou, E. (2012). mlbench: Machine learning benchmark problems. R package version 2.1-1. http://CRAN.R-project.org/package=mlbench.

Li, K. C. (1991). Sliced inverse regression for dimension reduction. *Journal of the American Statistical Association*, *86*, 316–327.

Mardia, K., Kent, J., & Bibby, J. (1979). *Multivariate analysis*. Academic.

Meyer, D., Dimitriadou, E., Hornik, K., Weingessel, A., Leisch, F., Chang, C.C., et al. (2019). e1071: Misc functions of the department of statistics (e1071). TU Wien, r package version 1.7-0.1. http://CRAN.R-project.org/package=e1071.

Pardoe, I., Yin, X., & Cook, R. D. (2006). Graphical tools for quadratic discriminant analysis. *Technometrics, 49*(2), 172–183.

Roever, C., Raabe, N., Luebke, K., Ligges, U., Szepannek, G., & Zentgraf, M. (2018). klaR: Classification and visualization. R package version 0.6-14. http://CRAN.R-project.org/package=klaR.

Schott, J. (1993). Dimension reduction in quadratic discriminant analysi. *Computational Statistics and Data Analysis, 16*, 161–174.

Szepannek, G. (2018). hda: Heteroscedastic discriminant analysis. R package version 0.2-14. http://CRAN.R-project.org/package=klaR.

Szepannek, G., Harczos, T., Klefenz, F., & Weihs, C. (2009). Extending features for automatic speech recognition by means of auditory modelling. In *Proceeding of European Speech and Signal Processing Conference (EUSIPCO), Glasgow* (pp. 1235–1239).

Weihs, C., Ligges, U., Luebke, K., & Raabe, N. (2005). klaR—*analyzing german business cycles* (pp. 225–343). Berlin: Springer.

Weisberg, S. (2002). Dimensionality reduction regression in R. *Journal of Statistical Software, 7*(1), 1–22.

Young, D., Marco, V., & Odell, P. (1987). Quadratic discrimination: Some results on optimal low-dimensional representation. *Journal of Statistical Planning and Inference, 17*, 307–319.

Chapter 7
Comprehensive Feature-Based Landscape Analysis of Continuous and Constrained Optimization Problems Using the R-Package Flacco

Pascal Kerschke and Heike Trautmann

Abstract Choosing the best-performing optimizer(s) out of a portfolio of optimization algorithms is usually a difficult and complex task. It gets even worse, if the underlying functions are unknown, i.e., so-called *black-box problems*, and function evaluations are considered to be expensive. In case of continuous single-objective optimization problems, *exploratory landscape analysis (ELA)*, a sophisticated and effective approach for characterizing the landscapes of such problems by means of numerical values before actually performing the optimization task itself, is advantageous. Unfortunately, until now it has been quite complicated to compute multiple ELA features simultaneously, as the corresponding code has been—if at all—spread across multiple platforms or at least across several packages within these platforms. This article presents a broad summary of existing ELA approaches and introduces flacco, an R-package for feature-based landscape analysis of continuous and constrained optimization problems. Although its functions neither solve the optimization problem itself nor the related *algorithm selection problem (ASP)*, it offers easy access to an essential ingredient of the ASP by providing a wide collection of ELA features on a single platform—even within a single package. In addition, flacco provides multiple visualization techniques, which enhance the understanding of some of these numerical features, and thereby make certain landscape properties more comprehensible. On top of that, we will introduce the package's built-in, as well as web-hosted and hence platform-independent, graphical user interface (GUI). It facilitates the usage of the package—especially for people who are not familiar with R—and thus makes flacco a very convenient toolbox when working towards algorithm selection of continuous single-objective optimization problems.

P. Kerschke (✉) · H. Trautmann
Information Systems and Statistics, University of Münster, Leonardo-Campus 3,
48149 Münster, Germany
e-mail: kerschke@uni-muenster.de

H. Trautmann
e-mail: trautmann@uni-muenster.de

© Springer Nature Switzerland AG 2019
N. Bauer et al. (eds.), *Applications in Statistical Computing*,
Studies in Classification, Data Analysis, and Knowledge Organization,
https://doi.org/10.1007/978-3-030-25147-5_7

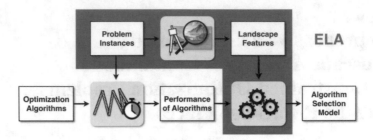

Fig. 7.1 Scheme for using ELA in the context of an algorithm selection problem

7.1 Introduction

Already a long time ago, Rice introduced the *algorithm selection problem (ASP)* (Rice 1976), which aims at finding the best algorithm A out of a set of algorithms $\mathscr{A}$ for a specific instance[1] $I \in \mathscr{I}$. Thus, one can say that the algorithm selection model m tries to find the best-performing algorithm A for a given set of problem instances $\mathscr{I}$.

As the performance of an optimization algorithm strongly depends on the structure of the underlying instance, it is crucial to have some a priori knowledge of the problem's landscape in order to pick a well-performing optimizer. One promising approach for such heads-up information are landscape features, which characterize certain landscape properties by means of numerical values. Figure 7.1 schematically shows how *exploratory landscape analysis (ELA)* can be integrated into the model-finding process of an ASP. In the context of continuous single-objective optimization, relevant problem characteristics cover information such as the degree of *multimodality*, its *separability* or its *variable scaling* (Mersmann et al. 2011). However, the majority of these so-called "high-level properties" have a few drawbacks: they (1) are categorical, (2) require expert knowledge (i.e., a decision maker), (3) might miss relevant information, and (4) require knowledge of the entire (often unknown) problem. Therefore, Bischl et al. (2012) introduced so-called "low-level properties", i.e., the landscape features, which characterize a problem instance by means of—not necessarily intuitively understandable—numerical values, based on a sample of observations from the underlying problem. Figure 7.2 shows which of their automatically computable low-level properties (shown as white ellipses) could potentially describe the expert-designed high-level properties (grey boxes).

In recent years, researchers all over the world have developed further (low-level) landscape features and implemented them in many different programming languages—mainly in MATLAB (2013), Python (VanRossum and The Python Development Team 2015), or R (R Core Team 2018). Most of these features are very cheap to compute as they only require an initial design, i.e., a small amount of

---

[1] In this context, an *instance* is the equivalent to an optimization problem, i.e., it maps the elements of the decision space $\mathscr{X}$ to the objective space $\mathscr{Y}$.

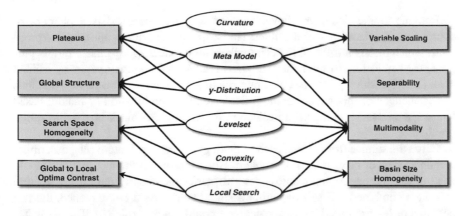

Fig. 7.2 Relationship between expert-based high-level properties (gray rectangles in the outer columns) and automatically computable low-level properties (white ellipses in the middle)

(often randomly chosen) exemplary observations, which are used as representatives of the original problem instance.

For instance, Jones and Forrest (1995) characterized problem landscapes using the *correlation* between the distances from a set of points to the global optimum and their respective fitness values. Later, Lunacek and Whitley (2006) introduced *dispersion* features, which compare pairwise distances of all points in the initial design with the pairwise distances of the best points in the initial design (according to the corresponding objective). Malan and Engelbrecht (2009) designed features that quantify the *ruggedness* of a continuous landscape and Müller and Sbalzarini (2011) characterized landscapes by performing *fitness-distance analyses*. In other works, features were constructed using the problem definition, hill climbing characteristics and a set of random points (Tinus et al. 2013), or based on a tour across the problem's landscape, using the change in the objective values of neighboring points to measure the landscape's *information content* (Muñoz et al. 2012, 2015a).

The fact that feature-based landscape analysis also exists in other domains— e.g., in discrete optimization (Daolio et al. 2016; Jones 1995; Ochoa et al. 2014) including its subdomain, the *traveling salesperson problem* (Mersmann et al. 2013; Hutter et al. 2014; Pihera and Musliu 2014)—also led to attempts to discretize the continuous problems and afterward use a so-called *cell mapping* approach (Kerschke et al. 2014), or compute *barrier trees* on the discretized landscapes (Flamm et al. 2002).

In recent years, Morgan and Gallagher (2015) analyzed optimization problems by means of *length scale* features, Kerschke et al. (2015, 2016) distinguished between so-called "funnel" and "non-funnel" structures with the help of *nearest better clustering* features, Malan et al. (2015) used *violation landscapes* to characterize constrained continuous optimization problems, and (Shirakawa and Nagao 2016) introduced the *bag of local landscape features*.

Looking at the list from above, one can see that there exists a plethora of approaches for characterizing the fitness landscapes of continuous optimization problems, which makes it difficult to keep track of all the new approaches. For this purpose, we would like to refer the interested reader to the survey paper by Muñoz et al. (2015b), which provides a nice introduction and overview on various methods and challenges of this whole research field, i.e., the combination of feature-based landscape analysis and algorithm selection for continuous black-box optimization problems.

Given the studies from above, each of them provided new feature sets, describing certain aspects or properties of a problem. However, due to the fact that one would have to run the experiments across several platforms, usually only a few feature sets were combined—if at all—within a single study. As a consequence, the full potential of feature-based landscape analysis could not be exploited. This issue has now been addressed with the implementation of the R-package flacco (Kerschke 2017), which is introduced within this paper.

While the general idea of flacco (Kerschke and Trautmann 2016), as well as its associated graphical user interface (Hanster and Kerschke 2017) were already published individually in earlier works, this paper combines them for the first time and—more importantly—clearly enhances them by providing a detailed description of the implemented landscape features.

The following section provides a very detailed overview of the integrated landscape features and the additional functionalities of flacco. Section 7.3 demonstrates the package usage based on a well-known optimization problem and Sect. 7.4 introduces the built-in visualization techniques. In Sect. 7.5, the functionalities of the package's graphical user interface are presented and Sect. 7.6 concludes this work.

7.2 Integrated ELA Features

The here presented R-package provides a unified interface to a collection of the majority of feature sets from above within a single package[2] and consequently simplifies their accessibility. In its default settings, flacco computes more than 300 different numerical landscape features, distributed across 17 so-called *feature sets* (further details on these sets are given on p. xx). Additionally, the total number of performed function evaluations, as well as the required runtime for the computation of a feature set, are automatically tracked per feature set.

Within flacco, basically all feature computations and visualizations are based on a *feature object*, which stores all the relevant information of the problem instance. It contains the initial design, i.e., a data frame of all the (exemplary) observations from the decision space along with their corresponding objective values and—if provided—the number of blocks per input dimension (e.g., required for the *cell mapping* approach, cf. Kerschke et al. 2014), as well as the exact (mathematical)

---

[2]The authors intend to extend flacco by any feature set that has not yet been integrated into it.

function definition, which is needed for those feature sets, which perform additional function evaluations, e.g., the *local search* features (which were already mentioned in Fig. 7.2).

Such a feature object can be created in different ways as shown in Fig. 7.3. First, one has to provide some input data by either passing (a) a matrix or data frame of the (sampled) observations from the decision space (denoted as X), as well as a vector with the corresponding objective values (y), or (b) a data frame of the initial design (init.design), which basically combines the two ingredients from (a), i.e., the decision space samples and the corresponding objective values, within a single data set. Then, the function createFeatureObject takes the provided input data and transforms it into a specific R object—the feature object. During its creation, this object can (optionally) be enhanced/specified by further details such as the exact function of the problem (fun), the number of blocks per dimension (blocks), or the upper (upper) and lower (lower) bounds of the problem's decision space. If the latter are not provided, the initial design's minimum and maximum (per input variable) will be used as boundaries.

The package's function createInitialSample allows to create a random sample of observations (denoted as X within Fig. 7.3) within the box-constraints $[0, 1]^d$ (with d being the number of input dimensions). This sample-generator also allows to configure the boundaries or use an *improved latin hypercube sample* (Beachkofski and Grandhi 2002; McKay et al. 2000) instead of a sample that is based on a random uniform distribution.

As one of the main purposes of *exploratory landscape analysis* is the description of landscapes when given a very restricted budget (of function evaluations), it is highly

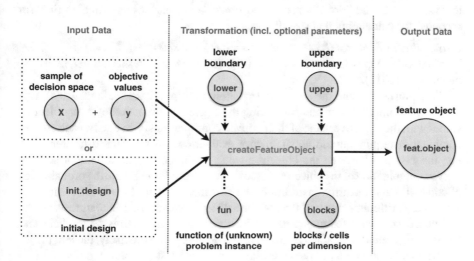

Fig. 7.3 Scheme for creating a feature object: (1) provide the input data, (2) call createFeatureObject and pass additional arguments if applicable, (3) receive the feature object, i.e., a specific R-object, which is the basis for all further computations and visualizations

recommended to keep the sample sizes small. Within recent work, Kerschke et al. (2016) have shown that initial designs consisting of $50 \cdot d$ observations—i.e., sampled points in the decision space—can be completely sufficient for describing certain high-level properties by means of numerical features. Also, one should note that the majority of research within the field of continuous single-objective optimization deals with at most 40-dimensional problems. Thus, unless one possesses a high budget of function evaluations, as well as the computational resources, it is not useful to compute features based on initial designs with more than 10,000 observations or more than 40 features. And even in case one is in possession of such resources, one should consider whether it is necessary to perform feature-based algorithm selection; depending on the problems, algorithms, and resources, it might as well be possible to simply run all optimization algorithms on all problem instances.

Given the aforementioned feature object, one has laid the basis for calculating specific feature sets (using the function `calculateFeatureSet`), computing all available feature sets at once (`calculateFeatures`) or visualizing certain characteristics (described in more detail in Sect. 7.4).

In order to get an overview of all currently implemented feature sets, one can call the function `listAvailableFeatureSets`. This function, as well as `calculateFeatures`, i.e., the function for calculating all feature sets at once, allows to turn off specific feature sets, such as those, which require additional function evaluations or follow the *cell mapping* approach. Turning off the *cell mapping* features could, for instance, be useful in case of higher dimensional problems due to the *curse of dimensionality*[3] (Friedman 1997).

In general, the current version of `flacco` consists of the feature sets that are described below. Note that the final two "features" per feature set always provide its costs, i.e., the additionally performed function evaluations and running time (in seconds) for calculating that specific set.

Classical ELA Features (83 features across 6 feature sets):
The first six groups of features are the "original" ELA features as defined by Mersmann et al. (2011).

Six features measure the *convexity* by taking multiple times (default: 1,000) two points of the initial design and comparing the convex combination of their objective values with the objective value of their convex combinations. Obviously, each convex combination requires an additional function evaluation. The set of these differences—i.e., the distance between the objective value of the convex combination and the convex combination of the objective values—is then used to estimate the probabilities of convexity and linearity. In addition, the arithmetic mean of these differences, as well as the arithmetic mean of the absolute differences are used as features as well.

The 26 *curvature* features measure information on the estimated gradient and Hessian of (a subset of) points from the initial design. More precisely, the lengths' of the gradients, the ratios between the biggest and smallest (absolute) gradient directions and the ratios of the biggest and smallest eigenvalues of the Hessian matrices

---

[3]In case of a 10-dimensional problem in which each input variable is discretized by three blocks, one already needs $3^{10} = 59{,}049$ observations to have one observation per cell—on average.

are aggregated using the minimum, first quartile ($= 25\%$-quantile), arithmetic mean, median, third quartile ($= 75\%$-quantile), maximum, standard deviation and number of missing values.[4] Note that estimating the derivatives also requires additional function evaluations (default: $100 \times d$ with search space dimensionality d).

The five *y-distribution* features compute the kurtosis, skewness and number of peaks based on a kernel-density estimation of the initial design's objective values.

For the next group, the initial design is divided into two groups (based on a configurable threshold for the objective values). Afterward the performances of various classification techniques, which are applied to that binary classification task, are used for computing 20 *levelset* features. Based on the classifiers and thresholds,[5] flacco uses the R-package mlr (Bischl et al. 2016) to compute the mean misclassification errors (per combination of classifier and threshold). Per default, the error rates are based on a tenfold cross-validation, but this can be changed to any other resampling-strategy implemented in mlr. In addition to the "pure" misclassification errors, flacco computes the ratio of misclassification errors per threshold and pair of classifiers.

The 16 *local search* features extract information from several local search runs (each of them starting in one of the points from the initial design). Per default, we run $50 \times d$ local searches with the quasi-Newton method L-BFGS-B (Byrd et al. 1995) (as implemented in R's optim-function) to find a local optimum. The found optima are then clustered (per default via single-linkage hierarchical clustering, see, e.g., Jobson 2012) in order to decide whether the found optima belong to the same basins. The resulting total number of found optima ($=$ clusters) is the first feature of this group, whereas the ratio of this number of optima and the conducted local searches form the second one. The next two features measure the ratio of the best and average objective values of the found clusters. As the clusters likely combine multiple local optima, the number of optima per cluster can be used as a representation for the basin sizes. These basin sizes are then averaged for the best, all non-best and the worst cluster(s). Finally, the number of required function evaluations across the different local search runs is aggregated using the minimum, first quartile ($= 25\%$-quantile), arithmetic mean, median, third quartile ($= 75\%$-quantile), maximum and standard deviation.

The remaining 11 *meta model* features extract information from linear and quadratic models (with and without interactions) that were fitted to the initial design. More precisely, these features compute the model fit ($= R_{adj}^2$) for each of the four aforementioned models, the intercept of the linear model without interactions, the condition (i.e., the ratio of largest and smallest absolute coefficient) of the quadratic model without interactions and the minimum, maximum as well as ratio of maximum

---

[4]If a point is a local optimum the gradient is zero for all dimensions of a sample point, then the ratio of biggest and smallest gradient obviously cannot be computed and therefore results in a missing value ($=$ NA).

[5]The default classifiers are linear (LDA), quadratic (QDA) and mixture discriminant analysis (MDA) and the default threshold for dividing the data set into two groups are the 10%-, 25%- and 50%-quantile of the objective values.

and minimum of all (absolute) coefficients—except for the intercept—of the linear model without interactions.

Cell Mapping Features (20/3):
This approach discretizes the continuous decision space using a predefined number of blocks (= cells) per input dimension. Then, the interaction of the cells, as well as the observations that are located within a cell, are used to compute the corresponding landscape features (Kerschke et al. 2014).

As shown in Fig. 7.4, the ten *angle* features extract information based on the location of the best and worst (available) observation of a cell w.r.t. the corresponding cell center. The distance from cell center to the best and worst point within the cell, as well as the angle between the three points are computed per cell and afterward aggregated across all cells using the arithmetic mean and standard deviation. The remaining features compute the difference between the best and worst objective value per cell, normalize these differences by the biggest span in objective values within the entire initial design and afterward aggregate these distances across all cells (using the arithmetic mean and the standard deviation).

The four *gradient homogeneity* features aggregate the cell-wise information on the gradients between each point of a cell and its corresponding nearest neighbor. Figure 7.5 illustrates the idea of this feature set. That is, for each point within a cell, we compute the gradient towards its nearest neighbor, normalize it, point it towards the better one of the two neighboring points and afterwards sum up all the normalized gradients (per cell). Then, the lengths of the resulting vectors are aggregated across all cells using the arithmetic mean and standard deviation.

The remaining six cell mapping features aggregate the (estimated) *convexity* based on representative observations of (all combinations of) three either horizontally, vertically or diagonally successive cells. Each cell is represented by the sample obser-

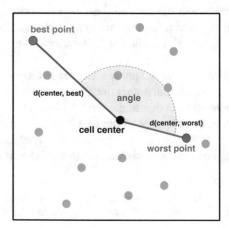

Fig. 7.4 Overview of the "ingredients" computing the *cell mapping angle features*: the location of the best and worst points within a cell

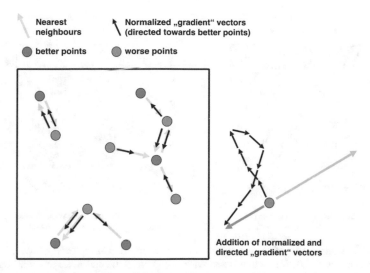

Fig. 7.5 General idea for computing the *cell mapping gradient homogeneity features*: (1) find the nearest neighbors, (2) compute and normalize the gradient vectors, (3) point them towards the better of the two points and (4) sum them up

vation that is located closest to the corresponding cell center. We then compare—for each combination of three neighboring cells—their objective values: if the objective value of the middle cell is below/above the mean of the objective values of the outer two cells, we have an indication for (soft) convexity/concavity. If the objective value of the middle cell even is the lowest/biggest value of the three neighboring cells, we have an indication for (hard) convexity/concavity. Averaging these values across all combinations of three neighboring cells, results in the estimated "probabilities" for (soft/hard) convexity or concavity.

Generalized Cell Mapping Features (75/1):
Analogously to the previous group of features, these features are also based on the block-wise discretized decision space. Here, each cell will be represented by exactly one of its observations—either its best or average value or the one that is located closest to the cell center—and each of the cells is then considered to be an absorbing Markov chain. That is, for each cell the transition probability for moving from one cell to one of its neighboring cells is computed. Based on the resulting transition probabilities, the cells are categorized into attractor, transient, periodic and uncertain cells. Figure 7.6 shows two exemplary cell mapping plots—each of them is based on the same feature object, but follows a different representation approach—which color the cells according to their category: attractor cells are depicted by black boxes, gray cells indicate uncertain cells, i.e., cells that are attracted by multiple attractors, and the remaining cells represent the certain cells, which form the basins of attractions. Furthermore, all non-attractor cells possess arrows that point toward their attractors and their length's represent the attraction probabilities. The different cell types, as

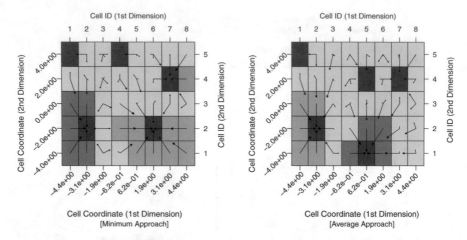

Fig. 7.6 Cell mappings of a two-dimensional version of *Gallagher's Gaussian 101-me Peaks* (Hansen et al. 2009) function based on a minimum (left) and average (right) approach. The black cells are the attractor cells, i.e., potential local optima. Each of those cells might come with a basin of attraction, i.e., the colored cells which point towards the attractor. All uncertain cells, i.e., cells that are attracted by multiple attractors, are shown in grey. The arrows show in the direction(s) of their attracting cell(s)

well as the accompanying probabilities are the foundation for the 25 *generalized cell mapping* (GCM) features (per approach):

1. total number of attractor cells and ratio of cells that are periodic (usually the attractor cells), transient (= nonperiodic cells) or uncertain cells,
2. aggregation (minimum, arithmetic mean, median, maximum and standard deviation) of the probabilities for reaching the different basins of attractions,
3. aggregation of basin sizes when the basins are only formed by the "certain" cells,
4. aggregation of basin sizes when the "uncertain" cells also count toward the basins (a cell which points towards multiple attractors contributes to each of them),
5. number and probability of finding the attractor cell with the best objective value.

Further details on the GCM-approach are given in Kerschke et al. (2014).

Barrier Tree Features (93/1):
Building on top of the transition probability representation and the three approaches from the GCM features, a so-called *barrier tree* (Flamm et al. 2002), as shown in Fig. 7.7, is constructed (per approach). The local optima of the problem landscape—i.e., the valleys in case of minimization problems—are represented by the leaves of the tree (indicated by filled circles within Fig. 7.7) and the branching nodes (depicted as non-filled diamonds) represent the ridges of the landscape, i.e., the locations where (at least) two neighboring valleys are connected. Based on these trees, the following 31 features can be computed for each of the three cell-representing approaches:

1. number of leaves (= filled circles) and levels (= branches of different colors), as well as tree depth, i.e., distance from root node (red triangle) to the lowest leaf,

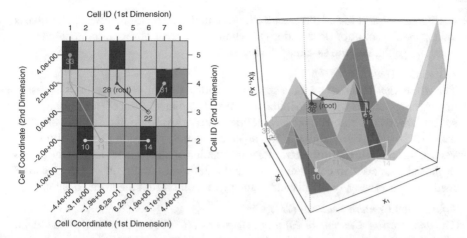

Fig. 7.7 Visualizations of a barrier tree in 2D (left) and 3D (right). Both versions of the tree belong to the same problem—a two-dimensional version of *Gallagher's Gaussian 101-me Peaks* function based on the minimum approach. The root of the tree is highlighted by a red triangle and the leaves of the tree, which usually lie in an attractor cell, are marked with filled circles. The saddle points, i.e., the points where two or more basins of attraction come together, are indicated by non-filled circles, as well as one incoming and two outgoing branches

2. ratio of the depth of the tree and the number of levels,
3. ratio of the number of levels and the number of non-root nodes,
4. aggregation (minimum, arithmetic mean, median, maximum and standard deviation) of the height differences between a node and its predecessor,
5. aggregation of the average height difference per level,
6. ratio of biggest and smallest basin of attraction (based on the number of cells counting towards the basin) for three different "definitions" of a basin: (a) based on the "certain" cells, (b) based on the "uncertain" cells or (c) based on the "certain" cells, plus the "uncertain" cells for which the probability towards the respective attractor is the highest,
7. proportion of cells that belong to the basin, which contains the attractor with the best objective value, and cells from the other basins—aggregated across the different basins,
8. widest range of a basin.

Nearest Better Clustering Features (7/1):
These features extract information based on the comparison of the sets of distances from (a) all observations towards their nearest neighbors and (b) their *nearest better neighbors*[6] (Kerschke et al. 2015). More precisely, these features measure the ratios of the standard deviations and the arithmetic means between the two sets, the correlation between the distances of the nearest neighbors and nearest better neighbors, the

---

[6]Here, the "nearest better neighbor" is the observation, which is the nearest neighbor among the set of all observations with a better objective value.

coefficient of variation of the distance ratios and the correlation between the fitness value, and the count of observations to whom the current observation is the nearest better neighbor, i.e., the so-called "indegree".

Dispersion Features (18/1):
The *dispersion features* by Lunacek and Whitley (2006) compare the dispersion among observations within the initial design and among a subset of these points. The subsets are created based on predefined thresholds, whose default values are the 2%-, 5%-, 10%- and 25%-quantile of the objective values. For each of these threshold values, we compute the arithmetic mean and median of all distances among the points of the subset, and then compare it—using the difference and ratio—to the mean or median, respectively, of the distances among all points from the initial design.

Information Content Features (7/1):
The *Information Content of Fitness Sequences (ICoFiS)* approach (Muñoz et al. 2015a) quantifies the so-called *information content* of a continuous landscape, i.e., smoothness, ruggedness, or neutrality. While similar methods already exist for the information content of discrete landscapes, this approach provides an adaptation to continuous landscapes that for instance accounts for variable step sizes in random walk sampling. This approach is based on a symbol sequence $\Phi = \{\phi_1, \ldots, \phi_{n-1}\}$, with

$$
\phi_i := \begin{cases} \bar{1} & \text{, if } \frac{y_{i+1} - y_i}{||\mathbf{x}_{i+1} - \mathbf{x}_i||} < -\varepsilon \\ 0 & \text{, if } \left| \frac{y_{i+1} - y_i}{||\mathbf{x}_{i+1} - \mathbf{x}_i||} \right| \leq \varepsilon \\ 1 & \text{, if } \frac{y_{i+1} - y_i}{||\mathbf{x}_{i+1} - \mathbf{x}_i||} > \varepsilon \end{cases}.
$$

This sequence is derived from the objective values $y_1, \ldots, y_n$ belonging to the n points $\mathbf{x}_1, \ldots, \mathbf{x}_n$ of a random walk across (the initial design of) the landscape and depends on the *information sensitivity* parameter $\varepsilon > 0$.

This symbol sequence Φ is aggregated by the *information content* $H(\varepsilon) := \sum_{i \neq j} p_{ij} \cdot \log_6 p_{ij}$, where p_{ij} is the probability of having the "block" $\phi_i \phi_j$, with $\phi_i, \phi_j \in \{\bar{1}, 0, 1\}$, within the sequence. Note that the base of the logarithm was set to six as this equals the number of possible blocks $\phi_i \phi_j$ for which $\phi_i \neq \phi_j$, i.e., $\phi_i \phi_j \in \{\bar{1}0, 0\bar{1}, \bar{1}1, 1\bar{1}, 10, 01\}$ (Muñoz et al. 2015a).

Another aggregation of the information is the so-called *partial information content* $M(\varepsilon) := |\Phi'|/(n-1)$, where Φ' is the symbol sequence of alternating 1's and $\bar{1}$'s, which is derived from Φ by removing all zeros and repeated symbols. Muñoz et al. (2015a) then suggest to use these two characteristics for computing the following five features:

1. *maximum information content* $H_{max} := \max_\varepsilon \{H(\varepsilon)\}$,
2. *settling sensitivity* $\varepsilon_s := \log_{10}(\min_\varepsilon \{\varepsilon : H(\varepsilon) < s\})$, with the default of s being 0.05 as suggested by Muñoz et al. (2015a),
3. $\varepsilon_{max} := \arg \max_\varepsilon \{H(\varepsilon)\}$,
4. *initial partial information* $M_0 := M(\varepsilon = 0)$,
5. *ratio of partial information sensitivity* $\varepsilon_r := \log_{10}(\max_\varepsilon \{\varepsilon : M(\varepsilon) > r \cdot M_0\})$, with the default of r being 0.5 as suggested by Muñoz et al. (2015a).

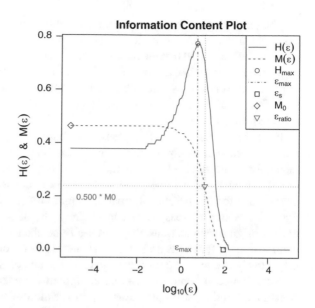

Fig. 7.8 *Information Content Plot* of a 2D version of *Gallagher's Gaussian 101-me Peaks* function. This approach analyzes the behavior of the tour along the points of the problem's initial design

The various characteristics and features described above can also be visualized within an *Information Content Plot* as exemplarily shown in Fig. 7.8.

Further Miscellaneous Features (40/3):
The remaining three feature sets provided within flacco are based on some very simple ideas. Note that there exists no (further) detailed description for these three feature sets as—according to the best knowledge of the author—none of them has been used for any (published) experiments. However, from our perspective they are worth to be included in future studies.

The simplest and quickest computable group of features are the 16 *basic* features, which provide rather obvious information of the initial design: (a) the number of observations and input variables, (b) the minimum and maximum of the lower and upper boundaries, objective values and number of blocks per dimension, (c) the number of filled cells and the total number of cells, and (d) a binary flag stating whether the objective function should be minimized.

The 14 *linear model* features fit a linear model—with the objective variable being the depending variable and all the remaining variables as explaining variables—within each cell of the feature object and aggregate the coefficient vectors across all the cells. That is, we compute (a) the length of the average coefficient vector, (b) the correlation of the coefficient vectors, (c) the ratio of maximum and minimum, as well as the arithmetic mean of the standard deviation of the (non-intercept) coefficients, (d) the previous features based on coefficient vectors that were a priori normalized per cell, (e) the arithmetic mean and standard deviation of the lengths of the coefficient vectors (across the cells), and (f) the arithmetic mean and standard deviation of the ratio of biggest and smallest (non-intercept) coefficients per cell.

The remaining ten features extract information from a *principal component* analysis, which is performed on the initial design—both, including and excluding the objective values—and that are based on the covariance, as well as correlation matrix, respectively. For each of these four combinations, the features measure the proportion of principal components that are needed to explain a predefined percentage (default: 0.9) of the data's variance, as well as the percentage of variation that is explained by the first principal component.

Further information on all implemented features, especially more detailed information on each of its configurable parameters, is given within the package's documentation. Note that in contrast to the original implementation of most of the feature sets the majority of parameters, e.g., the classifiers used by the level set features, or the ratio for the partial information sensitivity (information content), are configurable. While for reasons of conveniency, the default values are set according to their original publications, they can easily be changed using the `control` argument within the functions `calculateFeatureSet` or `calculateFeatures`, respectively. An example for this is given at the end of Sect. 7.3.

Note that each feature set can be computed for unconstrained, as well as box-constrained optimization problems. Upon creation of the feature object, R asserts that each point from the initial sample X lies within the defined `lower` and `upper` boundaries. Hence, if one intends to consider unconstrained problems, one should set the boundaries to `-Inf` or `Inf`, respectively. When dealing with box-constrained problems, one has to take a look at the feature sets themselves. Given that 14 out of the 17 feature sets do not perform additional function evaluations at all, none of them can violate any of the constraints. This also holds for the remaining three feature sets. The *convexity* features always evaluate (additional) points that are located in the center of two already existing (and thereby feasible) points. Consequently, each of them has to be located within the box-constraints as well. In case of the *local search* features, the box-constraints are also respected as long as one uses the default local search algorithm (L-BFGS-B), which was designed to optimize bound-constrained optimization problems. Finally, the *curvature* features do not violate the constraints either as the internal functions, which are used for estimating the gradient and Hessian of the function, were adapted in such a way that they always estimate them by evaluating points within the box-constraints—even in case the points are located close to or even exactly on the border of the box.

As described within the previous two sections, numerous researchers have already developed feature sets and many of them also shared their source code, enabling the creation of this wide collection of landscape features in the first place. By developing `flacco` on a publicly accessible platform,[7] other R-users may easily contribute (further feature sets) to the package. Aside from using the most recent (development) version of the package, one can also use its stable release from CRAN.[8] Moreover, the package repository on GitHub also provides a link to the corresponding online

---

[7]The development version is available on GitHub: https://github.com/kerschke/flacco.

[8]The stable release is published on CRAN: https://cran.r-project.org/package=flacco.

tutorial,[9] the platform-independent web-application[10] (further details are given in Sect. 7.5) and an issue tracker, where one could report bugs, provide feedback or suggest the integration of further feature sets.

7.3 Exemplary Feature Computation

In the following, the usage of `flacco` will exemplarily be presented on a well-known optimization problem, namely *Gallagher's Gaussian 101-me Peaks* (Hansen et al. 2009). It is the 21st out of the 24 artificially designed, continuous single-objective optimization problems from the *Black-Box Optimization Benchmark* (BBOB, Hansen et al. 2010). Within this benchmark, it belongs to a group of five multimodal problem instances with a weak global structure. Each of the 24 function classes from BBOB can be seen as a specific problem generator, whose problems are identical up to rotation, shifts and scaling. Therefore the exact instance has to be defined by a function ID (`fid`) and an instance ID (`iid`). For this exemplary presentation, we will use the second instance and generate it using `makeBBOBFunction` from the R-package `smoof` (Bossek 2017). The resulting landscape is depicted as a three-dimensional perspective plot in Fig. 7.9 and as a contour plot in Fig. 7.10.

In a first step, one needs to install the package along with all its dependencies, load it into the workspace and afterward generate the input data—i.e., a (hopefully) representative sample X of observations from the decision space along with their corresponding objective values y.

```
> install.packages("flacco", dependencies = TRUE)
> library("flacco")
> f = smoof::makeBBOBFunction(dimension = 2, fid = 21, iid = 2)
> ctrl = list(init_sample.lower = -5, init_sample.upper = 5)
> X = createInitialSample(n.obs = 100, dim = 2, control = ctrl)
> y = apply(X, 1, f)
```

The code above would also work without explicitly specifying the lower and upper boundaries (within `ctrl`), but in that case, the decision sample would automatically be generated within $[0, 1] \times [0, 1]$ rather than within the (for BBOB) common box-constraints of $[-5, 5] \times [-5, 5]$. In a next step, this input data is used (eventually in combination with further parameters such as the number of blocks per dimension) to create the feature object as already schematized in Fig. 7.3. The following R-code shows how this could be achieved.

```
> feat.object = createFeatureObject(X = X, y = y, fun = f,
+     blocks = c(8, 5))
```

As stated earlier, a feature object is the fundamental object for the majority of computations performed by `flacco` and thus, it obviously has to store quite a lot of

---

[9]Link to the package's tutorial: http://kerschke.github.io/flacco-tutorial/.

[10]Link to the package's GUI: https://flacco.shinyapps.io/flacco/.

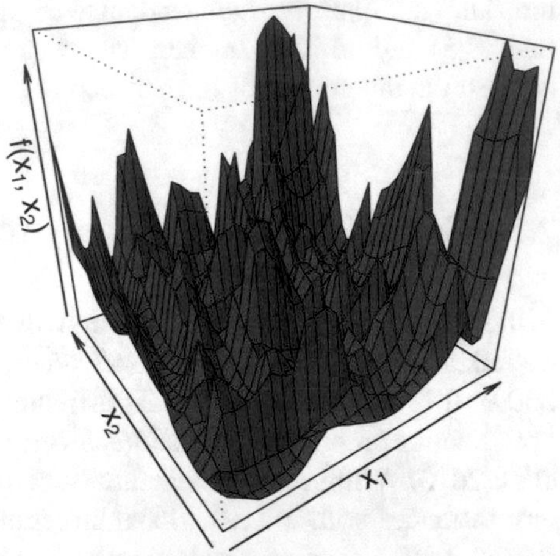

Fig. 7.9 3D-Perspective plot of *Gallagher's Gaussian 101-me Peaks* function

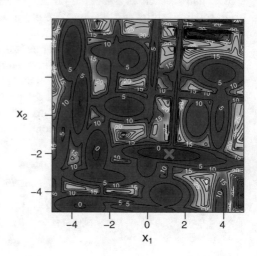

Fig. 7.10 Contour plot of *Gallagher's Gaussian 101-me Peaks* function. The red cross marks the global optimum and the coloring of the plot represents the objective values

information—such as the number of observations and (input) variables, the names and boundaries of the variables, the amount of empty and nonempty cells, etc. Parts of that information can easily be printed to the console by calling `print` on the feature object.[11]

---

[11]Note that the shown numbers, especially the ones for the number of observations per cell, might be different on your machine, as the initial sample is drawn randomly.

```
> print(feat.object)

Feature Object:
- Number of Observations: 100
- Number of Variables: 2
- Lower Boundaries: -5.00e+00, -5.00e+00
- Upper Boundaries: 5.00e+00, 5.00e+00
- Name of Variables: x1, x2
- Optimization Problem: minimize y
- Function to be Optimized: smoof-function (BBOB_2_21_2)
- Number of Cells per Dimension: 8, 5
- Size of Cells per Dimension: 1.25, 2.00
- Number of Cells:
  - total: 40
  - non-empty: 36 (90.00%)
  - empty: 4 (10.00%)
- Average Number of Observations per Cell:
  - total: 2.50
  - non empty: 2.78
```

Given the feature object, one can easily calculate any of the 17 feature sets that were introduced in Sect. 7.2. Specific feature sets can for instance—exemplarily shown for the *angle*, *dispersion* and *nearest better clustering* features—be computed as shown in the code below.

```
> angle.features = calculateFeatureSet(
+    feat.object = feat.object, set = "cm_angle")
> dispersion.features = calculateFeatureSet(
+    feat.object = feat.object, set = "disp")
> nbc.features = calculateFeatureSet(
+    feat.object = feat.object, set = "nbc")
```

Alternatively, one can compute all implemented features simultaneously via:

```
> all.features = calculateFeatures(feat.object)
```

Each of these calculations results in a list of (mostly numeric, i.e., real-valued) features. In order to avoid errors due to possibly similar feature names among different feature sets, each feature inherits the abbreviation of its feature set's name as a prefix, e.g., cm_angle for the *cell mapping angle* features or disp for the *dispersion* features. Using the results from the previous example, the output for the calculation of the seven *nearest better clustering* features could look like this:

```
> str(nbc.features)

  List of 7
  $ nbc.nn_nb.sd_ratio      : num 0.382
  $ nbc.nn_nb.mean_ratio    : num 0.59
  $ nbc.nn_nb.cor           : num 0.264
  $ nbc.dist_ratio.coeff_var: num 0.391
  $ nbc.nb_fitness.cor      : num -0.591
  $ nbc.costs_fun_evals     : int 0
  $ nbc.costs_runtime       : num 0.022
```

At this point, the authors would like to emphasize that **one should not try to interpret the numerical feature values** on their own as the majority of them simply are not intuitively understandable. Instead, these numbers should rather be used to distinguish between different problems—usually in an automated fashion, e.g., by means of a machine learning algorithm.

Also, it is important to note that features, which belong to the same feature set, are based on the same idea and only differ in the way they aggregate those ideas. For instance, the *nearest better clustering* features are based on the same distance sets—the distances to their nearest neighbors and nearest better neighbors. However, the features differ in the way they aggregate those distance sets, e.g., by computing the ratio of their (i) standard deviations or (ii) arithmetic means. As the underlying approaches are the bottleneck of each feature set, computing single features on their own would not be very beneficial and therefore has not been implemented in this package. Therefore, given the fact that multiple features share a common idea, which is usually the expensive part of the feature computation—for this group of features it would be the computation of the two distance sets—they will be computed together as an entire feature set instead of computing each feature on its own.

Figure 7.11 shows the results of a so-called "microbenchmark" (Mersmann 2014) on the chosen BBOB instance. Such a benchmark allows the comparison of the (logarithmic) runtimes across the different feature sets. As one can see, the time for calculating a specific feature set heavily depends on the chosen feature set. Although the majority of feature sets can be computed in less than half a second (the *basic* features even in less than a millisecond), three feature sets—namely *curvature, level set* and *local search*—usually needed between one and two seconds. And although two seconds might not sound that bad, it can have a huge impact on the required resources when computing all feature sets across thousands or even millions of instances. The amount of required time might even increase when considering the fact that some of the features are stochastic. In such a case it is strongly recommended to calculate each of those feature sets multiple times (on the same problem instance) in order to capture the variance of these features.

Note that the current definition of the feature object also contains the actual *Gallagher's Gaussian 101-me Peaks* function itself—it was added to the feature object by setting the parameter fun = f within the function createFeatureObject. Without that information it would have been impossible to calculate feature sets,

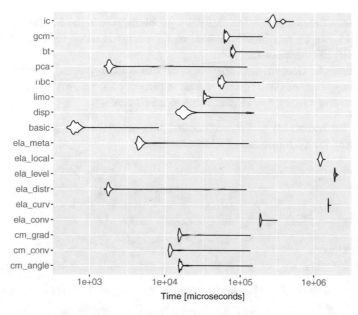

Fig. 7.11 Violin plots of a microbenchmark, measuring the logarithmic runtimes (in microseconds) of each feature set based on 1,000 replications on a two-dimensional version of the *Gallagher's Gaussian 101-me Peaks* function

which require additional function evaluations, i.e., the *convexity*, *curvature* and *local search* features of the classical ELA features. Similarly, the *barrier tree, cell mapping* and *general cell mapping* features strongly depend on the value of the `blocks` argument, which defines the number of blocks per dimension when representing the decision space as a grid of cells.[12]

As mentioned in Sect. 7.2, the majority of feature sets uses some parameters, all of them having certain default settings. However, as different landscapes/optimization problems might require different configurations of these parameters, the package allows its users to change the default parameters by setting their parameter configuration within the `control` argument of the functions `calculateFeatures` or `calculateFeatureSet`, respectively. Similarly to the naming convention of the features themselves, the names of parameters use the abbreviation of the corresponding feature set as a prefix. This way, one can store all the preferred configurations within a single control argument and does not have to deal with a separate control argument for each feature set. In order to illustrate the usage of the control argument, let's assume one wants to solve the following two tasks:

---

[12]The *barrier tree* features can only be computed if the total number of cells is at least two and the *cell mapping convexity* features require at least three blocks per dimension.

1. Calculate the *dispersion* features for a problem, where the Manhattan distance (also known as city-block distance or L_1-norm) is more reasonable than the (default) Euclidean distance.
2. Calculate the *General Cell Mapping* features for two instead of the usual three GCM approaches (i.e., `"min"` = best value per cell, `"mean"` = average value per cell and `"near"` = closest value to a cell center).

Due to the availability of a shared control parameter, these adaptations can easily be performed with a minimum of additional code:

```
> ctrl = list(
+    disp.dist_method = "manhattan",
+    gcm.approaches = c("min", "near")
+ )
> dispersion.features = calculateFeatureSet(
+    feat.object = feat.object, set = "disp", control = ctrl)
> gcm.features = calculateFeatureSet(
+    feat.object = feat.object, set = "gcm", control = ctrl)
```

This control parameter also works when computing all features simultaneously:

```
> all.features = calculateFeatures(feat.object, control = ctrl)
```

A detailed overview of all the configurable parameters is available within the documentation of the previous functions and can for instance be accessed via the command `?calculateFeatures`.

7.4 Visualization Techniques

In addition to the features themselves, `flacco` provides a handful of visualization techniques allowing the user to get a better understanding of the features. The function `plotCellMapping` produces a plot of the discretized grid of cells, representing the problem's landscape when using the cell mapping approach. Figure 7.6, which was already presented in Sect. 7.2, shows two of these cell mapping plots—the left one represents each cell by the smallest fitness value of its respective points, and the right plot represents each cell by the average of the fitness values of the respective cell's sample points. In each of those plots, the *attractor cells*, i.e., cells containing the sample's local optima, are filled black. The *uncertain cells*, i.e., cells that transition to multiple attractors, are colored gray, whereas the *certain cells* are colored according to their *basin of attraction*, i.e., all cells which transition towards the same attractor (and no other one) share the same color. Additionally, the figure highlights the attracting cells of each cell via arrows that point towards the corresponding attractor(s). Note that the length of the arrows is proportional to the transition probabilities.

Given the feature object from the example of the previous section, i.e., the one that is based on *Gallagher's Gaussian 101-me Peaks* function, the two cell mapping plots can be produced with the following code:

```
> plotCellMapping(feat.object,
+    control = list(gcm.approach = "min"))
> plotCellMapping(feat.object,
+    control = list(gcm.approach = "mean"))
```

Analog to the cell mapping plots, one can visualize the barrier trees of a problem. They can either be plotted in 2D as a layer on top of a cell mapping (per default without any arrows) or in 3D as a layer on top of a perspective/surface plot of the discretized decision space. Using the given feature object, the code for creating the plots from Fig. 7.7 (also already shown in Sect. 7.2) would look like this:

```
> plotBarrierTree2D(feat.object)
> plotBarrierTree3D(feat.object)
```

The representation of the tree itself is in both cases very similar. Each of the trees begins with its root, depicted by a red triangle. Starting from that point, there will usually be two or more branches, pointing either towards a saddle point (depicted by a non-filled diamond)—a node that connects multiple basins of attraction—or towards a leaf (filled circles) of the tree, indicating a local optimum. Branches belonging to the same level (i.e., having the same number of predecessors on their path up to the root) of the tree are colored identically. Note that all of the aforementioned points, i.e., root, saddle points and leaves, belong to distinct cells. The corresponding (unique) cell IDs are given by the numbers next to the points.

The last plot, which is directly related to the visualization of a feature set, is the so-called *information content plot*, which was shown in Fig. 7.8 in Sect. 7.2. It depicts the logarithm of the *information sensitivity* ε against the *(partial) information content*. Following these two curves—the solid red line representing the *information content* $H(\varepsilon)$, and the dashed blue line representing the *partial information content* $M(\varepsilon)$—one can (more or less) easily derive the features from the plot. The blue diamond at the very left represents the *initial partial information* M_0, the red circle on top of the solid, red line shows the *maximum information content* H_{max}, the green triangle indicates the *ratio of partial information sensitivity* $\varepsilon_{0.5}$ (with a ratio of $r = 0.5$) and the black square marks the *settling sensitivity* ε_s (with $s = 0.05$). Further details on the interpretation of this plot and the related features can be found in Muñoz et al. (2012, 2015a). In accordance to the code of the previous visualization techniques, such an information content plot can be created by the following command:

```
> plotInformationContent(feat.object)
```

In addition to the previously introduced visualization methods, flacco provides another plot function. However, in contrast to the previous ones, the *Feature Importance Plot* (cf. Fig. 7.12) does not visualize a specific feature set. Instead, it can be used to assess the importance of several features during a feature selection process of a machine learning (e.g., classification or regression) task. Given the high amount of features, provided by this package, it is very likely that many of them are redundant

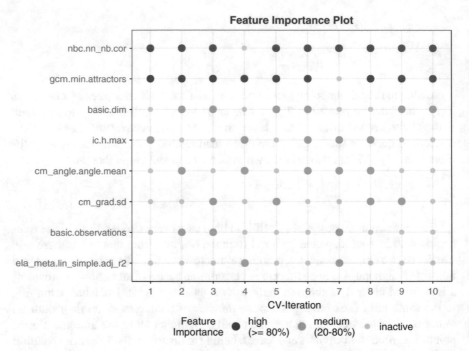

Fig. 7.12 Plot of the feature importance as derived from a 10-fold cross-validated feature selection

when training a (classification or regression) model and therefore, a feature selection strategy would be useful.

The scenario of Fig. 7.12 shows the (artificially created) result of such a feature selection after performing a tenfold cross-validation using the R-package `mlr` (Bischl et al. 2016).[13] Such a resampling strategy is useful in order to assess whether a certain feature was selected because of its importance to the model or more or less just by chance. The red points show whether an important feature—i.e., a feature, which has been selected in the majority of iterations (default: $\geq 80\%$)—was selected within a specific iteration (= fold of the cross-validation). Less important features are depicted as orange points. The only information that is required to create such a *Feature Importance Plot* is a list, whose elements (one per fold / iteration) are vectors of character strings. In the illustrated example, this list consists of ten elements (one per fold) with the first one being a character vector that contains exactly the three features `"gcm.min.attractors"`, `"nbc.nn_nb.cor"` and `"ic.h.max"`. The entire list that leads to the plot from Fig. 7.12 (denoted as `list.of.features`) would look like this:

---

[13] A more detailed step-by-step example can be found in the documentation of the respective `flacco`-function `plotFeatureImportancePlot`.

```
> str(list.of.features, vec.len = 2L)

List of 10
 $ : chr [1:3] "gcm.min.attractors" "ic.h.max" ...
 $ : chr [1:5] "gcm.min.attractors" "nbc.nn_nb.cor" ...
 $ : chr [1:5] "gcm.min.attractors" "nbc.nn_nb.cor" ...
 $ : chr [1:4] "gcm.min.attractors" "ic.h.max" ...
 $ : chr [1:4] "gcm.min.attractors" "nbc.nn_nb.cor" ...
 $ : chr [1:5] "gcm.min.attractors" "ic.h.max" ...
 $ : chr [1:5] "ic.h.max" "nbc.nn_nb.cor" ...
 $ : chr [1:5] "gcm.min.attractors" "ic.h.max" ...
 $ : chr [1:4] "gcm.min.attractors" "nbc.nn_nb.cor" ...
 $ : chr [1:3] "gcm.min.attractors" "nbc.nn_nb.cor" ...
```

Given such a list of features, the *Feature Importance Plot* can be generated with the following command:

```
> ggplotFeatureImportance(list.of.features)
```

Similarly to the feature computations, which were described in Sect. 7.3, each of the plots can be modified according to a user's needs by making use of the `control` parameter within each of the plot functions. Further details on the possible configurations of each of the visualization techniques can be found in their documentation.

7.5 Graphical User Interface

Within the previous sections, we have introduced a wide collection of feature sets that were previously implemented in different programming languages or at least in different packages. In addition, we have shown how one can use `flacco` to compute ELA features. While this is beneficial for researchers who are familiar with R, it comes with the drawback that researchers who are not familiar with R are left out. Therefore, Hanster and Kerschke (2017) implemented a graphical user interface (GUI) for the package, which can either be started from within R, or—which is probably more appealing to non-R-users—as a platform-independent web-application.

The GUI was implemented using the R-package `shiny` (Chang et al. 2016) and according to its developers, `shiny` *"makes it incredibly easy to build interactive web applications with* R*"* and its *"extensive prebuilt widgets make it possible to build beautiful, responsive, and powerful applications with minimal effort."*

If one wants to use the GUI-version that is integrated within `flacco` (version 1.5 or higher), one first needs to install `flacco` from CRAN (https://cran.r-project.org/package=flacco) and load it into the workspace of R.

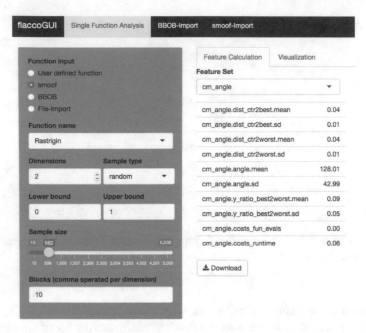

Fig. 7.13 Screenshot of the GUI, after computing the *cell mapping angle* features for a two-dimensional version of the *Rastrigin* function as implemented in `smoof`

```
> install.packages("flacco", dependencies = TRUE)
> library("flacco")
```

Afterwards it only takes the following line of code to start the application:

```
> runFlaccoGUI()
```

If one rather prefers to use the platform-independent GUI, one can use the web application, which is hosted on https://flacco.shinyapps.io/flacco/.

Once started, the application shows a bar on the top, where the user can select between the following three options (as depicted in Fig. 7.13): "Single Function Analysis", "BBOB-Import" and "Smoof-Import". Selecting the first tab, the application will show two windows: (1) a box (colored in gray) on the left, which inquires all the information that is needed for creating the feature object, i.e., the underlying function, the problem dimension (= number of input variables), the boundaries, the size and sampling type (latin hypercube or random uniform sampling) of the initial sample and the number of cells per dimension, and (2) a screen for the output on the right, where one can either compute the numerical landscape features under the tab "Feature Calculation" (as shown in Fig. 7.13) or create a related plot under the tab "Visualization" (as shown in Fig. 7.14).

For the definition of the underlying function (within the gray box), the user has the following options: (1) providing a user-defined function by entering the corresponding R expression, e.g., `sum(x^2)`, into a text box, (2) selecting one of the single-objective problems provided by `smoof` from a drop-down menu, (3) defining a BBOB function (Hansen et al. 2010) via its function ID (FID) and instance ID (IID), or (4) ignoring the function input and just provide the initial design, i.e., the input variables and the corresponding objective value, by importing a CSV-file.

Immediately after the necessary information for the feature object is provided, the application will automatically produce the selected output on the right half of the screen. If one does not select anything (on the right half), the application will try to compute the *cell mapping angle* features, as this is the default selection at the start of the GUI. However, the user can select any other feature set, as well as the option "all features", from a drop-down menu. Once the computation of the features is completed, one can download the results as a CSV-file by clicking on the "Download"-button underneath the feature values.

Aside from the computation of the landscape features, the GUI also provides functionalities for generating multiple visualizations of the given feature object. Aside from the *cell mapping plots* and two- and three-dimensional *barrier trees*, which can only be created for two-dimensional problems, it is also able to produce *information content plots* (cf. Sect. 7.4). Note that the latter currently is the only plotting function that is able to visualize problems with more than two input variables. In addition to the aforementioned plots, the application also enables the visualization of the function itself—if applicable—by means of an interactive *function plot* (for one-dimensional problems), as well as two-dimensional *contour* or three-dimensional *surface plots* (for two-dimensional problems). The interactivity of these plots, such as the one that is shown in the upper image of Fig. 7.14, was provided by the R-package `plotly` (Sievert et al. 2016) and (amongst others) enables to zoom in and out of the landscape, as well as to rotate or shift the latter.

While `flacco` itself allows the configuration of all of its parameters—i.e., the parameters used by any of the feature sets, as well as the ones related to the plotting functions—the GUI provides a simplified version of this. Hence, for the cell mapping and barrier tree plots, one can choose the approach for selecting the representative objective value per cell: (a) the best objective value among the samples from the cell (as shown in the lower image of Fig. 7.14), (b) the average of all objective values from the cell's sample points, or (c) the objective value of the observation that is located closest to the cell's center. In case of the information content plot, one can configure the range of the x-axis, i.e., the range of the (logarithmic) values of the information sensitivity parameter ε.

As already mentioned in Sect. 7.3, the authors highly recommend not to interpret single features on single optimization problems. Instead, one should rather compute several (sets of) landscape features across an entire benchmark of optimization problems. This way, they might be able to provide the information for distinguishing the benchmark's problems from each other and predicting a well-performing optimization algorithm per instance—cf. the *algorithm selection problem*, which was already mentioned as a potential use case in Sect. 7.1—or for adapting the parameters of such an optimizer to the underlying landscape.

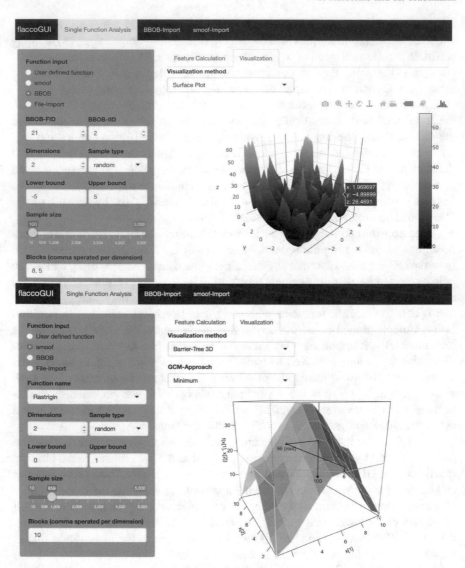

Fig. 7.14 Exemplary screenshots of two visualization techniques as provided by the GUI and shown for two different optimization problems. Top: An interactive surface plot of an instance of *Gallagher's 101-me Peaks* function (second instance of the 21st BBOB problem), which allows to zoom in and out of the landscape, as well as to rotate or shift it. Bottom: A 3D representation of the barrier tree for an instance of the *Rastrigin* function as implemented in `smoof`

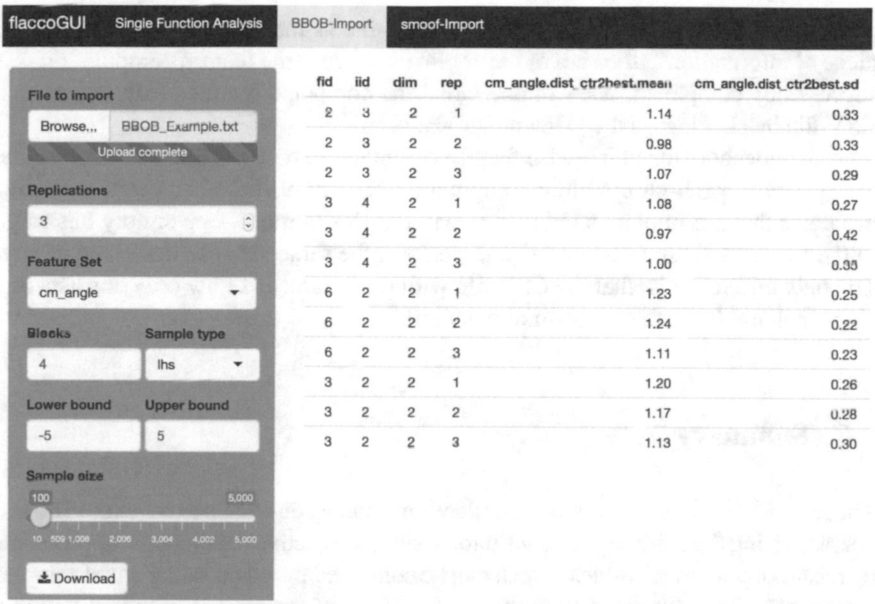
flaccoGUI
Single Function Analysis
BBOB-Import
smoof-Import
File to import
Browse...
BBOB_Example.txt
Upload complete
Replications
3
Feature Set
cm_angle
Blocks
4
Sample type
lhs
Lower bound
-5
Upper bound
5
Sample size
100
5,000
10 509 1,008 2,006 3,004 4,002 5,000
Download
fid iid dim rep cm_angle.dist_ctr2best.mean cm_angle.dist_ctr2best.sd
2 3 2 1 1.14 0.33
2 3 2 2 0.98 0.33
2 3 2 3 1.07 0.29
3 4 2 1 1.08 0.27
3 4 2 2 0.97 0.42
3 4 2 3 1.00 0.33
6 2 2 1 1.23 0.25
6 2 2 2 1.24 0.22
6 2 2 3 1.11 0.23
3 2 2 1 1.20 0.26
3 2 2 2 1.17 0.28
3 2 2 3 1.13 0.30

Fig. 7.15 Screenshot of the GUI, after computing the *cell mapping angle* features for four different BBOB instances with three replications each. Note that the remaining eight features from this set were computed as well and were simply cut off for better visibility

In order to facilitate such experiments, our application also provides functionalities for computing a specific feature set (or even all feature sets) simultaneously on multiple instances of a specific problem class—such as the BBOB functions or any of the other single-objective optimization problems provided by `smoof`. Figure 7.15 shows an exemplary screenshot of how such a batch computation could look like for the BBOB problems. In a first step, one has to upload a CSV-file consisting of three columns—the function ID (FID), instance ID (IID) and problem dimension (dim)—and one row per problem instance. Next, one defines the desired number of replications, i.e., how often should the features be computed per instance.[14] Afterward, one selects the feature set that is supposed to be computed, defines number of blocks and boundaries per dimension, as well as the desired sample type and size. Note that, depending on the size of the benchmark, i.e., the number of instances and replications, the number and dimensions of the initial sample, as well as the complexity of the chosen feature set, the computation of the features might take a while. Furthermore, please note that the entire application is reactive. That is, once all the required fields provide some information, the computation of the features will start and each time the user changes the information within a field, the computation will automatically restart. Therefore, the authors recommend to first configure

[14] As many of the features are stochastic, it is highly recommended to compute the features multiple(= at least 5 to 10) times.

all the parameters prior to uploading the CSV-file as the latter is the only missing piece of information at the start of the application. Once the feature computation was successfully completed, one can download the computed features/feature sets as a CSV-file by clicking on the "Download"-Button.

Analogously to the batch-wise feature computation on BBOB instances, one can also perform batch-wise feature computation for any of the other single-objective problems that are implemented in smoof. For this purpose, one simply has to go to the tab labeled "smoof-Import" and perform the same steps as described before. The only difference is, that the CSV-file with the parameters now only consists of a single column (with the problem dimension).

7.6 Summary

The R-package flacco provides a collection of numerous features for exploring and characterizing the landscape of continuous, single-objective (optimization) problems by means of numerical values, which can be computed based on rather small samples from the decision space. Having this wide variety of feature sets bundled within a single package simplifies further researches significantly—especially in the fields of algorithm selection and configuration—as one does not have to run experiments across multiple platforms. In addition, the package comes with different visualization strategies, which can be used for analyzing and illustrating certain feature sets, leading towards an improved understanding of the features.

This framework can be meaningfully combined with other R-packages such as mlr (Bischl et al. 2016) or smoof (Bossek 2017) and thereby enables the construction of better-performing optimization algorithms (which use the features to adapt to the problem at hand) or even algorithm selectors. Hence, flacco is a useful toolbox when (i) performing exploratory landscape analysis, and (ii) working towards algorithm selection models of single-objective, continuous optimization problems. Therefore, the package can be considered as a solid foundation for further researches—as for instance performed by members of the COSEAL network[15]—allowing to get a better understanding of algorithm selection scenarios and the employed optimization algorithms.

Lastly, flacco also comes with a graphical user interface, which can either be run from within R, or alternatively as a platform-independent standalone web-application. Therefore, even people who are not familiar with R have the chance to benefit of the functionalities of this package—especially from the computation of numerous landscape features—without the burden of interfacing or re-implementing identical features in a different programming language.[16]

---

[15]COSEAL is an international group of researchers with a focus on the **Co**nfiguration and **Se**lection of **Al**gorithms, cf. http://www.coseal.net/.

[16]The *European Center of Information Systems* (ERCIS) is an international network in the field of Information Systems, cf. https://www.ercis.org/.

Acknowledgements We acknowledge support by the ERCIS and thank Carlos Hernández (CIN-VESTAV, Mexico), Jan Dageförde, as well as Christian Hanster (University of Münster, Germany) for their valuable contributions to flacco and its GUI.

References

Abell, T., Malitsky, Y., & Tierney, K. (2013). Features for exploiting black-box optimization problem structure. In *Learning and intelligent optimization* (pp. 30–36). Berlin: Springer.

Beachkofski, B. K., & Grandhi, R. V. (2002). Improved distributed hypercube sampling. In *43rd AIAA/ASME/ASCE/AHS/ASC structures, structural dynamics, and materials conference*.

Bischl, B., Lang, M., Kotthoff, L., Schiffner, J., Richter, J., & Studerus, E., et al. (2016). mlr: Machine Learning in R. *Journal of Machine Learning Research, 17*(170), 1–5. R-package version 2.10.

Bischl, B., Mersmann, O., Trautmann, H., & Preuss, M. (2012). Algorithm selection based on exploratory landscape analysis and cost-sensitive learning. In *Proceedings of the 14th Annual Conference on Genetic and Evolutionary Computation, GECCO '12* (pp. 313–320). New York: ACM.

Bossek, J. (2017). smoof: Single-and multi-objective optimization test functions. *The R Journal*. https://journal.r-project.org/archive/2017/RJ-2017-004/index.html.

Byrd, R. H., Peihuang, L., Nocedal, J., & Zhu, C. (1995). A limited memory algorithm for bound constrained optimization. *SIAM Journal on Scientific Computing, 16*(5), 1190–1208.

Chang, W., Cheng, J., Allaire, J. J., Xie, Y., & McPherson, J. (2016). *Shiny: Web application framework for R*. R-package version 0.14.1.

Christoph, F., Hofacker, I. L., Stadler, P. F., & Wolfinger, M. T. (2002). Barrier trees of degenerate landscapes. *Zeitschrift für Physikalische Chemie International Journal of Research in Physical Chemistry and Chemical Physics 216*(2/2002), 155.

Daolio, F., Liefooghe, A., Verel, S., Aguirre, H., & Tanaka, K. (2016). Problem features versus algorithm performance on rugged multi-objective combinatorial fitness landscapes. *Evolutionary Computation*.

Friedman, J. H. (1997). On bias, variance, 0/1–loss, and the curse-of-dimensionality. *Data Mining and Knowledge Discovery, 1*(1), 55–77.

Guido VanRossum and The Python Development Team. (2015). *The Python Language Reference–Release 3.5.0*. Python Software Foundation.

Hansen, N., Auger, A., Finck, S., & Ros, R. (2010). Real-parameter black-box optimization benchmarking 2010: experimental setup. Technical Report RR-7215, INRIA.

Hansen, N., Finck, S., Ros, R., & Auger, A. (2009). Real-parameter black-box optimization benchmarking 2009: noiseless functions definitions. Technical Report RR-6829, INRIA.

Hanster, C., & Kerschke, P. (2017). Flaccogui: Exploratory landscape analysis for everyone.

Hutter, F., Lin, X., Hoos, H. H., & Leyton-Brown, K. (2014). Algorithm runtime prediction: Methods and evaluation. *Journal of Artificial Intelligence, 206*, 79–111.

Jobson, J. (2012). *Applied multivariate data analysis: Volume II: Categorical and multivariate methods*. Berlin: Springer.

Jones, T. (1995). *Evolutionary algorithms, fitness landscapes and search*. Ph.D. thesis, Citeseer.

Jones, T., & Forrest, S. (1995). Fitness distance correlation as a measure of problem difficulty for genetic algorithms. In *Proceedings of the 6th international conference on genetic algorithms* (pp. 184–192). Morgan Kaufmann Publishers Inc.

Kerschke, P. (2017). *Flacco: Feature-based landscape analysis of continuous and constrained optimization problems*. R-package version 1.6.

Kerschke, P., & Trautmann, H. (2016). The R-package FLACCO for exploratory landscape analysis with applications to multi-objective optimization problems. In *Proceedings of the IEEE congress on evolutionary computation (CEC)*. IEEE.

Kerschke, P., Preuss, M., Hernández, C., Schütze, O., Sun, J.- Q., Grimme, C., et al. (2014). Cell mapping techniques for exploratory landscape analysis. In *EVOLVE—A bridge between probability, set oriented numerics, and evolutionary computation V* (pp. 115–131). Berlin: Springer.

Kerschke, P., Preuss, M., Wessing, S., & Trautmann, H. (2015). Detecting funnel structures by means of exploratory landscape analysis. In *Proceedings of the 17th annual conference on genetic and evolutionary computation* (pp. 265–272). ACM.

Kerschke, P., Preuss, M., Wessing, S., & Trautmann, H. (2016). Low-budget exploratory landscape analysis on multiple peaks models. In *Proceedings of the 18th annual conference on genetic and evolutionary computation*. ACM.

Lunacek, M., & Whitley, D. (2006). The dispersion metric and the CMA evolution strategy. In *Proceedings of the 8th annual conference on genetic and evolutionary computation* (pp. 477–484). ACM.

Malan K. M., Oberholzer, J. F., & Engelbrecht, A. P. (2015). Characterising constrained continuous optimisation problems. In *Proceedings of the IEEE congress on evolutionary computation (CEC)* (pp. 1351–1358). IEEE.

Malan, K. M., & Engelbrecht, A. P. (2009). Quantifying ruggedness of continuous landscapes using entropy. In *Proceedings of the IEEE congress on evolutionary computation (CEC)* (pp. 1440–1447). IEEE.

MATLAB. (2013). *Version 8.2.0 (R2013b)*. The MathWorks Inc., Natick, Massachusetts.

McKay, M. D., Beckman, R. J., & Conover, W. J. (2000). A comparison of three methods for selecting values of input variables in the analysis of output from a computer code. *Technometrics*, *42*(1), 55–61.

Mersmann, O. (2014). *Microbenchmark: Accurate timing functions*. R-package version 1.4-2.

Mersmann, O., Bischl, B., Trautmann, H., Preuss, M., Weihs, C., & Rudolph, G. (2011). Exploratory landscape analysis. In *Proceedings of the 13th annual conference on genetic and evolutionary computation, GECCO '11* (pp. 829–836). New York: ACM.

Mersmann, O., Bischl, B., Trautmann, H., Wagner, M., Bossek, J., & Neumann, F. (2013). A novel feature-based approach to characterize algorithm performance for the traveling salesperson problem. *Annals of Mathematics and Artificial Intelligence*, *69*(2), 151–182.

Morgan, R., & Gallagher, M. (2015). Analysing and characterising optimization problems using length scale. *Soft Computing*, 1–18.

Müller, C. L., & Sbalzarini, I. F. (2011). Global characterization of the CEC 2005 fitness landscapes using fitness-distance analysis. In *Applications of evolutionary computation* (pp. 294–303). Berlin: Springer.

Muñoz, M. A., Kirley, M., & Halgamuge, S. K. (2012). Landscape characterization of numerical optimization problems using biased scattered data. In *2012 IEEE congress on evolutionary computation (CEC)* (pp. 1–8). IEEE.

Muñoz, M. A., Kirley, M., & Halgamuge, S. K. (2015a). Exploratory landscape analysis of continuous space optimization problems using information content. *IEEE transactions on evolutionary computation*, *19*(1), 74–87.

Muñoz, M. A., Sun, Y., Kirley, M., & Halgamuge, S. K. (2015b). Algorithm selection for black-box continuous optimization problems: A survey on methods and challenges. *Information Sciences*, *317*, 224–245.

Ochoa, G., Verel, S., Daolio, F., & Tomassini, M. (2014). Local optima networks: A new model of combinatorial fitness landscapes. In *Recent advances in the theory and application of fitness landscapes* (pp. 233–262). Berlin: Springer.

Pihera, J., & Musliu, N. (2014). Application of machine learning to algorithm selection for TSP. In *Proceedings of the IEEE 26th international conference on tools with artificial intelligence (ICTAI)*. IEEE press.

R Core Team. (2018). *R: A language and environment for statistical computing*. R foundation for statistical computing. Vienna, Austria.

Rice, J. R. (1976). The algorithm selection problem. *Advances in Computers*, *15*, 65–118.

Shirakawa, S., & Nagao, T. (2016). Bag of local landscape features for fitness landscape analysis. *Soft Computing, 20*(10), 3787–3802.

Sievert, C., Parmer, C., Hocking, T., Chamberlain, S., Ram, K., Corvellec, M., et al. (2016). *Plotly: Create interactive web graphics via 'plotly.js'*. R-package version 4.5.6.

## Part III

Perspectives on Statistics and Data Science

Chapter 8
A Note on Artificial Intelligence and Statistics

Katharina Morik

Abstract Now that data science receives a lot of attention, the three disciplines of data analysis, databases, and sciences are discussed with respect to the roles they play. In several discussions, I observed misunderstandings of Artificial Intelligence. Hence, it might be the right time to give a personal view of AI and the part of machine learning therein. Since the relation between machine learning and statistics is so close that sometimes the boundaries are blurred, explicit pointers to statistical research are made. Although not at all complete, the references are intended to support further interdisciplinary understanding of the fields.

8.1 Introduction

Data science has a long tradition in Dortmund. Long before the hype, an interdisciplinary curriculum between the faculties of statistics and computer science taught database, AI, and statistics in order to educate data scientists. Based on this curriculum, the faculty of statistics has had a Master's degree in Data Science since 2002. A common course on knowledge discovery in data is taught since 2003. Due to this collaboration, a convergence of notation and interest in research questions has been established between statistics and machine learning. A large common set of theoretical properties for the learned or statistical models exists. The additional properties concerning the resource consumption of algorithms, namely energy, communication, memory are intensively studied in machine learning, although also statistics is interested in implementations (Weihs et al. 2013). We may roughly characterise the machine learning and the statistics perspective as largely overlapping. Let us look at the different aspects or levels of machine learning. First, there is the general formal basis, expressed by a set of formulas. Often, this is common among mathematics, statistics, and computer science. Second, there is a learning task or model that specifies the particular approach to learning with the representation of what is learned,

K. Morik (✉)
Fakultät für Informatik, Technische Universität Dortmund, Dortmund, Germany
e-mail: katharina.morik@tu-dortmund.de

© Springer Nature Switzerland AG 2019
N. Bauer et al. (eds.), *Applications in Statistical Computing*,
Studies in Classification, Data Analysis, and Knowledge Organization,
https://doi.org/10.1007/978-3-030-25147-5_8

Fig. 8.1 The four aspects of machine learning and the curves of interest or research focus of statistics (solid line) and computer science (dotted line)

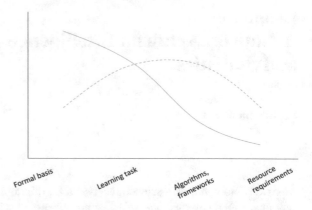

Formal basis   Learning task   Algorithms, frameworks   Resource requirements

from what it is learned, and what the quality conditions are, the latter including regularisation. Also, here are many common views in statistics and computer science. Third, there are several algorithms that are used by a model. Algorithm engineering is trained in computer science extensively and particular implementation schemes and frameworks are developed. Fourth, the algorithms are executed at some hardware. Research results in properties of the respective entities, i.e. platforms, algorithms, models. Novel machine learning research aims at making explicit the link from the restrictions of the model to those of constrained hardware.[1] Although, machine learning and statistics deal with all aspects, the distribution of their research activities is different. This is schematically shown in Fig. 8.1. Where statistics is more engaged in the first aspect and decreases activities towards the further aspects, machine learning focuses on the second and third with decreasing activities towards both ends. The broader field of AI has not yet been as extensively discussed with its links to statistics. This paper catches up and characterises machine learning as part of AI and then gives some references to data-driven sciences.

8.2 The Intelligence of Artificial Intelligence

John McCarthy, Marvin Minsky, Nathaniel Rochester and Claude Shannon organised the Summer Research Project on Artificial Intelligence in 1956. Their proposal used the term *Artificial Intelligence* for the first time and stated: *The study is to proceed on the basis of the conjecture that every aspect of learning or any other feature of intelligence can in principle be so precisely described that a machine can be made to simulate it.*[2] The Oxford dictionary defines artificial intelligence: *The theory and development of computer systems able to perform tasks normally requiring human intelligence...* This definition does not assume any similarity between the

---

[1] https://sfb876.tu-dortmund.de.

[2] http://www-formal.stanford.edu/jmc/history/dartmouth/dartmouth.html.

way in which humans and machines succeed in their performance nor does it refer to cognition. Implicitly, it refers to an observer who would call a human behaviour intelligent. It focuses on the task to be fulfilled and not on an internal function that is used to perform it. There are widespread misconceptions about AI. In the following, I list some issues which help to understand what AI is. The points reflect my opinion, unless indicated otherwise by references.

Dynamic concept of intelligence: Whether a task requires intelligence or not, is not a fixed statement but rather a proposition that is based on historical developments of human education. For centuries, writing and arithmetic were regarded as intelligent. With the regular school attendance of all, it was no longer classified as intelligent activities. The implicit assumption is that intelligence cannot be attributed to the majority but follows a kind of normal distribution with a large p quantile.

The human—machine competition: The capabilities of computers are sometimes considered non-intelligent by definition. That is, every task in which the computer outperforms humans, is no longer stated as one that demands intelligence. Hence, computation is thought of as a simple activity, because it could be automated. The definition of AI itself does not imply the exclusiveness of human intelligence. However, many people seem to feel themselves in competition with machines and make their self-image dependent on their superiority over machines.[3] Since winning chess games is most prestigious, the victory of DeepBlue (IBM) over world champion Garri Kasparov in two of six games in 1996 has been taken as evidence that machines can be intelligent. Similarly, playing Go counts as a very demanding task and the victory of AlphaGo over Lee Sedol in 2016 sparked an AI hype.

Different difficulties for humans and machines: The difficulty for computers to perform a task does not necessarily correspond to its difficulty for human performance. Computational complexity shows the difficulty of particular reasoning tasks independently of the reasoner. For body-related actions, the difficulty of a task is no longer comparable between man and machine. Tying shoelaces, for instance, is much easier for people than for machines. It is an open question, whether and in which way machine learning and human learning follow the same ranking of learning tasks with respect to their difficulty.

Tasks, not personalities: The definition of AI is about particular tasks, not about *all* capabilities of human beings. The first tasks which were modelled by classic AI were planning, reasoning, game playing, natural language processing, image understanding, and robotics. A high-performance chess playing system does not master any other of the human capabilities. Hence, the success of one particular AI system does not imply the overall intelligence of the system. This principle is also valid for systems that combine several capabilities. If the task is learning itself and the learning machine looks similar to humans, it is hard to distinguish the capabilities of humans and machines. This AI stimulates movies and novels most. Where research investigates the modelling of (interacting) tasks, fiction discusses whether many interacting

---

[3] A personal remark: the value of a person is not dependent on comparing him or her with any animal, machine, or even another person. Hence, there is no need to feel reduced, if a machine is stronger or more intelligent.

capabilities which learn from the environment and help each other's evolution will end-up in a being that cannot be differentiated from a human.

Acting descriptions: The distinction of the description and the described is one problem of understanding AI. Where, for example, everybody knows that a description on paper does not act, an AI description is executed by a program to act. The classical AI models tasks by their (logical) description. For instance, the rules that guide the path of an ant—follow the pheromone, perform the easiest move—can be executed in any environment by a general interpreter or solver. Given pheromone signals and obstacles on the ground as input, the path looks always different but is produced by a simple rule-based system. The rule-based ant robot acts like an ant. In this way, classical AI *explains* and performs problem solving at the same time. Theories about natural language syntax (Brill 1992) or natural language dialog (Hoeppner et al. 1986) or even medicine could be modelled and executed that way (Shortliffe and Buchanan 1990).

Machine learning: While early rule-based systems have been designed by hand, machine learning describes learning itself such that computers learn from failure, from examples, from observations. A rule-based system with its inference mechanism and interpreter is the AI system of the first level. Its inputs are the learning results. The learning system is the second level. Learning knowledge bases or knowledge discovery in data automatically builds systems that perform tasks which—when done by human beings—would require intelligence.

Reconstructing human learning formally by a program may help to understand, e.g. children's development of concepts (Morik and Muehlenbrock 1999).

Systems for automated scientific discovery (Nordhausen and Langley 1990; Zytkow 2002), or automated experimentation (Sparkes et al. 2010; Schneider 2017) characterise scientific tasks and execute them, e.g., in order to design new drugs. The never ending language learning project at Carnegie Mellon University (Yang and Mitchell 2016) learns to extract knowledge from web pages. Additionally, the never ending learning system NEQA enhances question-answering over knowledge bases by learning to map questions to formal queries (Abujabal et al. 2018). In this sense, machine learning automatically builds AI. Where Inductive Logic Programming (Muggleton et al. 1998; Kietz and Morik 1994) and statistical relational learning (Getoor and Taskar 2007; Kersting 2006) provide models that can be understood by non-experts with some training, the numerical approaches to learning are less easy to understand, in particular, if kernel functions or neural networks with many layers are used. Deep Learning with the convolutional or recurrent or generative adversarial neural networks results in high-performance models but they require additional components for their explanation (Goodfellow et al. 2016).

Many machine learning applications rely on other systems, to which they deliver input and make them smarter. On one hand, database and knowledge base systems, search engines, routing systems, model-driven control and many other systems are necessary to provide input to machine learning and execute the learned models. On the other hand, machine learning is the key mechanism that can make virtually any system smarter.

Machine learning also enhances itself. Even complete learning processes are used as input to other learning mechanisms such that the chains of operators are enhanced and learning for new applications is partially automated. RapidMiner started with what has been named modestly "parameter optimization", developed automatic enhancements of learning pipelines that are learned from learning processes, and finally came up with Auto Model to create predictive models in 4 clicks.[4] Similarly, the tool Keras, which runs on top of TensorFlow or Theano, provides functions to automatically search for architecture and hyperparameters of deep learning models. It claims to create a deep learning model in 4 lines of code.[5] The Auto-sklearn library replaces a scikit-learn estimator and optimises hyperparameters at scale (Falkner et al. 2018). Interestingly enough, it uses Bayesian optimization. Resource-aware model-based optimization of models has been investigated for R programs (Kotthaus et al. 2017). This learning to learn or self-training makes up for the exponential dynamics of AI.

The mind–body problem: The role of the body for cognitive capabilities is the philosophical problem of mind and body (Bieri 1981). This problem is highly relevant for AI. Embedding cognition into a physical object is investigated in AI robotics. The underlying assumption is, that task performance should not be programmed but is to be learned from interaction with an environment. Machine learning uses an acting body to learn, in other words: intelligent behaviour emerges from inter-actions (Steels 1990). Claiming a general intelligence which underlies bodily as well as mental actions investigates the automatic evolution or learning of sets of capabilities (Schmidhuber et al. 2011). Particularly Japanese scientists investigate androids which combine many cognitive capabilities with a human-like body. The robot behaviours emerge from interactions with the environment. The human-like robots of Hiroshi Ishiguro and the way in which they learn through communication investigates intelligence, be it human or artificial. The Japanese school of robotics does not separate the investigation of the mind from that of the body. It contributes to a better understanding of embodied intelligence and is put to use for complex services which require the combination of diverse capabilities (Ikegami et al. 2017; Keshmiri et al. 2018; Belkacem et al. 2018). Away from the Frankenstein metaphor, it offers a framework for the investigation of man–machine communication and the interaction of several learning tasks.

The Internet: The internet, originally a network of AI laboratories with email, search, and distributed programming services, has become a unique source of data. The level 1 tasks of indexing and search are performed by a level 2 learning system. Moreover, an ecosystem of level 3 systems enhances the learning. For instance, the results are structured for a better overview and content is combined to form a knowl-edge graph. Learning on knowledge graphs allows further services and extraction of facts (Nickel et al. 2011; Bordes et al. 2011). The users' reactions are analysed for better ranking (Joachims 2002; Swaminathan and Joachims 2015). In addition to

---

[4]https://rapidminer.com/products/auto-model/.

[5]https://www.kdnuggets.com/2018/08/auto-keras-create-deep-learning-model-4-lines-code.html.

ranking the search results, there is an auction for advertisements that will be shown to the user. This comes along with other learning tasks, e.g., enhancing a web site for winning the auction.

Investigating social networks (Holder et al. 2016), detecting communities and the influence of links (Barbieri et al. 2013; Kutzkov et al. 2013) or summarising graphs (Tsalouchidou 2018) has many facets. Of course, all these investigations develop methods of machine learning and at the same time exploit techniques from other fields of computer science.

AI and machine learning: AI is the pioneer of computer science. As soon as techniques developed in AI become standard, they disappear from the AI agenda. We have seen the importance of machine learning for AI. Of course, machine learning proves its value in many applications, also outside of AI, in sciences and engineering. We should not forget that the application that was the most important for Google in 2016 has been that for the cooling of its computing centre. Due to machine learning, the costs have been reduced by 40%.[6]

8.3   Data-Driven Sciences

Machine learning is analysing data. Hence, there is a close relation with statistics. We may illustrate it by the cross-industry standard process for data mining, CRISP (Wirth and Hipp 2000). It shows a business process (see Fig. 8.2). Originally, the first step was named *business understanding* and covered interactions between the analysts and the application specialists, finding the right questions and defining the tasks of the analysis. The cycle has not been closed then, but the deployment has been considered ending the cycle by finding the answers. I have changed this, here, because there is always some *process* that produces the data and which is to be enhanced by the deployment. In many commercial applications, the **process** is a production of some goods and the data are logged sensor data of the production. Advanced applications change the production on the basis of the analysis and, thus, produce new data, which are again analysed. For data science applications, the first step corresponds to the *scientific question or hypothesis* and to the experiments that produce the data. Astrophysics and particle physics create huge experiments for gathering data that could possibly verify or establish new hypotheses (Dorner et al. 2017; Aaij et al. 2017). Simulations are necessary, because the ground truth is not available (Bunse et al. 2017). It is similar to the steps *problem, plan, data collection* in the empirical enquiry scheme proposed by Wild and Pfannkuch (1999). The main difference is the central role of data and their processing steps in the CRISP model.

Let us look at the first step from the two perspectives, the statistical and the machine learning one. The statistical field of experiment design gives clear hypotheses and how they could be tested, e.g. by using multi-objective optimisation (Tillmann et al.

---

[6]Interview Vinton Cerf in ZEIT online 25.3.2017 https://www.zeit.de/digital/internet/2017-03/vinton-g-cerf-google-internet-kuenstliche-intelligenz-interview.

Fig. 8.2 The CRISP model

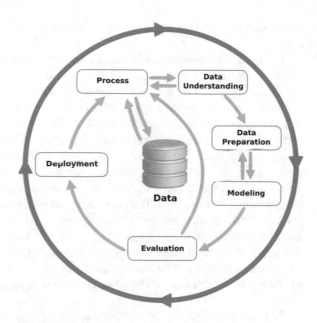

2010). Moreover, quality measures for the design of observational studies prepare the data analysis early on, Pumplün et al. (2005). Machine learning often thinks about the enumeration of all possible hypotheses. Of course, the hypothesis space then needs to be restricted. Many algorithms were developed for the combinatorial enumeration of frequent item-sets (Calders and Goethals 2002). A language bias allows to efficiently search in the space of logical formulas (Kietz and Wrobel 1992). We might also consider clustering a method of data exploration that leads to hypotheses. Ryszard Michalski already in 1986 created a conceptual clustering algorithm that moved beyond similarity and could be used for creating classification hypotheses (Michalski and Stepp 1986). Bayesian networks are directed acyclic graphs whose nodes represent variables that take on values according to the probabilities of values of their parent nodes. All nodes have assigned a conditional probability table for its states. Bayesian networks aim at causal modelling (Pearl 1996). Generative probabilistic models are not restricted to a certain hypothesis, but deliver the probabilities of all subsets of node states, given the remaining node states of a probabilistic graphical model (Wainwright and Jordan 2008).

The second step, **data understanding**, covers different tasks. First, we need a program for reading in the data and storing them. In statistics as well as machine learning toolboxes, correlations and distributions of feature values, diverse visualisations of the data, looking for typical or impossible feature values are the next steps to become acquainted with the data. Interacting with the scientific or business field, the role of the features can be determined, i.e., whether it is a target feature (label), an observed or already aggregated or constructed feature. It is also important to know, how costly the measurement of a feature is and how likely measurement failures are.

Anonymisation or other means of guaranteeing privacy are additional tasks of this step.

The **data preparation** step is sometimes considered a task of database engineering, only. However, the imputation of missing values is again performed by statistics, e.g. Surmann et al. (2018) and supported by machine learning tools. For instance, RapidMiner might replace missing values by the minimum, maximum, average value or learn to predict the missing values. Also the feature transformation, generation and selection is in itself a learning task. In astrophysics, the distribution of the relevant quantity cannot be accessed experimentally, but has to be reconstructed from distributions of correlated quantities that are measured, instead (Bunse et al. 2018). The automatic transformation of time series data can optimise learning performance (Mierswa and Morik 2005). The combinatorial generation of new features for time series in a steel factory, from which relevant ones are automatically selected, enabled successful further learning (Stolpe et al. 2016). Feature selection is the most important step in data preparation. There are very many algorithms that optimise the feature set. Among others, it can be handled as an experiment design (Pumplün et al. 2004), as finding the optimal representation, e.g. through regularisation (Zou and Hastie 2005).

Representation learning is often said to remove the need for feature engineering. One approach is to transfer hyperparameters from one learning application to another one. Deep learning claims to inherently optimise the representation for learning. Again, the investigations are undertaken in statistics (Shwartz-Ziv and Tishby 2017) and machine learning (Maennel et al. 2018), alike.

The **modelling step** is often regarded as the only domain of machine learning and statistics. We have seen, however, that the steps before are covered by these two fields, as well. And certainly, the *evaluation* cannot be separated from the modelling. A given classification or regression task defines a set of hypotheses, namely all correlated premises for the conclusion of the goal. The function $f(x) = \beta x + \beta_0$ expresses weights β for all features, i.e. components of the vector x. Data are used to determine β such that the error is minimised or the likelihood is maximised. Depending on further restrictions the regularised quality measure is optimised. A wealth of models and quality measures together with their properties (convergence, bias, variance, robustness, runtime) is provided by statistical research (Hastie et al. 2008). We might consider this task of learning and its evaluation the centre of statistics. It is also the key modelling step in the CRISP model.

End-to-end learning covers representation learning, model selection and validation schemes such that the overall pipeline or the automated parts of the CRISP model are formalised by a deep neural network. The general inference engine is then the gradient descent method. Limits and merits of end-to-end learning are discussed in a recent paper (Glasmachers 2017).

The **evaluation** of the learned models covers the comparison of diverse models. Model selection as algorithm selection has already been mentioned above, coined as Auto Modeling (Sect. 8.2). In statistics, the selection of algorithms is investigated, as well, see, e.g., Horn et al. (2016), Lang et al. (2014). The evaluation might also lead to the insight, that data are missing or the leading research question needs to be reformulated.

Deployment originally has been something that a company does on the basis of the evaluated model. For instance, mailing actions to selected customers were the deployment of customers' data analytics. In production applications, the learned model may change the production process and, hence, lead to changed data. In data science, the insight may lead to new hypotheses, enhanced simulations, or even changes of the experiment that gathers the data. Some might call the deployment of the model learned from data of a scientific experiment the scientific progress, or even the scientific result. Others demand the incorporation of empirical results into the theory of the discipline being an additional step, which could be named scientific deployment. Behind these different views is the debate about causality. Some are happy with useful predictions, others aim at causal inference (Peters et al. 2017; Pearl 2019). The controversial arguments are not between machine learning on the one hand and statistics on the other hand, but within each discipline, there is a lively dispute. We shall not move into that discussion here, but only state that, again, artificial intelligence and statistics share an interesting scientific topic.

We have shown how close statistics and machine learning actually are by following the CRISP model. However, we have also shown that taking into account computing architectures as well as algorithms and programming frameworks is the additional expertise of machine learning.

Acknowledgements This work builds upon research in the collaborative research centre SFB 876 *Providing Information by Resource Constrained Analysis* funded by the Deutsche Forschungsgemeinschaft (DFG), projects A1, B3, and C3—http://sfb876.tu-dortmund.de.

References

Abujabal, A., Saha Roy, R., Yahya, M., & Weikum, G. (2018). Never-ending learning for open-domain question answering over knowledge bases. In *Proceedings of the WWWW* (pp. 1053–1062). Switzerland: Republic and Canton of Geneva.

Barbieri, N., Manco, G., Ritacco, E., Carnuccio, M., & Bevacqua, A. (2013). Probabilistic topic models for sequence data. *Machine Learning, 93*(1), 5–29.

Belkacem, A.N., Nishio, S., Suzuki, T., Ishiguro, H., & Hirata, M. (2018). Neuromagnetic decoding of simultatenous bilateral hand movements for multidimensional brain-machine interfaces. *IEEE TNSRE, 26*(6), 1301–1310.

Bieri, P. (Ed.). (1981). *Analytische philosophie des Geistes*. Hain Verlag.

Bordes, A., Weston, J., Collobert, R. & Bengio, Y., et al. (2011). Learning structured embeddings of knowledge bases. In *AAAI* (Vol. 6).

Brill, E. (1992). A simple rule-based part of speech tagger. In *Proceedings of the 3rd Conference on Applied Natural Language Processing* (pp. 152–155). ACL.

Bunse, M., Bockermann, C., Buss, J., Morik, K., Rhode, W., & Ruhe, T. (2017). Smart control of monte carlo simulations for astroparticle physics. In *ADASS XXVII*.

Bunse, M., Piatkowski, N., Ruhe, T., Rhode, W., & Morik, K. (2018). Unification of deconvolution algorithms for cherenkov astronomy. In *5th IEEE DSAA*.

Calders, T., & Goethals, T. (2002). Mining all non-derivable frequent itemsets. In T. Elomaa, H. Mannil, & H. Toivonen (Eds.), *Proceedings of the 6th ECML PKDD* (Vol. 2431, pp. 74–85). Berlin: Springer

Dorner, D., Lauer, R.J., FACT Collaboration:, Adam, J., Ahnen, M., & Baack, D., et al. (2017). First study of combined blazar light curves with FACT and HAWC. In *Proceedings of 6th GAMMA* (Vol. 1792).

Falkner, S., Klein, A., & Hutter, F. (2018). BOHB: Robust and efficient hyperparameter optimization at scale. In *Proceedings of the 35th ICML* (pp. 1436–1445).

Getoor, L., & Taskar, B. (2007). *Introduction to statistical relational learning*. MIT Press.

Glasmachers, T. (2017). Limits of end-to-end learning. In *Proceedings of The 9th ACML* (pp. 17–32).

Goodfellow, I., Bengio, Y., & Courville, A. (2016). *Deep learning*. MIT Press.

Hastie, T., Tibshirani, R., & Friedman, R. (2008). *The elements of statistical learning: data mining, inference, and prediction*. Berlin: Springer.

Hoeppner, W., Morik, K., & Marburger, H. (1986). *Talking it over: The natural language dialog system HAM-ANS* (pp. 189–258). Berlin: Springer.

Holder, L. B., Caceres, R., Gleich, D. F., Riedy, J., Eliassi-Rad, T., et al. (2016). Current and future challenges in mining large networks: report on the second SDM workshop on mining networks and graphs. *KDD Explorations Newsletter*, *18*(1), 39–45.

Horn, D., Schork, K., & Wagner, T. (2016). Multi-objective selection of algorithm portfolios: Experimental validation. In J. Handl, E. Hart, P.R. Lewis, M. López-Ibáñez, G. Ochoa, & B. Paechter (Eds.), *Parallel Problem solving from nature – PPSN XIV* (pp. 421–430). Berlin: Springer.

Ikegami, T., Mototake, Y.I., Kobori, S., Oka, M., & Hashimoto, Y. (2017). Life as an emergent phenomenon: Studies from a large-scale boid simulation and web data. *Philosophical transactions of the royal society a: Mathematical, physical and engineering sciences*, *375*(2109).

Joachims, T. (2002). Optimizing search engines using clickthrough data. In *Proceedings of KDD* (pp. 133–142).

Kersting, K. (2006). An inductive logic programming approach to statistical relational learning. *AI Communications*, *19*(4), 389–390.

Keshmiri, S., Sumioka, H., Nakanishi, J., & Ishiguro, H. (2018). Bodily-contact communication medium induces relaxed mode of brain activity while increasing its dynamical complexity: A pilot study. *Frontiers in Psychology*, *9*, 7.

Kietz, J.-U., & Wrobel, S. (1992). Controlling the complexity of learning in logic through syntactic and task–Oriented models. In S. Muggleton (Ed.), *Inductive logic programming* (Vol. 38, Chap. 16, pp. 335–360). Academic.

Kietz, J.-U., & Morik, K. (1994). A polynomial approach to the constructive induction of structural knowledge. *Machine Learning Journal*, *14*(2), 193–217.

Kotthaus, H., Richter, J., Lang, A., Thomas, J., Bischl, B., & Marwedel, P., et al. (2017). Rambo: Resource-aware model-based optimization with scheduling for heterogeneous runtimes and a comparison with asynchronous model-based optimization. In *Proceedings of the 11th LION* (pp. 180–195).

Kutzkov, K., Bifet, A., Bonchi, F., & Gionis, A. (2013). Strip: Stream learning of influence probabilities. In *Proceedings of the 19th KDD* (pp. 275–283). ACM.

Lang, M., Kotthaus, H., Marwedel, P., Weihs, C., Rahnenführer, J., & Bischl, B. (2014). Automatic model selection for high-dimensional survival analysis. *Journal of Statistical Computation and Simulation*, *85*(1), 62–76.

LHCb Collaboration, Aaij, R., Schellenberg, M., Spaan, B., Stevens, H., & et al. (2017). Measurement of cp violation in $b^0 \rightarrow j/\psi k_s^0$ and $b^0 \rightarrow \psi(2s)k_S^0$ decays. *Journal of High Energy Physics*, *11*(170).

Maennel, H., Bousquet, O., & Gelly, S. (2018). Gradient descent quantizes relu network features. *arXiv e-prints* (p. 3). arXiv:1803.08367.

Michalski, R.S., Stepp, R.E. (1986). Conceptual clustering: Inventing goal-oriented classifications of structured objects. In R.S. Michalski, J.G. Carbonell, and T.M. Mitchell (Eds.), *Machine learning - an artificial intelligence approach* (Vol. II, pp. 471–498). Tioga Publishing Company.

Mierswa, I., & Morik, K. (2005). Automatic feature extraction for classifying audio data. *Machine Learning Journal, 58*, 127–149.

Morik K., & Muehlenbrock, M. (1999). *Conceptual change in the explanations of phenomena in astronomy* (pp. 138–167). Pergamon.

Muggleton, S., Srinivasan, A., King, R., & Sternberg, M. (1998). Biochemical knowledge discovery using inductive logic programming. In H. Motoda (Ed.), *Proceedings of the 1st International Conference on Discovery Science*. Berlin: Springer.

Nickel, M., Tresp, V., & Kriegel, H.-P. (2011). A three-way model for collective learning on multi-relational data. In *Proceedings of the 28th ICML*, (pp. 809–816).

Nordhausen, B., & Langley, P. (1990). An integrated approach to empirical discovery. In J. Shrager & P. Langley (Eds.), *Computational models of scientific discovery formation* (Chap. 4, pp. 97–128). Morgan Kaufmann.

Pearl, J. (1996). A casual calculus for statistical research. In D. Fisher & H.-J. Lenz (Eds.), *Learning from data* (Chap. 1, pp. 23–34). Berlin: Springer.

Pearl, J. (2019). The seven tools of causal inference, with reflections on machine learning. *Communications of the ACM, 62*(3), 54–60.

Peters, J., Janzing, D., & Schölkopf, B. (2017). *Elements of causal inference*. MIT Press.

Pumplün, C., Rüping, S., Morik, K., & Weihs, C. (2005). D-optimal plans in observational studies. Techreport, Sonderforschungsbereich 475 Komplexitätsreduktion in Multivariaten Datenstrukturen, Universität Dortmund.

Pumplün, C., Weihs, C., & Preusser, A. (2004). Experimental design for variable selection in data bases. In *Classification - the ubiquitous challenge, proceedings of the 28th annual conference of the gesellschaft für klassifikation e.V.* (pp. 192–199).

Schmidhuber, J., Thóisson, K.R., & Looks, M. (Eds.). (2011). *Proceedings of the 4th AGI*, (Vol. 6830). Berlin: Springer.

Schneider, G. (2017). Automating drug discovery. *Nature reviews drug discovery, 17*(2), 97–113.

Shortliffe, E.H., & Buchanan, B.G. (1990). A model of inexact reasoning in medicine. In G. Shafer & J. Pearl (Eds.), *Readings in uncertain reasoning* (Chapter V, pp. 259–273). Morgan Kaufmann.

Shwartz-Ziv, R., & Tishby, N. (2017). Opening the black box of deep neural networks via information, *3. arXiv e-prints*.

Sparkes, A., Aubrey, W., Byrne, E., Clare, A., Khan, M., Liakata, M., et al. (2010). Towards robot scientists for autonomous scientific discovery. *Automated experimentation, 2*(1).

Steels, L. (1990). Towards a theory of emergent functionality. In *Proceedings of the 1st SAB* (pp. 451–461). MIT Press: Cambridge, MA, USA.

Stolpe, M., Blom, H., & Morik, K. (2016). Sustainable industrial processes by embedded real-time quality prediction. In K. Kersting, J. Lässig, & K. Morik (Eds.), *Computational sustainability* (pp. 201–243). Berlin: Springer.

Surmann, D., Ligges, U., & Weihs, C. (2018). Predicting measurements at unobserved locations in an electrical transmission system. *Computational Statistics, 33*(3), 1159–1172.

Swaminathan, A., & Joachims, T. (2015). Counterfactual risk minimization: Learning from logged bandit feedback. In *Proceedings of the 32nd ICML* (pp. 814–823).

Tillmann, W., Vogli, E., Baumann, I., Kopp, G., & Weihs, C. (2010). Desirability-based multi-criteria optimization of HVOF spray experiments to manufacture fine structured wear-resistant $75cr_3c_2 - 25$ (nicr20) coatings. *Journal of thermal spray technology, 19*(1–2), 392–408.

Tsalouchidou, I., Bonchi, F., Morales, G., & Baeza-Yates, R. (2018). Scalable dynamic graph summarization. *IEEE TKDE*.

Wainwright, M. J., & Jordan, M. I. (2008). Graphical models, exponential families, and variational inference. *Foundations and Trends in Machine Learning, 1*(1–2), 1–305.

Weihs, C., Mersmann, O., & Ligges, U. (2013). *Foundations of statistical algorithms: With references to r packages*. Chapman and Hall/CRC.

Wild, C. J., & Pfannkuch, M. (1999). Statistical thinking in empirical enquiry. *International Statistical Review*, *67*(3), 97–113.

Wirth, R., & Hipp, J. (2000). Crisp-dm: Towards a standard process model for data mining. In *Proceedings of the 4th PADD* (pp. 29–39).

Yang, B., & Mitchell, T. (2016). Joint extraction of events and entities within a document context. In *Proceedings of the 15th NAACL*.

Zou, H., & Hastie, T. (2005). Regularization and variable selection via the elastic net. *Journal of the Royal Statistical Society B*, *67*, 301–320.

Zytkow, J.M. (2002). Automated scientific discovery. In W. Klösgen & J.M. Zytkow (Eds.), *Handbook of data mining and knowledge discovery* (pp. 671–679). Oxford University Press, Inc.

Chapter 9
Statistical Computing and Data Science in Introductory Statistics

Karsten Lübke, Matthias Gehrke and Norman Markgraf

Abstract In the last years, there is movement towards simulation-based inference (e.g. bootstrapping and randomization tests) in order to improve students' understanding of statistical reasoning as well as a call to introduce statistical computing and reproducible analysis within the curriculum. With the help of R mosaic and the concept of minimal R, we were able to include all this in an introductory statistics course for people studying while working a business-related major. Moreover, this also paves the road towards methods and concepts like data wrangling or algorithmic modelling, more related to data science than to classical statistics.

9.1 Introduction

Weihs and Ickstadt (2018), claim that

> we substantiate our premise that statistics is one of the most important disciplines to provide tools and methods to find structure in and to give deeper insight into data, and the most important discipline to analyse and quantify uncertainty. (Weihs and Ickstadt 2018)

This is generally acknowledged by the Guidelines for Assessment and Instruction in Statistics Education (GAISE) Reports (Wood et al. 2018) of the American Statistical Association or the discussion about the subject of statistics, see Wild et al. (2018) for a recent overview.

As already noted by Cobb in 2007 (more than 10 years ago!) (Cobb 2007), computers and statistical computing can help to achieve both: more relevance and better conceptual understanding.

K. Lübke (✉)
FOM University of Applied Sciences, Lissaboner Allee 7, 44269 Dortmund, Germany
e-mail: karsten.luebke@fom.de

M. Gehrke
FOM University of Applied Sciences, Franklinstr. 52, 60486 Frankfurt, Germany
e-mail: matthias.gehrke@fom.de

N. Markgraf
FOM University of Applied Sciences, Herkulesstr. 32, 45127 Essen, Germany
e-mail: nmarkgraf@hotmail.com

© Springer Nature Switzerland AG 2019
N. Bauer et al. (eds.), *Applications in Statistical Computing*,
Studies in Classification, Data Analysis, and Knowledge Organization,
https://doi.org/10.1007/978-3-030-25147-5_9

139

By the integration of statistical computing and elements of data science in introductory statistics courses, we can improve conceptual understanding of statistical reasoning, combine statistical and computational thinking as well as real-world relevance by 'flatten[ing] prerequisites' (Cobb 2015) for real-life data analysis. This is achieved by using the R (Core Team 2018) package `mosaic` (Pruim et al. 2017) with a special focus on modelling (Stigler and Son 2018). The data generating process usually cannot be observed directly. Therefore, we calculate a statistic from a sample used as estimation for the parameter of the data generating process or the population. By simulation and randomization, the distribution of this statistic can be shown. This leads to the basic idea of modelling: data are equal to a model plus an error or residual.

Moreover, integration of statistical computing can enable students to 'compute with data' (Nolan and Lang 2010) as computing literacy and programming are essential to statistical practice as well as what is now called data literacy:

> Data literacy is the ability to collect, manage, evaluate, and apply data, in a critical manner. (Ridsdale et al. 2015)

Furthermore, programming can facilitate reproducible research (Baumer et al. 2014), a problem also addressed by project TIER: Teaching Integrity in Empirical Research (https://www.projecttier.org/) and Open Science Framework (https://osf.io/). Statistical computing is also part of data science (see e.g. Donoho 2017 for an excellent discussion) which should be addressed by statistical education (Horton et al. 2014). As our students are studying while working, the bidirectional transfer between theory and practice, which is alleviated by working on real-life data, is of special importance (Schulte 2015).

9.2 R Package Mosaic

We think that the R package `mosaic` is of special use in teaching statistics and computing if the focus is on data exploration (Tukey 1977), data modelling (Kaplan 2011), and simulation-based inference (Chance et al. 2016).

With `mosaic` other packages are automatically loaded, like `dplyr` (Wickham et al. 2018) which help to preprocess the data. But more important, `mosaic` offers a consistent R formula interface Y~X for graphics (Wickham 2016; Kaplan and Pruim 2018), modelling, and inference. For the students, this makes the turn to typing interface and programming easier. They have to learn one template only:

```
my.analysis( my.y ~ my.x, options, data = my.data )
```

`my.analysis` then is the R function name of the statistical method applied to `my.data` using optional, additional `options`. This pattern is used for graphics like boxplots, descriptive statistics like means or proportions, statistical inference like t-tests or χ^2-tests, modelling techniques like linear or logistic regression, or algorithmic modelling techniques (Breiman 2001) like tree-based methods. The latter are sometimes more associated to data science (Donoho 2017) than to 'classical' statistics.

Moreover, `mosaic` offers an easy access point to simulation-based inference, like bootstrap and permutation tests. As pointed out by Wild et al. (2017):

> With the rapid, ongoing expansions in the world of data, we need to devise ways of getting more students much further, much faster. One of the choke points affecting both accessibility to a broad spectrum of students and faster progress is classical statistical inference based on normal theory. (Wild et al. 2017)

There is a growing body of evidence that simulation-based inference indeed enhances conceptual understanding in investigating uncertainty in inferential statistics (Chance et al. 2016; Tintle et al. 2018; Hildreth et al. 2018).

Permutation tests (see e.g. Ernst 2004) can be based on a Monte Carlo *shuffling* of the independent variable, expressed in `mosaic`:

```
do(often) * my.statistic( my.y ~ shuffle(my.x), options,
                          data = my.data )
```

Whereas Bootstrapping (see e.g. Hesterberg 2015) is based on a Monte Carlo *resampling* of data, expressed in `mosaic`:

```
do(often) * my.statistic( my.y ~ my.x, options,
                          data = resample(my.data) )
```

For both simulation-based inference methods, standard errors, confidence intervals, or p-values can be calculated and the resulting distributions can be visualized by means of histograms.

Here is a short example of using the CIA World Factbook. The first step is loading and data handling:

```
cia <- CIAdata(c("life", "fert")) # Get CIA World Factbook data
cia <- cia %>%
na.omit() # Delete rows with missing values
```

The second step is visual modelling (Fig. 9.1):

```
gf_point(life ~ fert, data = cia) # Scatterplot
```

Wheras the third step maybe a numerical summary:

```
cor(life ~ fert, data = cia) # Correlation Coefficient

## [1] -0.812093
```

Fig. 9.1 Scatterplot of total
fertility rate and Life
expectancy at birth. *Data
Source* CIA World Factbook

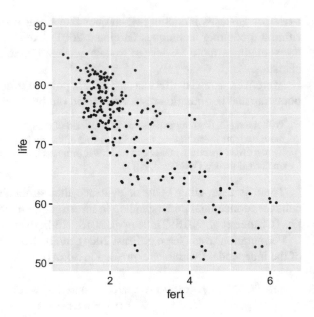

With consistent coding one can perform a (parametric) inference test for
$H_0 : \rho = 0$:

```
# (Parametric) Correlation Test
cor.test(life ~ fert, data = cia)

##
##  Pearson's product-moment correlation
##
## data:  life and fert
## t = -20.689, df = 221, p-value < 2.2e-16
## alternative hypothesis: true correlation is not equal to 0
## 95 percent confidence interval:
##   -0.8525149 -0.7620158
## sample estimates:
##        cor
## -0.812093
```

Simulation-based inference in `mosaic`: what if we had a different sample? So let
us build a Bootstrap distribution and look at the histogram (Fig. 9.2):

```
Bootdist <- do(1000) * cor(life ~ fert, data = resample(cia))
gf_histogram(~ cor, data = Bootdist) %>%
    gf_vline(xintercept = 0, color = "red")
```

Simulation-based inference in `mosaic`: what if we assume that the variables are
independent (Fig. 9.3)?

Fig. 9.2 Bootstrap
distribution of correlation
coefficient between total
fertility rate and life
expectancy at birth. *Data
Source* CIA World Factbook

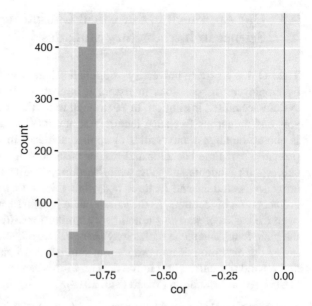

Fig. 9.3 Permutation
distribution of correlation
coefficient between Total
fertility rate and Life
expectancy at birth. *Data
Source* CIA World Factbook

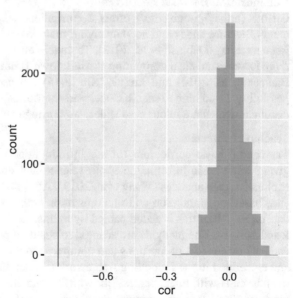

```
Nulldist <- do(1000) * cor(life ~ shuffle(fert), data = cia)
gf_histogram( ~ cor, data = Nulldist) %>%
  gf_vline(xintercept = -0.81, color = "red")
```

9.3 Our Approach to Statistical Computing and Data Science in Introductory Statistics

The FOM is a private university of applied sciences in Germany, founded in 1993 by employers' associations in Essen. It is a nonprofit-oriented, state-recognized, system-accredited institution. In 2018, with more than 40 000 students, 400 professors and numerous freelance lecturers at about 30 locations all over Germany, it is one of the largest universities of applied sciences in Germany. The FOM offers primarily part-time bachelor and master programs in mainly economically related subjects for students studying while working. Due to this structure, it has a heterogeneous student and lecturer body, but this setting provides opportunities for collaboration of lecturers, as the same course is offered at different locations. Moreover, the basics of statistics remain very similar for different courses of study; only the main focus and some topics may change. The relaunch of the Bachelor in Business Administration degree program in summer 2016 gave us the chance to rethink and rebuild our curriculum in the different introductory statistics courses along the GAISE recommendation (Wood et al. 2018).

Collaboration Between Lecturers
GitHub (https://github.com/) offers a convenient way for collaborative development of lecture slides and accompanying materials along with version control and issue tracking (Dabbish et al. 2012). To enable customization, modularization and reproducible statistical computing R Markdown (`rmarkdown`) (Allaire et al. 2018; Baumer et al. 2014) and `knitr` (Xie 2015) are used. R Markdown offers the possibility to combine text, R Code, and the corresponding output in one single document which are in our case slides, supporting exercises, case studies, etc.

Teaching Materials
The central basic curriculum for our introductionary statistics course starts after giving some basic information about science and quantitative data analysis with explorative data analysis (Wang et al. 2017). After that, we directly jump into inference based on simulation and randomization adapted to categorial and numerical data types. All of this is accompanied by letting the students do it themselves using R and `mosaic`. The methods are adapted to standard procedures like testing (difference of) means and proportions and compared to the 'classical' tests like t-tests—and embedded in the modelling framework (Stigler and Son 2018). The basic curriculum ends with linear regression, which is the also the central and main part of the curriculum. Inference in regression again is performed with simulation and randomization versus the standard approach using `lm` and the included t- and F-tests. The latter is still required for the students to read scientific papers.

Frequent quizzes like 'is the observed statistic unlikely under the permutation distribution' are integrated in the lecture slides in order to give feedback both to lecturers and students and also to enhance engagement and learning (McGowan and Gunderson 2010). To further support student achievement or attitudes in statistics, fun elements like cartoons, videos, and songs are included or linked (Lesser et al. 2016). This is to be accomplished by relevant case studies (i.e. small research projects for 180 min.) relevant for the individual students (Neumann et al. 2013; Sole and Weinberg 2017) which also emphasize the whole process from problem to solution (Weihs and Ickstadt 2018; Wild and Pfannkuch 1999; Tintle et al. 2016).

Of course, this shift towards statistical computing comes with a cost: *mathematical* (not to be confused with *theoretical*) statistics is de-emphasized. While sampling, experiment, measurement, variation as well as inference and modelling are still central, manual drawing and calculating, i.e. plugging numbers into equations, are shortened or cut. Also theoretical distributions and asymptotic and approximate procedures are reduced. Equations and code are given to foster understanding—and are simultaneously applied.

9.4 Preliminary Results

The general acceptance of the integration of statistical computing within introductory statistics was rather high among our lecturers: in a survey conducted at the end of the first term after the change (March 2018) more than two-thirds ($n = 37$) find the integration of computational thinking very useful. Additionally, two more surveys among our students were evaluated: a quasi-experiment and pre-/post-test survey.

Quasi-Experiment
Due to a change in the course of study in summer term 2017 introductionary statistics was teached simultaneously following the classical approach as well as following the new approach. The *new* approach uses, as described above, randomization and simulation-based inference for teaching inference. The setting of having two different curricula at the some time enables a quasi-experimental setup in some of the master courses.

To catch the acceptance of R, we used altogether 14 items based on the technology acceptance model (Venkatesh et al. 2003) on a multi-item Likert scale.[1] The scales on each item ranged from 1 (absolute agreement) to 7 (absolute disagreement). Eight additional items regarding statistics in general, using 7 point scales on each item as well, were included to capture perceived interest in and power of statistics. The items for perceived interest were adopted from the SATS-36 questionnaire (Schau 2017), whereas perceived power was build from items like

- statistics allows analysis of dependencies,
- modelling is a main task of statistics,
- statistics allows description of variation.

The four components were rescaled afterwards, so that higher values mean higher usefulness, usability, interest or power.

---

[1] This questionnaire was used in our first surveys 2015 and 2016 already.

Table 9.1 Results quasi-experiment. The first four items run from 1 to 7, the last two from 0 to 5

	Classical approach (second-term students)	New approach (first-term students)	p-value two-sided t-test
Usefulness	3.79	3.57	0.3667
Usability	3.16	2.98	0.4726
Interest in statistics	4.52	5.07	0.0617
Perceived power of statistics	5.41	5.52	0.5776
Understanding p-value	3.01	3.25	0.3842
Understanding confidence interval	1.79	2.00	0.4565

Finally, five questions (true/false) each in regard understanding p-value and confidence interval were included to obtain conceptional understanding. To use them in our analysis, we calculated scores (1 point for each correct answer, maximum of 5 each).

The survey was conducted between June 10, 2017 and June 24, 2017 as supervised questionnaire during lecture time. Target were Master of Sciences programs in their first or second term: first-term students were taught the *new* approach and second-term students the classical approach. Both the course in the first term and the course in the second term were given by the same lecturers. Total number of questionnaires used in evaluation was 125, 28 from first-term courses (3) and 97 from second-term courses (3). For results of this survey, see Table 9.1.

Overall, the results show that the acceptance of the technology (R) was just the same with coding as with the point-and-click interface of the R Commander (Rcmdr, Fox 2017)—but with a higher interest in statistics and a better understanding of statistical concepts, although not significantly. The ability to use R for statistical analysis does not depend on the user interface.

Pre-/Post-Test Survey

From a statistical educational point of view, the most important result is the conceptional understanding, measured using an adopted version of the Comprehensive Assessment of Outcomes in Statistics (CAOS, Delmas et al. 2007; Chance et al. 2016). From this item set, we used a subset of eight questions with focus on sampling, inference, causality, variation, graphical interpretation, etc. All were single-choice questions. Therefore, we counted the number of right answers as score.

This survey[2] took place in summer term 2018. Pre-test was performed at the beginning of the course after the students learned the first steps in R, post-test at the end, both as supervised questionnaires during lecture time. Total response was 600 completed questionnaires (pre-test: 408, 308 of them bachelor, 94 master; post-test: 192, 111 of them bachelor, 77 master, each remaining the selection of study

---

[2]Additionally, to the CAOS items, we used a multi-item scale based on the Unified Theory of Acceptance and Use of Technology (UTAUT2 Venkatesh et al. 2012) which is not part of the investigation referred to in this paper.

Table 9.2 Results CAOS pre- and post-test. Scores representing conceptional understanding of statistics run from 1 to 8

	Pre-test	Post-test	p-value one-sided t-test
All	2.80	3.24	0.0008
Bachelor	2.70	3.02	0.0383
Master	3.22	3.60	0.0595

program is missing). Due to missingness of data, the comparison of the CAOS part in the questionnaire was done on the full set of 600 completed questionnaires (not pairwise only).

As shown in Table 9.2, the students show a significant improvement during the term. Overall, students improved over time, the master students had in average a higher score in the beginning and at the end. The absolute improvement of master students was somewhat higher compared to bachelor students, but relatively to the starting value slightly lower.

More Lessons Learned
But there are also some other lessons learned: Many students struggle with the installation of R, RStudio, and mosaic. Maybe, a cloud-based service like https://rstudio.cloud/ can help. Furthermore, some students have difficulties with basics like entering a file path or distinguishing between different file formats which, at least for us, is basic computing literacy. They struggle in unpacking a zip folder and many students are not familiar in typing without auto-correction. On the other hand, also some lecturers need persuading for a detour of the traditional consensus curriculum despite their positive attitude towards the integration of computational thinking—and they need training. Of course, we are still learning and there is a lot of trial-and-error in the integration of statistical computing and data science in introductory statistics. But it is definitely worth it, as simulation-based inference frees up space for advanced topics like text mining which is important in a sense of general data literacy (Hardin et al. 2015; Kim and Escobedo-Land 2015).

9.5 Outlook

In the focus of (multivariate) modelling, confounding and bias as well as data acquisition and enrichment (Weihs and Ickstadt 2018) are of crucial importance. We will address these topics by the integration of basic concepts of causal inference in the curricula—a shift proposed also by Pearl (2009), Ridgway (2016), Angrist and Pischke (2017) and Kaplan (2018).

Currently, the additional use of more interactive tools like shiny (Chang et al. 2018) is underdevelopment. These tools can further help student comprehension (Doi et al. 2016). More specific, least squares line fitting and sampling, resampling, and re-allocation can be made visible like in Wild et al. (2017).

By interactive R tutorials based on `learnr` (Schloerke et al. 2018), we hope to foster self-controlled learning of the programming part within the curriculum.

As computational and statistical thinking as well as data understanding most often requires judgement (De Veaux and Velleman 2008), experience is needed. We hope to integrate this data literacy in the whole course of the study.

In the end, as Wild et al. (2018) wrote:

> For educational purposes, statistics needs to be defined by the ends it pursues rather than the means statisticians have most often used to pursue them in the past. Changing capabilities, like those provided by advancing technology, can change the preferred means for pursuing goals over time, but the fundamental goals themselves will remain the same. (Wild et al. 2018)

These goals are, as Weihs and Ickstadt (2018), pointed out

> to provide tools and methods to find structure in and to give deeper insight into data, and [...] analyse and quantify uncertainty. (Weihs and Ickstadt 2018)

We agree with Weihs and Ickstadt (2018) that 'only a balanced interplay of all sciences involved will lead to successful solution' but not only for data science but also for introductory statistics education.

Acknowledgements We thank Oliver Gansser, Bianca Krol, Sebastian Sauer, and numerous other colleagues for their contribution in the proposed change of the curriculum and for helpful comments in order to improve the teaching materials. Also, we thank Nathan Tintle for his support with the CAOS inventory. The remarks of Nicholas Horton, Randall Pruim, and two reviewers helped to improve this paper a lot. We gratefully acknowledge that our work was supported by an internal teaching innovation grant by our institution.

References

Allaire, J., Xie, Y., McPherson, J., Luraschi, J., Ushey, K., Atkins, A., et al. (2018). rmarkdown: Dynamic documents for R. https://CRAN.R-project.org/package=rmarkdown; R package version 1.10.

Angrist, J. D., & Pischke, J. S. (2017). Undergraduate econometrics instruction: Through our classes, darkly. *Journal of Economic Perspectives, 31*(2), 125–44. https://doi.org/10.1257/jep.31.2.125..

Baumer, B., Cetinkaya-Rundel, M., Bray, A., Loi, L., & Horton, N. J. (2014). R markdown: Integrating a reproducible analysis tool into introductory statistics. *Technological Innovations in Statistics Education, 8*(1).

Breiman, L. (2001). Statistical modeling: The two cultures (with comments and a rejoinder by the author). *Statistical Science, 16*(3), 199–231.

Chance, B., Wong, J., & Tintle, N. (2016). Student performance in curricula centered on simulation-based inference: A preliminary report. *Journal of Statistics Education, 24*(3), 114–126. https://doi.org/10.1080/10691898.2016.1223529..

Chang, W., Cheng, J., Allaire, J., Xie, Y., & McPherson, J. (2018). Shiny: Web application framework for R. https://CRAN.R-project.org/package=shiny; R package version 1.1.0.

Cobb, G. (2015). Mere renovation is too little too late: We need to rethink our undergraduate curriculum from the ground up. *The American Statistician, 69*(4), 266–282. https://doi.org/10.1080/00031305.2015.1093029..

Cobb, G. W. (2007). The introductory statistics course: A ptolemaic curriculum? *Technology Innovations in Statistics Education, 1*(1).

Dabbish, L., Stuart, C., Tsay, J., & Herbsleb, J. (2012). Social coding in github: transparency and collaboration in an open software repository. In *Proceedings of the ACM 2012 Conference on Computer Supported Cooperative Work* (pp. 1277–1286). ACM

De Veaux, R. D., & Velleman, P. F. (2008). Math is music; statistics is literature (or, why are there no six-year-old novelists?). *Amstat News, 375*, 54–58.

Delmas, G., Joan, G., Ooms, A., & Chance, B. (2007). Assessing students' conceptual understanding after a first course in statistics. *Statistics Education Research Journal, 6*(2).

Doi, J., Potter, G., Wong, J., Alcaraz, I., & Chi, P. (2016). Web application teaching tools for statistics using R and shiny. *Technology Innovations in Statistics Education, 9*(1).

Donoho, D. (2017). 50 years of data science. *Journal of Computational and Graphical Statistics, 26*(4), 745–766. https://doi.org/10.1080/10618600.2017.1384734..

Ernst, M. D. (2004). Permutation methods: A basis for exact inference. *Statistical Science, 19*(4), 676–685.

Fox, J. (2017). *Using the R commander: A point-and-click interface for R*. Boca Raton FL: Chapman and Hall/CRC Press. http://socserv.mcmaster.ca/jfox/Books/RCommander/

Hardin, J., Hoerl, R., Horton, N. J., Nolan, D., Baumer, B., Hall-Holt, O., et al. (2015). Data science in statistics curricula: Preparing students to "think with data". *The American Statistician, 69*(4), 343–353. https://doi.org/10.1080/00031305.2015.1077729..

Hesterberg, T. C. (2015). What teachers should know about the bootstrap: Resampling in the undergraduate statistics curriculum. *The American Statistician, 69*(4), 371–386.

Hildreth, L. A., Robison-Cox, J., & Schmidt, J. (2018). Comparing student success and understanding in introductory statistics under consensus and simulation based curricula. *Statistics Education Research Journal, 17*(1).

Horton, N. J., Baumer, B. S., Wickham, H. (2014). Teaching precursors to data science in introductory and second courses in statistics. arXiv:1401.3269

Kaplan, D.: *Statistical modeling: A fresh approach* (2 edn). Project MOSAIC Books.

Kaplan, D. (2018). Teaching stats for data science. *The American Statistician, 72*(1), 89–96. https://doi.org/10.1080/00031305.2017.1398107..

Kaplan, D., & Pruim, R. (2018). ggformula: Formula Interface to the Grammar of Graphics. https://CRAN.R-project.org/package=ggformula; R package version 0.9.0.

Kim, A. Y., Escobedo-Land, A. (2015). Okcupid data for introductory statistics and data science courses. *Journal of Statistics Education, 23*(2), null (2015). https://doi.org/10.1080/10691898.2015.11889737.

Lesser, L. M., Pearl, D. K., III, J. J. W. (2016). Assessing fun items' effectiveness in increasing learning of college introductory statistics students: Results of a randomized experiment. *Journal of Statistics Education, 24*(2), 54–62 (2016). https://doi.org/10.1080/10691898.2016.1190190, https://doi.org/10.1080/10691898.2016.1190190.

McGowan, H. M., & Gunderson, B. K. (2010). A randomized experiment exploring how certain features of clicker use effect undergraduate students' engagement and learning in statistics. *Technology Innovations in Statistics Education, 4*(1).

Neumann, D. L., Hood, M., & Neumann, M. M. (2013). Using real-life data when teaching statistics: student perceptions of this strategy in an introductory statistics course. *Statistics Education Research Journal, 12*(2).

Nolan, D., & Lang, D. T. (2010). Computing in the statistics curricula. *The American Statistician, 64*(2), 97–107. https://doi.org/10.1198/tast.2010.09132..

Pearl, J. (2009). Causal inference in statistics: An overview. *Statistics surveys, 3*, 96–146.

Pruim, R., Kaplan, D.T., & Horton, N.J. (2017). The mosaic package: Helping students to 'think with data' using R. *The R Journal, 9*(1), 77–102 (2017). https://journal.r-project.org/archive/2017/RJ-2017-024/index.html.

R Core Team. (2018). R: A language and environment for statistical computing. R foundation for statistical computing, Vienna, Austria. https://www.R-project.org/

Ridgway, J. (2016). Implications of the data revolution for statistics education. *International Statistical Review*, *84*(3), 528–549. https://doi.org/10.1111/insr.12110..

Ridsdale, C., Rothwell, J., Smit, M., Ali-Hassan, H., Bliemel, M., Irvine, D., et al. (2015). Strategies and best practices for data literacy education: Knowledge synthesis report (2015). https://doi.org/10.13140/RG.2.1.1922.5044

Schau, C.: Survey of attitudes toward statistiks (2017). https://www.evaluationandstatistics.com/

Schloerke, B., Allaire, J., & Borges, B. (2018). learnr: Interactive Tutorials for R. https://CRAN.R-project.org/package=learnr; R package version 0.9.2.1.

Schulte, F. P. (2015). Die Bedeutung und Erfassung des Erwerbs von Theorie-Praxis-/Praxis-Theorie-Transferkompetenz im Rahmen eines dualen Studiums. http://stifterverband.de/pdf/hds-essen-transferkompetenz.pdf

Sole, M. A., & Weinberg, S. L. (2017). What's brewing? A statistics education discovery project. *Journal of Statistics Education*, *25*(3), 137–144. https://doi.org/10.1080/10691898.2017.1395302..

Stigler, J. W., Son, J. Y. (2018). Modeling first: A modeling approach to teaching introductory statistics. In M. A. Sorto, A. White, L. Guyot (Eds.), *Looking back, looking forward. Proceedings of the Tenth International Conference on Teaching Statistics*. https://iase-web.org/icots/10/proceedings/pdfs/ICOTS10_9C3.pdf.

Tintle, N., Chance, B. L., Cobb, G. W., Rossman, A.J., Roy, S., Swanson, T., et al. (2016). *Introduction to statistical investigations*. Wiley Online Library.

Tintle, N., Clark, J., Fischer, K., Chance, B., Cobb, G., Roy, S., et al. (2018). Assessing the association between precourse metrics of student preparation and student performance in introductory statistics: Results from early data on simulation-based inference vs. nonsimulation-based inference. *Journal of Statistics Education*, *26*(2), 103–109 (2018). https://doi.org/10.1080/10691898.2018.1473061, https://doi.org/10.1080/10691898.2018.1473061.

Tukey, J. W. (1977). *Exploratory data analysis*. Mass: Reading.

Venkatesh, V., Morris, M. G., Davis, G. B., & Davis, F. D. (2003). User acceptance of information technology: Toward a unified view. *MIS Quarterly*, *27*(3), 425–478.

Venkatesh, V., Thong, J. Y., & Xu, X. (2012). Consumer acceptance and use of information technology: extending the unified theory of acceptance and use of technology. *MIS Quarterly* (pp. 157–178).

Wang, X., Rush, C., & Horton, N. J. (2017). Data visualization on day one: Bringing big ideas into intro stats early and often. *Technology Innovations in Statistics Education*, *10*(1).

Weihs, C., & Ickstadt, K. (2018). Data science: The impact of statistics. *International Journal of Data Science and Analytics*,. https://doi.org/10.1007/s41060-018-0102-5..

Wickham, H. (2016). *ggplot2: Elegant Graphics for data analysis*. New York: Springer. http://ggplot2.org

Wickham, H., François, R., Henry, L., Müller, K. (2018). dplyr: A grammar of data manipulation. https://CRAN.R-project.org/package=dplyr; R package version 0.7.6.

Wild, C. J., & Pfannkuch, M. (1999). Statistical thinking in empirical enquiry. *International Statistical Review*, *67*(3), 223–248. https://doi.org/10.1111/j.1751-5823.1999.tb00442.x..

Wild, C. J., Pfannkuch, M., Regan, M., & Parsonage, R. (2017). Accessible conceptions of statistical inference: Pulling ourselves up by the bootstraps. *International Statistical Review*, *85*(1), 84–107. https://doi.org/10.1111/insr.12117..

Wild, C. J., Utts, J. M., Horton, N. J. (2018). What Is statistics? (pp. 5–36). Cham: Springer International Publishing. https://doi.org/10.1007/978-3-319-66195-7_1, https://doi.org/10.1007/978-3-319-66195-7_1

Wood, B. L., Mocko, M., Everson, M., Horton, N. J., & Velleman, P. (2018). Updated guidelines, updated curriculum: The gaise college report and introductory statistics for the modern student. *CHANCE*, *31*(2), 53–59. https://doi.org/10.1080/09332480.2018.1467642..

Xie, Y. (2015) *Dynamic Documents with R and knitr* (2nd Edn). Chapman and Hall/CRC, Boca Raton, Florida. https://yihui.name/knitr/. ISBN 978-1498716963

Chapter 10
Approaching Ethical Guidelines for Data Scientists

Ursula Garczarek and Detlef Steuer

Abstract The goal of this article is to inspire data scientists to participate in the debate on the impact that their professional work has on society, and to become active in public debates on the digital world as data science professionals. How do ethical principles (e.g. fairness, justice, beneficence and non-maleficence) relate to actual situations in our professional lives? What lies in our responsibility as professionals by our expertise in the field? More specifically, this article makes an appeal to statisticians that may neither consider themselves data scientists, nor what they do data science, to join that debate, and to be part of the community that establishes data science as a proper profession in the sense of Airaksinen (2009), a philosopher working on professional ethics. As we will argue, data science has one of its roots in statistics but also contains additional tasks and features that extend it. To shape the future of statistics, and to take responsibility for the statistical contributions to data science, statisticians should actively engage in the discussions. In Sect. 10.1, the term data science is defined, and the technical changes that have led to a strong influence of data science on society are outlined. In Sect. 10.2, the systematic approach from Commission Nationale Informatique & Liberte (2018) is introduced. Along the lines of that approach, prominent examples are given for ethical issues arising from the work of data scientists. In Sect. 10.3, we provide reasons why data scientists should engage in shaping morality around data science and to formulate codes of conduct and codes of practice for data science professionals. In Sect. 10.4, we present established ethical guidelines for the related fields of statistics and computing science. Section 10.5 describes the necessary steps in the community to develop professional ethics for data science. Finally in Sect. 10.6, we motivate our own engagement and give our starting statement for the debate: *Data science is in the focal point of cur-*

U. Garczarek
Cytel Inc, Clinical Research Services ICC, Route de Pré-Bois 20 C.P. 1839,
1215 Geneva 15, Geneva, Switzerland
e-mail: Ursula.Garczarek@cytel.com

D. Steuer (✉)
Helmut-Schmidt-Universität, Universität der Bundeswehr Hamburg, Holstenhofweg 85,
22043 Hamburg, Germany
e-mail: steuer@hsu-hh.de

© Springer Nature Switzerland AG 2019   151
N. Bauer et al. (eds.), *Applications in Statistical Computing*,
Studies in Classification, Data Analysis, and Knowledge Organization,
https://doi.org/10.1007/978-3-030-25147-5_10

rent societal development. Without becoming a profession with professional ethics, data science will fail in building trust in its interaction with and its much-needed contributions to society!

10.1 Definition of Data Science, the Roots and the Changing Role

We start with the definition of data science as given by Donoho which we find very useful. We will describe how data science relates to statistics and machine learning and why the role of a data scientists in society is becoming increasingly important.

10.1.1 Definition of Data Science

There is currently no generally agreed definition of data science. Here we use the definition of Donoho (2017) of *greater* data science:

Data science is the science of learning from data; it studies the methods involved in the analysis and processing of data and proposes technology to improve methods in an evidence-based manner. The scope and impact of this science will expand enormously in coming decades as scientific data and data about science itself become ubiquitously available.

Donoho also provides a classification of the related activities into six divisions:

1. Data gathering, preparation and exploration,
2. data representation and transformation,
3. computing with data,
4. data modelling,
5. data visualization and presentation,
6. science about data science.

Items 1–5 describe the work of a data scientist, item 6 differentiates what he calls *greater data science* from data science.

10.1.2 Relation of Data Science, Statistics and Artificial Intelligence

The lack of an agreed definition of data science is a symptom of a larger problem: it is not (yet) a profession of its own. Some see it as subdivision of machine learning, and thus a subdivision of artificial intelligence, others as subdivision of statistics, that is exploratory statistics, and many see it as a collection of methods from both statistics and machine learning, used by people of different professional backgrounds, or people with no actual professional background only trained in the application of those methods, without the necessary formal scientific education. By starting with the definition of Donoho (Sect. 10.1.1) we already make two statements:

1. Data science should become a profession in the sense of Airaksinen (2009), with a definition, a grounding in science and a task and responsibility in society.
2. Exploratory statistics is a historical predecessor of data science.

With respect to the second point, we do not claim exploratory statistics to be the only predecessor of data science. With the same right, people from the artificial intelligence community can see machine learning as a historical predecessor of data science. Therefore, we want the machine learning and artificial intelligence community to work together with the statistics community on the first point.

10.1.3 The Changes of the Societal Impact of Data Science

The biggest, relatively recent changes in practical data science are the availability of vast amount of data together with the increase in computational power. Technically speaking this enables fast, low-cost processing of ever-changing large databases by algorithms to derive continuously updated, highly condensed and aggregated data, i.e. results.

These results can be fed into human decision-making, that is based on the interpretation and understanding of the results, or they can be used in rules for automatic decision-making. Whether or not, at least interim, the decisions are made with human understanding of the results and how they were generated distinguishes black-box algorithms from other algorithms.

Focus of this article is the consequences of processing and analysing vast amounts of data about humans and human behaviour. Today's possibilities in these respects change human interaction and thus society directly and fundamentally. Examples for this broad claim will be given in subsequent sections.

As data science is the focal point of these developments, the role of data scientists in society becomes more influential and important. With increased influence and importance comes increased responsibility.

10.2 Ethical Issues in Data Science

The awareness that data science and its algorithms have an increased and fundamental impact on society is vivid around the world. There are ongoing or starting discussions in many countries and organizations in legal and political context, actually too many to cite. Instead, we refer to any search in news portals, social media and internet with terms as *algorithm, impact, society*.

Actually, such considerations are not really new. To our knowledge, the first data science application recognized to have a large impact on societal processes are election forecasts and polls on voting behaviour. Many countries have thus regulations on what is allowed to publish when in context of an upcoming or ongoing election. An overview over such regulations is given in Wissenschaftlicher Dienst des Bundestags Fachbereich WD 3 (2018).

In the next section, we want to give examples of ethical issues to make them tangible and to highlight those that we consider relevant for data scientists. Among the various initiatives and their publications (see also Sect. 10.3), here we chose the systematic approach as established by CNIL (Commission nationale de l'informatique et des libertés) in 2017 Commission Nationale Informatique & Liberte (2017, 2018) for exemplifying ethical issues in data science.

That report is the result of a public debate organized by the French data protection authority. We will follow its structure and give examples for each of the given categories of ethical issues to make them tangible. The main points relevant for consideration by data scientists are identified.

10.2.1 Six Main Ethical Issues According to CNIL

In the debate, six main issues were identified. Citations referencing (Commission Nationale Informatique & Liberte 2017) are given in front of each of the following sections. These citations are set in italics to be easily identifiable.

10.2.1.1 Autonomous Machines: A Threat to Free Will and Responsibility?

Delegation of complex and critical decisions and tasks to machines increases the human capacity to act and poses a threat to human autonomy and free will and may water down responsibilities.

The most widely discussed application of this type are autonomous vehicles. Autonomous vehicles have the potential to increase traffic safety, but who is responsible for remaining accidents? Will it be possible to overrule a machine's decision on *lowest* or *allowable* risk, i.e. in case of an emergency.

On a more abstract level, any sufficiently complex system may be called an autonomous machine.

Already today many Kafkaesque situations arise due to complex semi-automatic regulations, i.e. the story of a man who was released from his job by an algorithm due to an error, and no human was able to stop that procedure (Diallo 2018) after the lay-off was triggered.

It must be noted, that in these settings the data scientist is typically not involved directly. Maybe he or she built some model in preparation to steer the machine, but from the model to the actual steering of the machine there are typically many adaptations and rounds of testing the implementation which are done by technicians and engineers following some definition of the machine's intended purpose and specification. In that case, to the authors' opinion, those involved defining the machine's specification together with those implementing the steering take the actual responsibility, and the data scientist may or may not be part of that team, and may even not have any knowledge about the application of the original model for that purpose.

10.2.1.2   Bias, Discrimination and Exclusion

Algorithms and artificial intelligence can create biases, discrimination or even exclusion towards individuals and groups of people.

General Remarks
This issue is one where data science expertise is very important for understanding the extent of the problem. We start stressing one point that is often overlooked, when algorithmic bias is discussed. The very nature of the most commonly applied algorithms, called pattern recognition or classification and clustering, if applied to humans, is applying *prejudice*. In statistical language, they form a prior belief on an individual generated by experience with other individuals assigned to the same group. The goal of these algorithms is the assignment of a new object, in this case, a person, according to some measured characteristics of this person into some group. Judgements and predictions on, e.g. future behaviour or reactions to medical treatment for the individual are then made according to previously observed behaviours or reactions of the others in the group. Obviously, if this leads to an improved medical decision-making, this is to the benefit to the individual and the society at large.

In many examples, though, there is a possible benefit to some and a negative impact on others. In those cases, questions of fairness and justice are touched by the use of these algorithms for judgement, prediction and decision-making in general.

Any of their use constitute *bias*, if the measured characteristics, that lead to the assignment into the group, are only correlated but not causally related to the features that are judged about. *Bias* here is defined as the difference between the probabilities that the model assigns to the unknown (future) features of a person as compared to the true underlying probabilities for this (future) feature of the person. One way to understand it is that any such model must be of high quality both on the explanatory dimension of describing causal relationships and the predictive dimension on allowing some quality of prediction even on the individual level (Shmueli 2010). For those familiar with the theory of causal graphs, the formal reason is that the relationship between what is predicted or judged about for the individual and the measured characteristic of the individual is conditionally independent given the individual at a certain point in time. Note that this bias is created independently on whether or not the underlying database is representative for the larger population for the measured characteristics. The bias is created by applying an approach (= data + method) that is suitable for correlational analyses only for judgements that require causal reasoning on individual level.

Practically, this is not different from humans basing their judgement on a person on experiences (= data) they have made with other people that are alike based on some arbitrary (that is bearing no causal relationship) assessment on similarity.

If this is implemented by an algorithm the impact to an individual can be as severe or even more severe. More severe, if there is no escape from the algorithms' decision, while it would have been possible to go to another decision maker, as, e.g. in medical decision-making on getting access to a specific treatment when asking for a second opinion. From a societal perspective, implementing an algorithm that systematically applies a prejudice that is objective in that it is not based on emotion and sentiment but subjective and arbitrary in the choices of characteristics used to make a decision (bearing of a causal relationship and thus arbitrary choice among all characteristics that define an individual human being) raises ethical issues for justice and equality.

Combined with the observable concentration of data ownership, like currently for social media or search data, and with the scalability of computing power such a systematic bias can easily become a universal norm. Where the algorithm uses characteristics that include or are related to protected characteristics by anti-discrimination laws (mostly race, sexual orientation, religion or belief, age and disability) any judgement and any decision based on the algorithm constitute instances of *discrimination*, when they result in one person being treated less favourably than another in a comparable situation.

This does not happen only in badly designed or malfunctioning systems. It is in the core of all classification applied to people.

Another, practically incurable, drawback of those algorithms is that they infer from data of the past, on the members of the group and/or the individual on which one wants to judge, and human behaviour on an individual level and their patterns do change over time.

Examples
The probably most famous example is COMPAS (Correctional Offender Management Profiling for Alternative Sanctions) a software used in the US judicial system to classify the probability of defendants' recidivism. A good discussion of the approach can be found in O'Neil (2016). It was shown in a detailed analysis (Angwin et al. 2016a, b) that the privately owned algorithm used in the juridical system gave far better prognoses for white than for black people, thus it discriminated implicitly based on colour. The machine-generated prognosis was intended just to help the judges, but in interviews it could be seen, that it played a crucial rule in the judgements. Especially decisions by the judges whether defendants could get out on parole or had to go to jail were strongly influenced by the algorithm's output and discriminated against black people.

It must be stressed, that this bias in application was not intentional as far as it is known. The bias most probably was introduced through available data on prisoners in conjunction with the above-described fundamental misunderstanding that observed correlations would be good enough to make decisions that require causal reasoning.

Examples of the application of algorithms are not restricted to the US. In Europe, for example there is a recent initiative in Austria to classify unemployed people in one of the three possible groups: *bad (<= 25%)*, *mediocre* or *good chances (>=66%)* to be employed for at least 6 months in 24 months from now Fanta (2018). The idea is to spend money to bring people back into the workforce more on target. Controversial is the stated goal to spend *less* money on those in the lowest group. It is reported that age and nationality increase one's probability to be put in the lowest group. Both points seem to be openly discriminating. The official stance is that the algorithm does not decide but only helps a human to decide and therefore no discrimination would happen (Futurezone 2018). This is ignoring to the large influence that those supportive systems have, when there is a shortage of money: decision makers typically need to justify, if they deviate from the algorithmic choices, but not if they follow the machine's decision. The default mode of operation may change through the use of such a simple helper algorithm. The ethical issue arises from the intention of economic optimization through the implementation of the algorithm. So what is the responsibility of the data scientist that provides a dedicated algorithm for this and knows that its decision- making is not objective and prejudice free?

A very similar system is already in use in Poland Panoptykon Foundation (2015).

In the examples given, in addition to generating bias, the automatic classifiers act like self-fulfilling prophecies. The automatic, even secret, classification of an individual will influence his or her future life, in the direction the chosen algorithm determines. At the same time, it becomes impossible to assess the algorithm's performance in the future, as the future of the individual's life is changed based on the algorithm's outcome and there is no control group.

Also, the algorithms act very similar to ancient oracles. For an outsider, it is impossible to find out which characteristics of a person exactly have led to the given classification. They are black-box algorithms, a feature shared by many of the algorithms from the artificial intelligence community. There only is the spell of the oracle, no reasoning, and no possible recourse. For that reason, black-box algorithms will always be problematic for usage in any juridical system or for any scoring implying a value judgement of an individual, i.e. credit scoring. Of course, the machine learning community is well aware of the problems following from applying black-box algorithms to real-life problems. There is a lot of research going on make these models interpretable and therefore easier to trust.

The given applications are examples for applications, where some people have a benefit and others negative consequences from the application of the algorithm. It is accepted, that the application may be not in the interest of the individual that is judged.

Of course, this is not a drawback inherent in using algorithmic decision-making. It is possible to set up procedures with no intention to inflict negative consequences on some to the benefit of others, if care is given to transparency and possible discriminating behaviours. For example, in Germany, there exists a program RADAR-iTE (Regelbasierte Analyse potentiell destruktiver Täter zur Einschätzung des akuten Risikos - islamistischer Terrorismus) (Bundeskriminalamt 2018) where an algorithm is used to try identifying the more dangerous people in a group of people already under investigation by law enforcement.

Decisions are based on a set of 72 questions, which are transparent for anybody involved. Because those under inspection by RADAR-iTE already are under investigation, the most important aspect of its application is resource allocation by law enforcement. There is no additional negative effect on those individuals that are judged to be high risk beyond being under investigation already. Publicized numbers (FAZ 2018) give around half (96 of 205) of the suspects are considered *low risk* after classification by RADAR-iTE, only around 40% (82 of 205) are considered *high risk*. Transparency of all steps seems guaranteed throughout all decisions performed with respect to algorithmic classifications.

In this case, those applying the algorithms and those being judged share in some sense the goal to reduce the number of individuals that are observed. The application of the algorithm has the potential to help an individual by being removed from the group of *high risk* people.

The implications of a similar algorithm if it was applied to screen the overall population would lead to a completely different assessment. Technically, there is no barrier to such use. It can only be prevented by *morality* and *law*.

10.2.1.3 Algorithmic Profiling

Personalizing versus collective benefits: Individuals have gained a great deal from profiling and ever finer segmentation. This mindset of personalizing can affect the key collective principles like democratic and cultural pluralism and risk sharing in the realm of insurance.

The most discussed form of personalizing in the age of the Internet is the so-called *filter bubble* (Pariser 2012). The scandal around Cambridge Analytica using Facebook data for micro-targeting a very specific subset of the public with the aim to influence the US elections in 2016 made the dangers of highly personal news and marketing feeds obvious (Confessore 2018; FAZ 2018).

As a reaction, the legislative started to formulate laws to reduce the risks of such personalized targeting with fabricated news, i.e. in Germany the 'Netzwerkdurch-setzungsgesetz' (Bundesministerium der Justiz und für Verbraucherschutz 2018). Facebook restricted the admission to personal data for third parties in the aftermath of that scandal (Schroepfer 2018).

A data scientists role, if implementing schemes for targeting specific sub-population identified by profiling with the help of the vast amount of information available on each active person in the Internet, should at least be to warn of possible misuse. She or he should understand the *dangers for society* and only help to implement *lawful or ethical* algorithms.

A nice example of the second point on risk sharing is telemetry data collected by the so-called smart devices and transmitted to insurance companies. Since the beginning of 2018, each new automobile in the EU has to record telemetry data in a system called eCall (European Commission 2018). While that system will only transfer data in case of an emergency, there are systems that collect lots of information about all aspects of car usage, down to location and the music the driver listens to Seeger (2016). First, there are obvious problems with privacy, if there can be unlawful information sharing. The second problem here are insurance companies who try to give personalized policy premiums based on level of data sharing a car owner accepts. Probably even more problematic are health data, which can be accessed by insurance companies (Forbes 2015).

While at first, nothing seems at stake if an unhealthy living style is punished with higher policy costs, a second look reveals that the fundamental principle of an insurance, namely risk sharing among a large group, is eroded. In addition, there is a direct conflict of personalized insurance policies and personal freedom. Big monetary pressure on customers to live a good life *in the sense of the insurance companies* must be expected.

10.2.1.4   Preventing Massive Files While Enhancing AI: Seeking a New Balance

Artificial intelligence by being based on advanced techniques of machine learning requires a significant amount of data. Still, data protection laws are rooted in the belief that individuals' rights regarding their personal data must be protected and thus prevent the creation of massive files. AI brings up many hopes: to what extent the balance chosen by the lawmaker and applied until now should be renegotiated?

A field of research that is already very experienced and advanced in using large databases on humans and trying to find ways to make that balance is the medical field. Thus, the following two examples are able to illustrate the benefits of the availability of collected personal data and how the risks for individuals regarding their privacy or for the society regarding fair access to information were mitigated.

In July 2018, some valsartan products were discovered to have been contaminated with N-nitrosodimethylamine (NDMA). In September 2018, an expedited assessment of cancer risk associated with exposure to NDMA through contaminated valsartan products could be published Pottegård et al. (2018), providing reassuring interim evidence that the short- term overall risk of cancer in users of valsartan contaminated with NDMA was not markedly increased. This fast assessment in a relatively large cohort (5150 Danish patients) was possible by linking data from four official Danish registries on individual level thus collecting information on prescriptions, cancer diagnosis hospital admissions, mortality and migration. Privacy was implemented by a process where officials from the registries perform the linking, derive the important information, and then de-identify the data before it is sent to the scientists.

In 2018, the German health insurance company DAK Gesundheit in cooperation with scientists from the University of Bielefeld published a report on the health status and the health costs of children and adolescents based on the claims database from the people insured with the DAK Gesundheit (Greiner et al. 2018). Next to some general overview on the health status, a key topic was the investigation of the influence of socio-economic status and education of the parents on the health and induced health costs of the children. The main conclusion is that education is a stronger influencing factor than socio-economic status and that important preventive measures consist of giving children good health education. In the same report, and by guest authors

(Kuntz et al. 2018), also the results from the KiGGS study (Kurth et al. 2008) are discussed. That study puts its emphasis more on the principle of equal opportunity and the influence of socio-economic status on general health and specifically mental health. Publishing this together shows sensitivity of the topic in the political debate and the role that an open scientific environment has to play.

Both, the valsartan case and the DAK study show that there are true benefits for public health that can be generated from using large medical databases. When balancing these benefits with the risk for privacy violations for the people whose data is used, in the valsartan case, we want to highlight the high trust from the citizens that is given to officials: if data on any medical problem one encounters in life can be linked to the home address, citizens need to trust the government that this data is not accessible or made accessible to anyone that uses this information with other than the best intentions. With the DAK study, we want to highlight another important aspect of balancing benefit–risk: the ownership of data, and fair access to data. Data is the *new oil*, and evidence generation shapes how benefit is defined and how it is implemented. Thus, if risk is shared by people of all political opinions, then fairness requires that evidence generation is possible for people from different political opinions.

In general, an important measure for respecting privacy is to de-identify data in the databases, and making them non-identifiable. Guidelines exist for de-identification processes (e.g. the Safe Harbor method U.S. Department of Health and Human Services 2002), yet, with growing databases through social media use and genetic and biomarker research, non-identifiability is a moving target. A good countermeasure is implemented in the process for requesting access in the so-called MIMIC-III database (Goldberger et al. 2000; Pollard and Johnson 2016) on critical care unit patients. In addition to a required training on data privacy, and a strict de-identification of the data, all scientists accessing the data have to submit a data use agreement with 10 points, among which there is one requiring the scientists take immediate action should they realize that there is a way to de-identify data. This is acknowledging the fact that de-identification is no guarantee to de-identifiability at all times by installing a process to monitor de-identifiability by those who have the expertise and knowledge, namely the data scientists, holding them responsible for it and giving them, as a community, a general credit of trust.

10.2.1.5 Quality, Quantity, Relevance: The Challenges of Data Curated for AI

The acceptance of the existence of potential bias in datasets curated to train algorithms is of paramount importance.

Even if implemented in best of mind, there may be unexpected bias in the training data going beyond what has already been said about bias in Sect. 10.2.1.2. There are many examples to find, we want to give two.

One famous example of algorithmic training going wrong was Microsoft's Twitter bot *Tay* (Perez 2016). Tay was implemented to act on Twitter as a regular user. The bot should learn from the comments by others how to perform common Twitter conversations. In less than a day, the humans had learned how to manipulate the learning algorithm in such a way that *Tay* started to speak out fascistic and racist paroles. Microsoft decided to take *Tay* offline less than a day after it started learning.

A recent example of a similar event is an AI system at Amazon. That system should help to find the most qualified applicants in their huge stream of applications. The experiment had to be stopped, when it was noted that the algorithm systematically downgraded applications of women. In Dastin (2018), some probable causes for that behaviour are given. The training data contained mostly applications of men, so most of the successful applicants were men. There are not too many details, but as a consequence any appearance of the word *woman* reduced the chances of that applications.

Finally, the whole project was stopped, even after the developing team tried to correct for known shortcomings, because *there was no guarantee the machine would not devise ways to discriminate in other ways* (Dastin 2018).

The important observation in both cases is, that these black-box algorithms couldn't be improved. They had to be taken offline and completely replaced. As an obvious consequence such algorithms should not be used, where such a replacement is complicated or dangerous.

10.2.1.6   Human Identity Before the Challenge of Artificial Intelligence

Hybridization between humans and machines challenges the notion of our human uniqueness. How should we view the new class of objects, humanoid robots, which are likely to arouse emotional responses and attachment in humans?

This point from the debate in France run by CNIL is given only for the sake of completeness. At the moment, we do not believe that this is an ethical issue where data scientists have a special responsibility due to their expertise.

10.2.2   Conclusion from CNIL's Report

The given examples show the multitude of complex ethical issues that arise from a data scientist's work. In the next section, we argue that ethical guidelines for data scientists are one means to help them taking their responsibility.

10.3   Guidance for Data Science

The call for more guidance for digital technologies in media, in general, is loud and all across the globe, leading to various initiatives and groups engaging in discussions around ethical rules for developing and implementing those technologies. For an overview on initiatives and ethical values in the technical field, visit the website of the think tank doteveryone (James 2018) or the blog of Erickson (2018). There is a long history of computer scientists discussing the ethics of algorithms. A good starting point is the website http://fatml.org. Here, FATML is an acronym for *Fairness, Accountability, and Transparency in Machine Learning* and stands for a series of conferences. For the German-speaking communities, we recommend the slides to the 1-day workshop *Ethische Leitlinien wissenschaftlicher Fachgesellschaften* of the Deutschen Gesellschaft für Medizinische Informatik, Biometrie und Epidemiologie (GMDS) (Deutschen Gesellschaft für Medizinische Informatik 2017) or the Algorithmic Accountability Lab (AAL) at the University of Kaiserslautern http://aalab.informatik.uni-kl.de. AAL provides a good source for current discussions not specific for data scientists but about the use of algorithms in general with some hints towards data science.

This article is in that sense, one contribution among many. Its main purpose is to broaden the audience and increase the number of participants in the discussions, and to foster the development of *morality*, a *set of deeply held, widely shared and relatively stable values* (Horner 2003) on data science within and around the data science community. As any ethical guidance, be it in form of codes, oaths and even law, only has the intended impact, if people are willing to follow it, and the chance for that is high, if the underlying norms and values are in accordance with, in this case, the data science community's own morality.

10.3.1 Do We Really Need More Ethical Guidelines?

Not everyone would agree that data scientists need more guidance on how to make moral decisions in their professional life: many do work in companies with codes of conduct, work for institutions that require some oath, or are members of scientific societies that give ethical guidelines to their members, or have religious beliefs that give guidance to wrong or right in their life, and there is the fundamentally sceptical view that paper does not blush. Also, we are all obliged to obey to law. So what does a special set of ethical rules for the profession of data science add? In the authors' opinion, there are at least four obvious rationales:

1. For the individual data scientist, the translation from very general ethical principles from common morality, law or religion to an ethical issue at work can be quite difficult. Especially since most issues are not about intentions, but about the consequences of one's work. Those consequences are often not very easy to judge upon. Having some reference to well-thought-out and well-reasoned guidelines in that sense is neither more nor less than having publications on specific methods: it helps to avoid re-inventing the wheel ever so often. In addition, it can be helpful to have such a reference along with the reasoning for justification, if the consequences of an ethical decision increase the workload for a colleague or costs for an employer or client.

2. For data scientists as a community, having formulated codes of conduct or some service ideal makes the difference of acting as professionals or merely having

a job that does data crunching. In sociology, a profession is defined by means of professionalism. This implies that a profession has a certain degree of autonomy in society, its members' expertise is based on science, and the professional work exemplifies a service ideal (Airaksinen 2009). In other words: without a service ideal, there is no professionalism and without professionalism, there is no profession.

3. For data scientists as members of society, for their clients, employers and colleagues, written rules of conduct for data science services can help to establish a relationship of trust. If written clearly, they give lay people some mean to know what to expect from a data scientist, to compare what they are getting against that standard, and finally to gain trust if the expectations are met. Being trusted as a professional increases social status, reputation and possibly the money that is paid for the service.

   A code of conduct or ethical guidelines may even be the start of a well-defined job definition for data scientist!

4. In case of conflicts of interests, an ethical guideline under the maintainership of some professional society may offer an arbitration process between different interests.

10.4 Existing Guidelines and Codes

In the previous section, we provided references to ongoing efforts to develop ethical guidelines to data science itself and connected scientific or technical fields. Here, we want to give more details on three main guidelines from the fields of statistics and computer sciences from some of the largest and oldest established associations for those communities. If one could establish additional sub-guidelines that filled the gaps with respect to data science aspects, the audience would immediately be very large, and there would be no need to establish a new association. Both, ACM and ASA, acknowledge data science as an important field in their domains.

10.4.1 ASA: Ethical Guideline for Statistical Practice

The American Statistical Association was founded in Boston in 1839 and has more than 19000 members worldwide. The current Ethical Guidelines (American Statistical Association 2018) have been updated and approved by the ASA Board in April 2018. The guideline has eight sections, six of which describe the responsibilities towards individuals and groups of people to which the statistical work may matter:

- Professional integrity and accountability,
- integrity of data and methods,
- responsibilities to science/public/funder/client,
- responsibilities to research subjects,

- responsibilities to research team colleagues,
- responsibilities to other statisticians or statistics practitioners,
- responsibilities regarding allegations of misconduct,
- responsibilities of employers, including organizations, individuals, attorneys or other clients employing statistical practitioners.

Checking which of the ethical issues discussed in Sect. 10.2 are covered, one recognizes, that implicitly, it is a clear call for human responsibility addressing the issue raised on autonomous machines (Sect. 10.2.1.1). It only touches very briefly on the risk, that information presented as aggregates on groups may lead to bias, discrimination and exclusion (Sect. 10.2.1.2). It sets high standards for privacy and respecting data confidentiality (Sect. 10.2.1.4). With the integrity of data and methods section and throughout almost any other point, it gives clear guidance on quality, quantity and relevance of data, and to a general notion of scientific honesty. It also addresses ethical issues specific to human studies, not covered in Sect. 10.2, but relevant to all scientists working in that field. The guidelines have gaps concerning those ethical issues that result from the implementation of *statistical procedures* into daily practice. Missing are discussions on all ethical issues that can arise from implementing algorithmic results without further human interaction into automatic decision-making.

10.4.2 ACM: Code of Ethics and Professional Conduct

The Association of Computing Machinery (ACM) was founded in 1947 and has more than 100.000 members worldwide. The ACM has ethical guidelines for a long time. *The Code* (Association for Computing Machinery 2018) as it is named, has just been updated and adopted by ACM in June 2018. It has a preamble and four sections:

- General ethical principles,
- professional responsibilities,
- professional leadership responsibilities and
- compliance with the code.

On a general level, *The Code* addresses all ethical issues that we present in Sect. 10.2. Yet, the Code is not a code for data science, and it is not providing the constructive guidance ASA gives on the integrity of data and methods related to scientific honesty and on responsibilities to research subjects.

10.4.3 Ethical Guidelines of the German Informatics Society

The German Informatics Society (GI) has a long history of its ethical guidelines (Gesellschaft für Informatik 2018). The latest update was in June 2018. These guidelines are concise and consist of a preamble and 12 very short sections.

- Sections 1–4 concentrate on aspects of the professional competence of computer scientists,
- Sections 5 and 6 are about individual working conditions,
- Sections 7 and 8 are about teaching and researching in the field of computer science.
- Very interesting are Sects. 9, 10 and 11 which clearly state the societal responsibilities of computer scientists. We see some intercept with the work of data scientists there.
- Finally, Sect. 13 defines a mediating role of the German Informatics Society in case of conflicts stemming from these guidelines.

There are no data science specific sections in these guidelines, nevertheless many important aspects are touched. We think the structure of the ethical guidelines of the GI can be a good skeleton to develop ethical guidelines for data science.

10.4.4 Conclusion from Examining Existing Guidelines

The ethical guidelines for statisticians from the ASA are constructive and detailed for the ethical issues of statisticians and data scientists in the sense of Donoho (Sect. 10.1) that work in research and the special responsibilities towards participants in human studies. *The Code* of the ACM covers the area of using data from and about humans outside from human studies and issues that arise from implementing algorithms from data science for repeated use and that have an impact on individuals and communities. What we have in mind is a combination of those aspects, may be structured as in the guidelines of the GI, as data scientists work on data from all sources and across all those areas.

10.5 Development of Ethical Guidelines for Data Science

There are hurdles to overcome before a meaningful guideline can be established. In our view, the main ones are the lack of a sense of community and a lack of communication on ethics.

10.5.1 Data Scientists Have to Perceive Themselves as a Community

At the moment the term *data scientist* in not a protected professional title. Data scientists can have academic training in statistics, or computer science, as their main fields of professional training, but also engineering, psychology, business management, or they can be trained programmers or only have been following a 3-month course on data science learning Python, Julia or R. In that sense, data science today is not a profession but only occupation. Airaksinen (2009). Between the data scientists from statistics and computer science, on the ground, there is not much tension, but there are many turf battles on academic levels. So the first step would be to realize that ethical guidelines are a shared interest and to then start discussing the content within data science related societies, at conferences, in University courses, at work with colleagues.

Being a community does not mean that there is a need for a new association. A good option would be to add data science specific guidelines to those of the ACM, the ASA and the GI. Such an approach would have the big advantage, that it would not require to first establish a new data science association. Of course, the authors would like to see the European statistics societies embracing ethical issues in their agenda. It would seem natural the *Gesellschaft für Klassifikation – Data Science Society* took a leading role in discussing and defining such guidelines.

10.5.2   Data Scientists Have to Overcome Shyness or Ignorance to Discuss Ethics and Own Moral Views Related to Data Science

In the perception of the authors, it is uncommon for data scientists in industry to express any moral view on the work they do or on the impact their work may have for fellow people and the society at large. That might be, because only recently society and data scientists themselves have realized how much impact data science services have on individuals and communities. Maybe that is because the very nature of this impact is, to be de-personalized and it is easy to overlook one's own responsibility. Maybe it is because most people in data sciences are coming from a mathematical, technical or computer science background and are in general less vocal on anything outside hard science.

The places to change such culture fundamentally should be universities and colleges where data science is taught. Ethics and professional ethics should be part of the curriculum, just as inspiring critical thinking and expressing one's views. First steps in this direction happen just now, see, for example Veaux et al. (2016), where curriculum guidelines for undergraduate data science courses are described and ethics at least is mentioned as part of the education.

In the meantime, every data scientist can work towards that goal within her or his environment. Crucial is taking part in discussions at work in critical projects or within any community when there are, e.g. discussions on the so-called digital revolution, the influence of social media, or algorithms in health care or the criminal justice system.

Talking about ethical questions must become natural for any data scientist.

10.6   Conclusion

We wrote this article for most parts without assuming that our views are generally shared views, or that anyone has to agree that any given specific application is good or bad. Underlying, there is an understanding that the morality of the data science community is evolving and that it is a shared task to develop it, which in turn needs open discussions. Yet, there is at least one fundamental basic moral conviction of the authors, which we have taken as a generally agreed moral principle: as a human being one has to think about possible consequences of one's actions. That responsibility for the consequences grows with the knowledge and the potential one has to think about consequences.

Finally, we want to start the debate with a first statement:

Data science is in the focal point of current societal development. To build trust in data science and its interaction with society and to empower data science to take its responsibility for its contributions to society, data science must develop professional ethics and become a clearly defined profession!

References

Airaksinen, T. (2009). The Philosophy of Professional Ethics. In R.C. Elliot (Ed.), INSTITU-
    TIONAL ISSUES INVOLVING ETHICS AND JUSTICE, (Vol. 1, pp. 201) in Encyclopedia of
    life support systems (EOLSS), Developed under the Auspices of the UNESCO, Paris, France:
    Eolss Publishers. http://www.eolss.net.
American Statistical Association, Ethical Guidelines for Statistical Practice, Approved April
    2018.   https://www.amstat.org/ASA/Your-Career/Ethical-Guidelines-for-Statistical-Practice.
    aspx. Cited 9 Nov 2018.
Angwin, J., Larson, J., Mattu, S., & Kirchner, L. (2016a). ProPublica: Machine bias,
    May 23, 2016. https://www.propublica.org/article/machine-bias-risk-assessments-in-criminal-
    sentencing. Cited 24 Oct 2018.
Angwin, J., Larson, J., Mattu, S., & Kirchner, L. (2016b). ProPublica: How we analyzed the COM-
    PAS recidivism algorithm, May 23, 2016. https://www.propublica.org/article/how-we-analyzed-
    the-compas-recidivism-algorithm. Cited 24 Oct 2018.
Association for Computing Machinery (ACM), ACM Code of Ethics and Professional Conduct,
    Approved June 2018. https://www.acm.org/code-of-ethics. Cited 9 Nov 2018.
Bundeskriminalamt, Presseinformation: Neues Instrument zur Risikobewertung von potentiellen
    Gewaltstraftätern, RADAR-iTE (Regelbasierte Analyse potentiell destruktiver Täter zur Ein-
    schätzung des akuten Risikos - islamistischer Terrorismus), 2 Feb 2017. https://www.bka.de/DE/
    Presse/Listenseite_Pressemitteilungen/2017/Presse2017/170202_Radar.html Cited 6 Nov 2018.
Bundesministerium der Justiz und für Verbraucherschutz, Gesetz zur Verbesserung der Rechtsdurch-
    setzung in sozialen Netzwerken (Netzwerkdurchsetzungsgesetz - NetzDG), 1.9.2017, https://
    www.gesetze-im-internet.de/netzdg/BJNR335210017.html. Cited 6 Nov 2018.
Commission Nationale Informatique & Liberte, Algorithms and artificial intelligence: CNIL's report
    on the ethical issues, 25 May 2018. https://www.cnil.fr/en/algorithms-and-artificial-intelligence-
    cnils-report-ethical-issues. Cited 2 Nov 2018.
Commission Nationale Informatique & Liberte, HOW CAN HUMANS KEEP THE
    UPPER HAND? The ethical matters raised by algorithms and artificial intelligence, Dec

2017. https://www.cnil.fr/en/how-can-humans-keep-upper-hand-report-ethical-matters-raised-algorithms-and-artificial-intelligence. Cited 2 Nov 2018.

Confessore, N. (2018). New York times, Cambridge analytica and facebook: The scandal and the fallout so far, 4 Apr 2018. https://www.nytimes.com/2018/04/04/us/politics/cambridge-analytica-scandal-fallout.html. Cited 6 Nov 2018.

Dastin, J. (2018). Amazon scraps secret AI recruiting tool that showed bias against women, Reuters news, 22 Oct 2018, 5:12 am. https://www.reuters.com/article/us-amazon-com-jobs-automation-insight/amazon-scraps-secret-ai-recruiting-tool-that-showed-bias-against-women-idUSKCN1MK08G. Cited 1 Nov 2018.

De Veaux, R.D., Agarwal, M., Averett, M., Baumer, B.S., Bray, A., & Bressoud, T.C., et al. (2016). Curriculum guidelines for undergraduate programs in data science, 2016, Annual review of statistics and its application, (Vol. 4, no. 1, pp. 15-30), 2017. https://doi.org/10.1146/annurev-statistics-060116-053930.

U.S. Department of Health and Human Services. (2002). Standards for privacy of individually identifiable health information, final rule. Federal Register, 45(CFR), 160–164.

Deutschen Gesellschaft für Medizinische Informatik, Biometrie und Epidemiologie (GMDS), Manzeschke, A., Winter, A., Isele, C., Deserno, T., Pallas, F., Weber, K., & Niederlag, W. (2017). Workshop: Ethische Leitlinien wissenschaftlicher Fachgesellschaften, 4 May 2017. https://gmds.de/ueber-uns/organisation/praesidiumskommissionen/ethische-fragen-in-der-medizinischen-informatik-biometrie-und-epidemiologie/. Cited 6 Nov 2018.

Diallo, I. (2018). The machine fired me, 17 June 2018. https://idiallo.com/blog/when-a-machine-fired-me. Cited 2 Nov 2018.

Donoho, D. (2017). 50 Years of Data Science. Journal of Computational and Graphical Statistics, 26(4), 745–766. https://doi.org/10.1080/10618600.2017.1384734.

Erickson, L.C., Harris, N.E., & Lee, M.M. (2018). It's time to talk about data ethics, 26 Mar 2018. https://www.datascience.com/blog/data-ethics-for-data-scientists. Cited 6 Nov 2018.

European Commission, Cybersecurity & Digital Privacy Policy (Unit H.2), eCall: Time saved = lives saved, 14 Feb 2018. https://ec.europa.eu/digital-single-market/en/ecall-time-saved-lives-saved. Cited 7 Nov 2018.

FAZ, Jeder zweite Gefährder hat das Potential zum Terroristen, 18.12.2017. http://www.faz.net/aktuell/politik/inland/terror-gefahr-jeder-zweite-radikale-islamist-hochgefaehrlich-15347580.html. Cited 6 Nov 2018.

Fanta, A. (2018). Österreichs Jobcenter richten künftig mit Hilfe von Software über Arbeitslose, 13 Oct 2018. https://netzpolitik.org/2018/oesterreichs-jobcenter-richten-kuenftig-mit-hilfe-von-software-ueber-arbeitslose/. Cited 2 Nov 2018.

FAZ, Wir dachten, wir tun etwas völlig Normales, 21 Mar 2018. http://www.faz.net/aktuell/wirtschaft/diginomics/skandal-um-cambridge-analytica-dachten-wir-tun-etwas-voellig-normales-15506137.html.

Futurezone, AMS-Chef: Mitarbeiter schätzen Jobchancen pessimistischer ein als der Algorithmus, 12 Oct 2018. https://futurezone.at/netzpolitik/ams-chef-mitarbeiter-schaetzen-jobchancen-pessimistischer-ein-als-der-algorithmus/400143839. Cited 6 Nov 2018.

Galit, S. (2010). To explain or to predict? Statistical science, 25(3), 289–310. https://doi.org/10.1214/10-STS330, https://projecteuclid.org/euclid.ss/1294167961.

Gesellschaft für Informatik, Ethical Guidelines of the German Informatics Society, 29 June 2018. https://gi.de/ethicalguidelines/. Cited 6 Nov 2018.

Goldberger, A.L., Amaral LAN, Glass, L., Hausdorff, J.M., Ivanov, P.C., & Mark, R.G., et al. (2000). PhysioBank, PhysioToolkit, and PhysioNet: Components of a new research resource complex physiologic signals. Circulation 101(23), e215–e220. [Circulation Electronic Pages. http://circ.ahajournals.org/content/101/23/e215.full].

Greiner, W., Batram, M., Damm, O., Scholz, S., & Witte, J. (2018). Kinder- und Jugendreport 2018, Beiträge zur Gesundheitsökonomie und Versorgungsforschung (Band 23), Andreas Storm (Herausgeber), DAK-Gesundheit.

Horner, J. (2003). Morality, Ethics, and Law: Introductory concepts. *SEMINARS IN SPEECH AND LANGUAGE, 24*(4), 263–274.

James, L. (2018). Oaths, pledges and manifestos: A master list of ethical tech values, doteveryone. https://doteveryone.org.uk, 7 Mar 2018. https://medium.com/doteveryone/oaths-pledges-and-manifestos-a-master-list-of-ethical-tech-values-26e2672e161c. Cited 6 Nov 2018.

Kuntz, B., Mauz, M., & Lampert, T. (2018) Die KiGGS-Studie des Robert Koch-Instituts: Studiendesign, Erhebungsinhalte und Ergebnisse zur gesundheitlichen Ungleichheit im Kindes- und Jugendalter – Robert Koch-Institut, Berlin, Kinder- und Jugendreport 2018, Beiträge zur Gesundheitsökonomie und Versorgungsforschung (Band 23), Andreas Storm (Herausgeber), DAK-Gesundheit.

Kurth, B.M., Kamtsiuris, P., Hölling, H., Schlaud, M., Dölle, R., & Ellert, U., et al. (2008) The challenge of comprehensively mapping children's health in a nation-wide health survey: Design of the German KiGGS-Study BMC Public Health 20088:196. https://doi.org/10.1186/1471-2458-8-196 Kurth et al. licensee BioMed Central Ltd.

Marr, B. (2015). Forbes, How big data is changing insurance forever, 16 Dec 2015. https://www.forbes.com/sites/bernardmarr/2015/12/16/how-big-data-is-changing-the-insurance-industry-forever/.

O'Neil, C. (2016). *Weapons of math destruction: How big data increases inequality and threatens democracy.* New York: Crown Publishing Group.

Panoptykon Foundation, Niklas, J., Sztandar-Sztanderska, K., & Szymielewicz, K. (2015). Warsaw. https://panoptykon.org/sites/default/files/leadimage-biblioteka/panoptykon_profiling_report_final.pdf. Cited 2 Nov 2018.

Pariser, E. (2012). The filter bubble ; how the new personalized web is changing what we read and how we think, Penguin Books, New York. ISBN: 0143121235.

Perez, S. (2016). Microsoft silences its new A.I. bot Tay, after Twitter users teach it racism. https://techcrunch.com/2016/03/24/microsoft-silences-its-new-a-i-bot-tay-after-twitter-users-teach-it-racism/. Cited 1 Nov 2018.

Pollard, T.J., Johnson, A.E.W. (2016). The MIMIC-III clinical database. https://doi.org/10.13026/C2XW26.

Pottegård, A., Kristensen, K.B., Ernst, M.T., Johansen, N.B., Quartarolo, P., Hallas, J. (2018). Use of N-nitrosodimethylamine (NDMA) contaminated valsartan products and risk of cancer: Danish nationwide cohort study, (Vol. 362). BMJ Publishing Group Ltd. https://doi.org/10.1136/bmj.k3851, https://www.bmj.com/content/362/bmj.k3851.

Schroepfer, M. (2018). CTO facebook, An update on our plans to restrict data access on facebook, 4 Apr 2018. https://newsroom.fb.com/news/2018/04/restricting-data-access/. Cited 6 Nov 2018.

Seeger, J. (2016). ADAC-Untersuchung: Autohersteller sammeln Daten in großem Stil, 4 June 2016, https://www.heise.de/newsticker/meldung/ADAC-Untersuchung-Autohersteller-sammeln-Daten-in-grossem-Stil-3227102.html. Cited 6 Nov 2018.

Wissenschaftlicher Dienst des Bundestags Fachbereich WD 3: Verfassung und Verwaltung, Veröffentlichung der Ergebnisse von Umfragen vor Wahlen (Deutschland und Mitgliedstaaten der EU), Aktenzeichen WD 3 - 3000 - 058/18, 2018. https://www.bundestag.de/blob/556748/ea25753e1c4a357a2c2c1c4791d4c4a8/wd-3-058-18-pdf-data.pdf. Cited 2 Nov 2018.

Part IV
Statistics in Econometric Applications

Chapter 11
Dating Lower Turning Points of Business Cycles – A Multivariate Linear Discriminant Analysis for Germany 1974 to 2009

Ullrich Heilemann and Heinz Josef Münch

Abstract This paper examines with multivariate linear discriminant analysis (LDA) the accuracy of dating lower turning point phases (LTP) of a four-phase classification scheme of German business cycles from 1974 to 2009, in particular why the start of the Great Recession (GR) is dated so late. Based on quarterly year-to-year rates of change, 85% of LTP periods are classified correctly, which is the lowest rate of the four phases but is not different from the results for previous periods. The misclassification of the start of the GR as an upper turning point phase (UTP) and thereby skipping the DOWN phase seems to be the result of a rare coincidence of outliers of influential classifiers. The re-estimation of the classification functions with quarter-to-quarter rates of change showed promising results; however, they also failed to classify the start of the GR correctly.

11.1 Introduction

The Great Recession (GR)—the global financial and economic crisis from 2007 to 2009—has raised many questions about its causes, its quick international expansion, its late diagnosis, or, more generally, about the predictability of crises. The dissemination, technically defined as two succeeding quarters of shrinking real GDP (henceforth: GDP[1]), started in Germany in the second quarter of 2008 (2008-II)

---

[1] Adjusted for seasonal fluctuations (with Census X-12-ARIMA).

We are indebted to Roland Döhrn, Roland Schuhr and two referees for helpful comments and suggestions.

U. Heilemann (✉)
Universität Leipzig, Grimmaische Straße 12-14, 04105 Leipzig, Germany
e-mail: heilemann@wifa-leipzig.de

H. J. Münch
Schacht Jakob 5, 45259 Essen, Germany
e-mail: heinz.muench@gmx.de

© Springer Nature Switzerland AG 2019
N. Bauer et al. (eds.), *Applications in Statistical Computing*,
Studies in Classification, Data Analysis, and Knowledge Organization,
https://doi.org/10.1007/978-3-030-25147-5_11

173

but forecasters realised this only after the publication of official GDP data in mid-November when the monetary and fiscal policy had already started fighting the crisis. Though there is some evidence that the recession could have started in the summer of 2008 (Heilemann and Schnorr-Bäcker 2017), the exact reasons for their late diagnosis are still unclear. Macro-econometric model studies of these faults do not exist and for forecasters usually employing combinations of theories, methods and data ('informal GDP model'), detailed analyses of forecast errors are laborious, if possible at all in retrospect.

A straightforward way to detect at least some reasons for the late perception of the crisis offers multivariate linear discriminant analysis (LDA). Based on a standard four-phase classification of the business cycle and a (small) number of macroeconomic variables (predictors), LDA tries to match a period's a priori classification, or, in terms of classical statistics, the a priori classification stands for y and the a posteriori classification stands for $\hat{y}$. Deviations of a posteriori from a priori classifications can be analysed and help to identify the predictors responsible for these faults; for example, for failing to identify a crisis.

This paper examines the dating of the five lower turning point (LTP) phases between 1974-II and 2009-IV, in particular why the GR had been diagnosed so late. Building on previous LDA-based studies of the German business cycle, we examine the sensitivity of LTP classifications for data and specification, and more specifically, the use of rates of change against the previous quarter (*qtq* rates) data instead of rates of change against the previous year (*yty* rates) (each in percentage) as used in previous studies. A fuller use of LDA techniques was beyond the possibilities of this paper, but already the shortcuts taken here offer some new and interesting findings.

Section 11.2 presents the classification scheme and the LDA results; Sect. 11.3 examines the accuracy of LTP classifications. Section 11.4 analyses the influence of outliers in predictor data and the use of qtq instead of yty rates of change data. Section 11.5 has the summary and conclusions.

11.2   Classification Scheme and LDA Results

The purpose and the method of our LDA approach have been presented before,[2] so we will be brief on this here. The necessary a priori classification was the classical four-phase taxonomy of the cycle, comprising the phases 'lower turning point' (LTP), 'upswing' (UP), 'upper turning point' (UTP) and 'downswing' (DOWN); however, not all cycles must entail every phase. While for the US, this was several times the case (Zarnowitz 2006, p. 19), and for Germany, it did not occur until 2008 (see below).

The a priori dating of phases was taken from the upswing/downswing classification of industrial production by the Deutsche Bundesbank, and when it abandoned it in the 1990s, it was continued by us. From this, we derived a four-phase scheme

---

[2]See, for example, Heilemann and Münch (1999) and Heilemann and Weihs (2003). For a recent comparison of various concepts for dating business cycles and their phases, see Heilemann (2018).

from classifying the two quarters before and after their upper and lower turning points as UTP and LTP and in-between phases as UP and DOWN. This was tested by LDA estimates,[3] and, after being accordingly modified, served as (final) a priori classification.

Predictors were selected based on economic theory and knowledge about the German business cycle. We excluded survey data and leading indicators because we did not intend to use LDA results for forecasting. After several tests, we selected 12 variables usually employed in macroeconomic analyses and reported forecasts as sufficiently reliable and stable predictors, most of which expressed in *yty* rates.[4] To improve practical use, in 2005, the number of predictors was reduced to eight (Döhrn et al. 2005, p. 26 f.) and this set[5] will be used in the following. The ensuing 'stylized facts' of the cycle are as follows:

– The LTP phase is characterised by stagnating demand, negative investment in machinery, shrinking employment, declining unit labour cost (compared to the DOWN phase) and short-term interest rates decreasing;
– The UP phase is the stage in which demand strongly expands, inflation is low, short-term interest rates decline (compared to the LTP phase) and employment begins to increase;
– The UTP phase is marked by a further expansion of domestic and international demand, rising employment, increasing unit labour cost and prices and declining short-term and real long-term interest rates;
– Finally, in the DOWN phase demand declines, unit labour cost and prices continue to increase, employment begins to fall and real long-term interest rates continue to decline, while short-term interest rates increase.

As in previous studies, we estimated three discriminant functions. The sample started in 1974-II and ended in 2009-IV, both being LTP phases. The data are first vintage data as of spring 2010.

Table 11.1 (A) has the estimation results. Compared with previous findings for various sample periods, the present set of predictors is still successful in separating the phases. The explanatory power of the first function, measured by the eigenvalue ratio is, at 73%, fairly high and much greater than that of the second (25%).[6] The third function is considerably lower (2%), but Wilk's λ and the χ^2 value indicate that it still contributes to the separation of the phases. All F-values are significant.

---

[3] The classifications were made with the routine DISCRIMINANT of IBM SPSS 24, which also has a description of LDA and the various statistics employed here.

[4] The use of *yty* rates is sometimes associated with the growth cycle concept, *qtq* with the business concept. For a detailed discussion of the two concepts (cf. Zarnowitz 2006). We dispense with this distinction, and employ *yty* rates, which in German short-term analysis and forecasting play a much larger role than *qtq* rates.

[5] The eight variables are number of wage and salary earners, real GDP, real investment in machinery, net exports in percent of GDP, GDP price deflator, unit labour cost, short-term interest rates and real long-term interest rates. All are expressed in *yty* rates, except for net exports and interest rates.

[6] The first function can be thought of as discriminating between UP and DOWN, the second as discriminating between LTP and UTP (Heilemann and Münch 1999).

Table 11.1 Standardized canonical discriminant functions, 1974-II–2009-IV

Predictor	Coefficients Discriminant functions				F-Value[a]	Significance[b]
		1	2	3		
Wages and salary earners	A	0.63	−0.30	0.32	53.51	0.00
	B	0.27	0.77	−0.06	17.94	0.00
Real GDP	A	−0.12	−0.80	−0.07	41.32	0.00
	B	−0.32	0.22	0.76	5.70	0.00
Real investment in machinery	A	0.76	0.26	−0.22	57.98	0.00
	B	0.50	0.31	−0.16	10.15	0.00
Net exports as percent of GDP	A	0.48	0.51	0.76	11.60	0.00
	B	0.71	−0.47	0.43	14.08	0.00
GDP price deflator	A	1.08	1.41	−0.16	13.13	0.00
	B	0.33	−0.17	0.10	7.38	0.00
Unit labour cost	A	−0.13	−0.29	0.51	20.81	0.00
	B	0.09	0.07	1.05	3.96	0.01

(continued)

Table 11.1 (continued)

Predictor		Coefficients Discriminant functions			F-Value[a]	Significance[b]
		1	2	3		
Short-term interest rate	A	−1.59	−1.21	0.07	33.63	0.00
	B	−1.72	0.38	0.15	33.62	0.00
Real long-term interest rate	A	1.11	0.73	0.18	3.86	0.01
	B	1.25	−0.55	−0.45	6.53	0.00

Summary statistics

Discriminant function		Eigenvalue	Percent of variance	Canonical correlation	Test of function s	Wilks' λ	χ²	DF	Significance[b]
1	A	4.25	73.1	0.90	1–3	0.07	361.0	24	0.00
	B	2.05	80.1	0.82		0.22	208.7		0.00
2	A	1.46	25.1	0.77	2–3	0.37	135.6	14	0.00
	B	0.48	18.8	0.57		0.66	57.0		0.00
3	A	0.10	1.8	0.31	3	0.91	13.24	6	0.04
	B	0.03	1.0	0.16		0.97	3.54		0.74

Authors' computations A: estimated with yty rates of change data, B: estimated with qtq rates of change data.—[a]F-to-enter-statistic.—[b]p-value

Though interpretation of the parameters of discriminant functions must be cautious, the most important variables are employment (number of wage and salary earners), real investment, real GDP and short-term interest rates—very much the same as in the preceding studies.[7] The discriminant functions appear as quite robust; the recession classifications hardly change when re-estimated by successively leaving out each of the five LTP phases. When leaving out the period 2008-II–2009-IV, the beginning of the LTP phase shifts to 2009-I, but a DOWN phase is still missing.

Table 11.2 shows the classification results. Not surprisingly, the duration of upswings and downswings is in line with post-World War II experience: upswings— defined as LTP/UP or UP/UTP—and downswings last, on average, about 25 quarters and ten quarters, respectively.

The highest accuracy is to register for the UTP phase—from an economic perspective much surprising—, and the lower is for the LTP phase (Table 11.3). Misclassifications are confined to the phases' boundaries,[8] with one exception (1984-II).

11.3   Classifications of Lower Turning Point Phases

Economic cycles and their phases are statistical constructs, and their definitions vary considerably. Before more closely examining our findings, a brief look at the results of two other, widely used concepts to date turning points/recessions seems useful. Compared with the technical definition of recessions (henceforth: GDP dating), LDA classifications miss the second oil crisis (1980) and the fiscal consolidation crisis of 1993 (Tables 11.2 and 11.5 (Appendix)). The multivariate-based classification of business cycles by the Economic Cycle Research Institute (ECRI)[9] does not see a (separate) fiscal consolidation crisis in 1982 or 1993, but does see the second oil crisis in 1980 and the GR as having started in 2008-I. The duration of a recessions or their counterparts is seen differently: three quarters in the case of GDP dating, nine quarters with ECRI, and five quarters with LDA. On average, ECRI and LDA date the beginning of recessions three and one quarter, respectively, earlier than the GDP recession and its ending one or two quarters, respectively, later. As for the longer duration of recessions by LDA and ECRI, it should be noticed that on average the pre-crisis level of GDP is reached only four quarters after the end of GDP recessions.

The later beginning of the GR in the LDA classification, however, is much in line with assessments by the business community, politics (the ECB had raised interest

---

[7] All data and results not shown here are available upon request from the authors.

[8] Röhl (1998) did not allow within-phase errors, which, of course, reduced their number (Heilemann and Weihs 2003, p. 137).

[9] For the definition of recession by ECRI (and a critical appraise of GDP dating), cf. Layton and Banerji (2003). Recently, the Sachverständigenrat (2017, p. 134 f.) presented a two-phase classification of the German business cycle for 1950–2016. He is not very specific about his method but broadly refers to that used by the National Bureau of Economic Research, the Centre of Economic Policy Research, and expert knowledge.

Table 11.2 Dating and quantity of German business cycles, 1974-II–2009-IV[a]

Nr.	Cycle		Phases								
			LTP		UP		UTP		DOWN	
7	1974-II–1982-I	(32)	1974-II	(7)	1976-I	(13)	1979-II	(4)	1980-II	(8)
8	1982-II–1994-I	(48)	1982-II	(6)	1983-IV	(27)	1990-III	(6)	1992-I	(9)
9	1994-II–2001-IV	(31)	1994-II	(1)	1994-III	(23)	2000-II	(5)	2001-III	(2)
10	2002-I–2008-III	(27)	2002-I	(7)	2003-IV	(17)	2008-I	(3)	–	(0)
11	2008-IV–2009-IV	(5)	2008-IV	(5)	2010-I	(…)	…	(…)	…	(…)
All		(143)		(26)		(80)		(18)		(19)
Ø		(29)		(5)		(20)		(5)		(5)

Author's computations.(…): cycle-/phase-length.—[a]Until 1991: West Germany

Table 11.3 Accuracy of phase classifications, 1974-II–2009-IV

A priori classifications[a]	Number of cases		A posteriori classifications							
			LTP	In per cent	UP	In per cent	UTP	In per cent	DOWN	In per cent
LTP	26	A	22	84	2	8	0	0	2	8
		B	17	65	7	27	0	0	2	8
UP	80	A	1	1	72	90	7	9	0	0
		B	10	13	64	80	6	7	0	0
UTP	18	A	0	0	0	0	18	100	0	0
		B	0	0	4	22	11	61	3	17
DOWN	19	A	1	5	0	0	0	0	18	95
		B	1	5	0	0	3	16	15	79
Total correct classifications, in percent		A	91							
		B	75							
Total hit rates by phases	143	A	24	17	74	52	25	17	20	14
		B	28	20	75	52	20	14	20	14

Authors' computations A: results from discriminant functions estimated with *yty* rates of change data; B: results from discriminant functions estimated with *qtq* rates of change data.—[a]See Table 11.2

rates in July when Germany and other European countries were already in technical recessions), and forecasters widely shared positive assessment of this period; Even in October 2008, the German economy looked for them sufficiently healthy to withstand the international repercussions of the financial crises that had started in 2007 in the United States (Heilemann and Schnorr-Bäcker 2017).

LDA a posteriori results record four clear misclassifications for LTP (Table 11.4 (A)); two each because the beginning or end of the LTP phase was seen too early or too late and all in just one cycle (8). As to the GR, its start is seen only in 2008-IV, when GDP was already two quarters in decline. LDA still classifies these two quarters as UTP phases. The DOWN phase is missing—as noted above: the first time that a phase was skipped since our classification had started (with 1955).

Table 11.4 Classification errors in lower turning point phases, 1975–2009

Cycle	Period	Data	Phase		A posteriori phase probabilities[a]			
			A priori	Predicted	LTP	UP	UTP	DOWN
7	1975-IV	A	LTP	UP	22	78	0	0
		B		UP	33	67	0	0
8	1982-II	A	LTP	DOWN	3	0	0	97
		B		DOWN	3	1	36	60
	1982-III	A	LTP	DOWN	4	0	0	96
		B		DOWN	14	1	17	68
	1983-II	A	LTP	LTP	85	15	0	0
		B		UP	25	72	3	0
	1983-III	A	LTP	UP	37	63	0	0
		B		UP	40	51	9	0
9	1994-II	A	LTP	LTP	49	45	4	2
		B		UP	33	55	11	1
10	2002-II	A	LTP	LTP	95	5	0	0
		B		UP	49	50	1	0
	2003-III	A	LTP	LTP	82	18	0	0
		B		UP	36	62	2	0
Memory items								
11	2008-II	A	UTP	UTP	0	35	65	0
		B		UP	28	53	18	0
	2008-III	A	UTP	UP	28	53	18	0
		B		UP	37	47	16	0
	2009-III	A	LTP	LTP	100	0	0	0
		B		UP	47	53	0	0

Authors' computations A: results from discriminant functions estimated with yty rates of change data; B: results from discriminant functions estimated with qtq rates of change data
a In per cent

11.4 Explanations

Leaving aside the stochastic nature of LDA, misclassifications can arise for many reasons, such as misspecifications of the discriminant functions, structural changes within the sample period, or simply data anomalies or outliers. The second possibility has been explored before, so we limit ourselves here to outlier and specification errors, i.e. the use of *qtq* instead of *yty* data.

There are various methods to identify outliers in a single-equation context (Belsley et al. 1980, pp. 6 ff.),[10] but these are hard to employ in a multi-equation framework as LDA is and results are hard to objectify (ibid., p. 27). Here, we opted for a more straightforward, case-oriented, direct approach. We calculated for each predictor in each misclassified LTP period (and for the missing DOWN phase), the difference to the particular phase average in the a priori classification, weighting it with the F-values and re-classified these periods.

For 1975-IV, it appears that the misclassification (as a UP phase) is the result of unusually high values for three of the four most important predictors, particularly investment, where employment is the exception.

For 1982-II and 1982-III, the failures are difficult to explain. While GDP, employment and investment suggest an LTP phase, this is more than equalised by the other predictors and leads to clear classifications of these periods as DOWN phases, shifting the beginning of the LTP phase to 1982-IV.

For 1983-III, the results are clear, and no disturbances by outliers are found.

Given our scheme, the late start of the LTP phase compared to GDP dating is not unusual, but the short UTP phase and especially the absence of the DOWN phase are. The most important cause is that for almost all predictors, the data differ considerably from their averages (even including that of the current cycle).

To evaluate the influence of *yty* data, discriminant functions were re-estimated with qtq data (for five of the eight predictors), ignoring the lead of *qtq* rates. To facilitate comparison, we did not adjust a priori classifications as had been the case with *yty* estimates. The explanatory power of the discriminant functions differs from that of *yty* functions. For the first function, it is much higher, for the second and third much smaller (Table 11.1 (B)). However, F-values for the most important predictors show much conformity. Major differences from *yty* results register with short-term interest rates and net exports, which are now the first and the third-most-important predictors, replacing investment and GDP. Overall, classification hit rates are somewhat lower, which is mainly due to the lower accuracy of classifications of UTP and LTP phases (Table 11.3 (B)), as was to be expected due to the summary statistics for the second function.

The reasons for the misclassification of 1975-IV with *qtq* data are much the same as with *yty* data. The lower a posteriori probability for LTP and the higher for UTP appear to be mostly due to the now lower value for investment. The errors for 1982-II/1982-III and for 1983-III (Table 11.4 (B)) seem to have the same causes; 1994-II,

---

[10]Previous leave-one-out tests have shown that, given the sample period, the accuracy of the classifications is not much affected.

difficult to classify because of the short duration, is classified as UP primarily due to high GDP and long-term interest rates. The misclassifications for 2002-II are again difficult to explain, as the probabilities for the UP and LTP phases differ little, similar as with the *yty* classifications.

The UP classification of 2003-III seems to be the result of the high expansion of employment and high short-term interest rates.

The small number of within-cycle errors aside, all in all the classifications of the two specifications differ very little. A pattern in the deviations from the a priori classifications cannot be detected; there is no general lead or lag regarding the beginning or end of LTP. *qtq* functions also fail to classify 2008-II and 2008-III as DOWN phases, and the UP phase lasts until 2008-III. The reasons for *yty* functions to miss DOWN phases are that in 2008-II, GDP, investment, net exports and wage cost led to a UTP phase classification; except for GDP, they also did so for 2008-III, though to a lesser degree as the probabilities reveal. With *qtq* functions, in both quarters only GDP and short-term interest rates signal an LTP phase and only two predictors suggest DOWN phases, most of all because of employment and investment.

11.5 Summary and Conclusions

Based on a four-phase scheme for the German business cycle, we used linear multivariate discriminant analysis to classify the period 1974-II–2009-IV. We concentrated on the accuracy of lower turning point phase classifications, and in particular, the dating of the beginning of the Great Recession. Although 22 of the 26 LTP were seen correctly, this is the lowest proportion of any of the four phases; however, this is not different from previous experience. Misclassifications occur, with one exception only at the edges of phases; in three of four cases, they are caused by outliers, most of all with data of investment, employment and GDP.

The start of the Great Recession was seen as late as in 2008-IV when Germany was already two quarters into a GDP recession. While this is not different from previous experience with the two concepts, the skipping of the DOWN phase and the immediate transition from a UTP to an LTP phase, however, is new. The main reason for this is that in these quarters the data of almost all predictors are unusually high and hence clearly classified them as UTP phases. The re-estimation of the discriminant functions with *qtq* data—given their limitations—and their classification led to comparatively good results. However, even with *qtq* data, the DOWN phase was skipped.

The results illustrate not only the role of concepts of dating business cycles. Concerning the Great Recession, they demonstrate the main difficulty of business cycle analysis that data within or between a multivariate setting like the LDA approach or the informal GDP model may deliver different and wrong messages. Although a large number of indicators had suggested since the summer of 2008 that the German economy was in a downturn or in a GDP recession, macroeconomic data used here provided a rosier picture—as in Germany has often been the case with crises caused by contagion (Reinhart and Rogoff 2009, pp. 230 ff.)—which was owed mainly to outliers as the present analysis should have shown.

The present results give little cause to change or modify the LDA approach in general, although new data, new ideas for a priori classifications or predictors always lead to rethink our scheme. The results themselves are, given our straightforward procedure, of course, preliminary and the data sensitivity of the classification accuracy should be checked with more sophisticated techniques. In any case, however, the economic analysis should take a more critical look at the data used than is often the case and appears to have been the case in the Great Recession.

Appendix

See Table 11.5.

Table 11.5 Dating and duration of recessions and lower turning point phases, 1974–2009

Cycle	ECRI		LDA-based scheme		GDP dating	
	Dating	Length[a,b]	Dating	Length[a]	Dating	Length[a]
First oil crisis 1974/75						
6	1973-03–1975-07	10	1974-III–1975-IV	7	1974-II–1975-II	5
Second oil crisis 1980						
7	1980-01–1982-10	11	–	–	1980-II–1980-IV	3
Fiscal consolidation crisis 1982						
8	–	–	1982-II–1983-III	6	1982-II–1982-III	2
Unification crisis 1991						
9	1991-01–1994-4	13	1994-II	1	1991-II–1991-III	2
Fiscal consolidation crisis 1993						
10	–		–		1992-II–1993-I	4
High-tech crisis 2002/03						
11	2001-01–2003-08	10	2002-I–2003-IV	8	2002-IV–2003-I	2

(continued)

Table 11.5 (continued)

Cycle	ECRI		LDA-based scheme		GDP dating	
	Dating	Length[a,b]	Dating	Length[a]	Dating	Length[a]
Finance and economic crisis 2008/09						
12	2008-04–2009-01	3	2008-IV–2009-IV	5	2008-II–2009-I	4
Ø		9.4		5		3.1
σ		3.4		2.4		1.1

Sources Economic Cycle Research Institute (ECRI) (2010), Heilemann (2019), Döhrn et al. (2005) and Statistical Office (real terms, with Census X-12-ARIMA adjusted for seasonal and calendar effects, May 2018) and authors' calculations—[a]In quarters—[b]Formerly in months

References

Belsley, D. A., Kuh, E., & Welch, R. E. (1980). *Regression diagnostics*. New York, NY: Wiley.

Döhrn, R., Barabas, G., Gebhardt, H., Middendorf, T., Milton, A. R., Münch, H. J., et al. (2005). Die wirtschaftliche Entwicklung im Inland: Weiterhin kein kräftiger Aufschwung. *RWI-Konjunkturberichte, 56*, 23–58.

Economic Cycle Research Institute (ECRI) (Ed.). (2010), *Business cycle peak and trough Dates, 21 Countries, 1948–2009*. Retrieved September 25, 2018, from http://www.businesscycle.com/resources/cycles/.

Heilemann, U. (2019). Rezessionen in der Bundesrepublik Deutschland von 1966 bis 2013. *In Wirtschaftsdienst, 99*, 1–7.

Heilemann, U., & Münch, H. J. (1999). Classification of West German business cycles 1955–1994. *Jahrbücher für Nationalökonomie und Statistik, 219*, 632–656.

Heilemann, U., & Schnorr-Bäcker, S. (2017). Could the start of the German recession 2008–2009 have been foreseen? Evidence from real-time data. *Jahrbücher für Nationalökonomie und Statistik, 237*, 29–62.

Heilemann, U., & Weihs, C. (2003). Ergebnisse Teilprojekt B3 "Multivariate Bestimmung und Untersuchung von Konjunkturzyklen". In Sonderforschungsbereich 475 "Komplexitätsreduktion in multivariaten Datenstrukturen" an der Universität Dortmund (Hrsg.), Dortmund (pp. 135–156).

Layton, A. P., & Banerji, A. (2003). What is a recession?: A reprise. *Applied Economics, 35*(16), 1789–1797. https://doi.org/10.1080/0003684032000152853.

Reinhart, C. M., & Rogoff, K. S. (2009). *This time is different—Eight centuries of financial folly*. Princeton, NJ: Princeton University Press.

Röhl, M. C. (1998). *Computerintensive Dimensionsreduktion in der Klassifikation*. Dortmunder dissertation. Lohmar: Eul Verlag.

Sachverständigenrat zur Begutachtung der gesamtwirtschaftlichen Entwicklung. (2017). *Jahresgutachten 2017/18*. Wiesbaden: Statistisches Bundesamt.

Zarnowitz, V. (2006). Phases and stages of recent U.S. business cycles. In U. Heilemann & C. Weihs (Eds.), *Classification and clustering in business cycle analysis* (RWI Schriften, 79.) (pp. 19–32). Essen: RWI.

Chapter 12
Partial Orderings of Default Predictions

Walter Krämer and Peter N. Posch

Abstract We compare and generalize various partial orderings of probability fore-casters according to the quality of their predictions. It appears that the calibration requirement is quite at odds with the possibility of some such ordering. However, if the requirements of calibration and identical sets of debtors are relaxed, compa-rability obtains more easily. Taking default predictions in the credit rating industry as an example, we show for a database of 5333 (Moody's) and 6505 10-year default predictions (S&P), that Moody's and S&P cannot be ordered neither according to their grade distributions given default nor non-default or to their Gini- curves, but Moody's dominate S&P with respect to the ROC-criterion.

12.1 Introduction

In what follows, we touch upon the classification issue, which has been part of Claus Weihs' scientific interests throughout his career. See Weihs et al. (2007) or Bischl et al. (2013) for two more recent examples. The classification we are interested in here is sorting borrowers into classes according to the probability of repaying their debts. From an abstract viewpoint, this can be viewed as forecasting a Bernoulli variable with values 0 and 1. A long-standing example is weather forecasts, where statements like: 'The probability of rain in Dortmund tomorrow is 20%' have been common for quite a while. While the production of such forecasts has been heavily discussed both in the statistics and economics literature, much less is known about evaluating their relative performance. Section 12.2 summarizes previous results and suggests various extensions. It appears that the concept of calibration (DeGroot and Fienberg 1983) is a rather tough requirement which prevents most probability forecasters from being

W. Krämer (✉)
Fakultät Statistik, Technische Universität Dortmund, Dortmund, Germany
e-mail: walterk@statistik.tu-dortmund.de

P. N. Posch
Fakultät WiSo, Technische Universität Dortmund, Dortmund, Germany
e-mail: peter.posch@tu-dortmund.de

© Springer Nature Switzerland AG 2019   187
N. Bauer et al. (eds.), *Applications in Statistical Computing*,
Studies in Classification, Data Analysis, and Knowledge Organization,
https://doi.org/10.1007/978-3-030-25147-5_12

unequivocally comparable. Section 12.3 extends previous comparability results to non-identical sets of debtors and Sect. 12.4 applies our results to default prediction made by the leading rating agencies Moody's and S&P.

12.2 Partial Orderings of Probability Forecasts

There are two ways of comparing probability forecasters A and B. First, by computing scoring rules such as the well-known Brier Score,

$$S = \frac{1}{n} \sum_{i=1}^{n} (\theta_i - p_i)^2, \qquad (12.1)$$

where n is the number of forecasts made, $\theta_i \in \{0, 1\}$ denotes whether the event in question has occurred ($\theta = 1$) or not ($\theta = 0$), and where p_i is the forecasted probability of the event in trial no. i (Winkler 1996). As the Brier-Score is the mean squared error of the probability forecasts and the observed events, a forecaster wants to minimize its value. Second, by comparing complete distributions of forecasts, as will be done below. This allows statements such as 'A is better than B irrespective of any scoring rule from some class of scoring rules' (Krämer 2006), but has the disadvantage that certain pairs of forecasters cannot be ranked (e.g. this method provides only a partial ordering). It can be shown that it is then possible to find sensible (in the sense of 'proper', see Krämer 2006) scoring rules such that A is better than B according to rule no. 1 and B is better than A according to rule no. 2. Krämer and Güttler (2008) provide an example where, for an identical set of private enterprises, Moody's provides better default forecasts than S&P for some scoring rules, while S&P is better for others.

Let $0 = a_1 < a_2 < \ldots < a_k = 1$ be the probabilities that are available as forecasts for the future event in question. We take k to be small and finite here (for generalizations see Schervish 1989). For instance, the US National Weather Service has only multiples of 10% as predicted probabilities of rain, so $k = 11$ (including 0 and 100%). In the credit rating industry, the major agencies have 7 different rating categories with up to three rating modifiers each resulting in a maximum of 21 different grades including a default category. Typically, the worst 4 categories are combined into one, so there are usually 17 categories.

We are not concerned here with the methods by which default forecasts are produced. Rather, we take forecasts as given and take forecasters to be defined by the discrete bivariate probability function $r(\theta_i, a_j)$, $i = 1, 2$, $j = 1, \ldots, k$, resulting from some such method, whichever it may be, with $\theta = 1$ indicating default and $\theta = 0$ indicating non-default.

Of course, the true bivariate probability function $r(\theta_i, a_j)$ is known only after infinitely many trials. In practice, we take its empirical counterpart as a suitable approximation. As this paper is focused on fundamental problems of comparability, this sampling issue is ignored below (see however Krämer and Güttler 2008 for some discussion of the statistical significance of observed differences in ranking quality).

Without loss of generality, the set $\mathscr{A} = \{a_1, ..., a_k\}$ of available probabilities can be taken as identical for all forecasters involved. If not, define $\mathscr{A} := \mathscr{A}^A \cup \mathscr{A}^B$. Following Vardeman and Meeden (1983), the following additional notation will be used:

$p(1) := \sum_j r(1, a_j)$ = overall probability of default (PD).

$p(0) := \sum_j r(0, a_j)$ = overall probability of no default (probability of survival PS).

$q(a_j) :=$ probability with which default forecast a_j is made.

$p(1|a_j) := \frac{r(1,a_j)}{q(a_j)}$ = conditional probability of default given probability forecast a_j.

$p(0|a_j) := \frac{r(0,a_j)}{q(a_j)}$ = conditional probability of survival given probability forecast a_j.

$q(a_j|1) := \frac{r(1,a_j)}{p(1)}$ = conditional probability of prediction a_j given default.

$q(a_j|0) := \frac{r(0,a_j)}{p(0)}$ = conditional probability of prediction a_j given no default.

The problem is: Given two forecasters A and B, characterized by their respective bivariate probability functions $r^A(\theta_i, a_j)$ and $r^B(\theta_i, a_j)$, which one is 'better'?

One sensible requirement is that among borrowers with predicted default probability a_j, the relative percentage of defaults is equal to a_j. Formally:

$$a_j \stackrel{!}{=} p(1|a_j) = \frac{r(1, a_j)}{q(a_j)} \tag{12.2}$$

whenever $q(a_j) > 0$. Such forecasters are called 'well calibrated' (DeGroot and Fienberg 1983).

This calibration requirement has various consequences for the bivariate probability function $r(\theta_i, a_j)$. For instance, it is obvious from (12.2) that given the probabilities a_j of the future event and the $r(1, a_j)$, the marginal frequencies $q(a_j)$ and therefore also the $q(a_j|0)$ and the $r(0, a_j)$ are fixed. These limited degrees of freedom for obtaining a bivariate probability function $r(\theta_i, a_j)$ that is compatible with calibration are further reduced by the requirement that

$$\sum q(a_j) = q(0) + \sum_{j=2}^{k} \frac{r(1, a_j)}{a_j} = r(0, 0) + \underbrace{r(1, 0)}_{=0} + \sum_{j=2}^{k} \frac{r(1, a_j)}{a_j} = 1. \tag{12.3}$$

These relationships are probably best clarified via a numerical example. Assume that among the a_j's, there is 0.2 and 0.4 and that $r(1, 0.2) = 0.1$ and $r(1, 0.4) = 0.2$, as in Table 12.1:

Table 12.1 A well-calibrated probability forecaster

a_j	...	0.2	...	0.4	...
$r(1, a_j)$	0	0.1	0	0.2	0
$q(a_j)$	0	0.5	0	0.5	0
$r(0, a_j)$	0	0.4	0	0.3	0
$q(a_j|1)$	0	1/3	0	2/3	0
$q(a_j|0)$	0	4/7	0	3/7	0

Then the first four entries in the table completely determine the rest, via $a_j r(0, a_j) = (1 - a_j) r(1, a_j)$, which follows from (12.2). In particular, there can only be zeros in the columns where $a_j \notin \{0.2, 0.4\}$. Also, the marginal probabilities $p(1) = 0.3$ and $p(0) = 0.7$ follow immediately from

$$r(1, 0.2) = 0.1 \quad \text{and} \quad r(1, 0.4) = 0.2.$$

These restrictions will be vital in establishing various relationships between partial orderings below.

The first such partial ordering relies on 'refinement' (DeGroot and Fienberg 1983). We say that A is *more refined* than B, in symbols $A \geq_R B$, if there exists a $k \times k$ Markov matrix M (i.e. a matrix with nonnegative entries whose columns sum to unity) such that

$$q^B(a_i) = \sum_{j=1}^{k} M_{ij} q^A(a_j), \text{ and} \tag{12.4}$$

$$a_i q^B(a_i) = \sum_{j=1}^{k} M_{ij} a_j q^A(a_j), \quad i = 1, \ldots, k. \tag{12.5}$$

Equation (12.4) means that given A's forecast a_j, an additional independent randomization is applied according to the conditional distribution M_{ij}, $j = 1, \ldots, k$ which produces forecasts with the same probability function as that of B. Condition (12.5) ensures that the resulting forecast is again well calibrated.

The concept of refinement easily extends to forecasters which are not necessarily well calibrated. Again following DeGroot and Fienberg (1983), we say that A is sufficient for B - in symbols: $A \geq_S B$ - if, for some Markov matrix M,

$$q^B(a_i|\theta) = \sum_{j=1}^{k} M_{ij} q^A(a_j|\theta), \quad i = 1, \ldots, k, \ \theta = 0, 1. \tag{12.6}$$

Vardeman and Meeden (1983) suggest to alternatively order probability forecasters according to the concentration of defaults in the 'bad' grades. This will here be

Table 12.2 Two well-calibrated forecasters

	Forecaster A				Forecaster B			
a_j	0.2	0.25	0.4	1	0.2	0.25	0.4	1
$r^A(1, a_j)$	0.1	0	0.2	0	0.1	0.05	0.1	0.05
$q^A(a_j)$	0.5	0	0.5	0	0.5	0.2	0.25	0.05
$r^A(0, a_j)$	0.4	0	0.3	0	0.4	0.15	0.15	0
$q^A(a_j|1)$	1/3	0	2/3	0	1/3	1/6	1/3	1/6
$q^A(a_j|0)$	4/7	0	3/7	0	8/14	3/14	3/14	0

called the VM-default order. Formally:

$$A \geq_{VM(d)} B \quad :\Longleftrightarrow \quad \sum_{i=1}^{j} q^A(a_i|1) \leq \sum_{i=1}^{j} q^B(a_i|1), \quad j = 1, \ldots, k. \qquad (12.7)$$

Or to put this differently: A dominates B in the Vardeman–Meeden default ordering if its conditional distribution, given default, first-order stochastically dominates that of B.

The same can be done for the non-defaults. A is better than B in the VM-non-default sense if non-defaults are more frequent in the 'good' grades. Formally:

$$A \geq_{VM(nd)} B \quad :\Longleftrightarrow \quad \sum_{i=1}^{j} q^A(a_i|0) \geq \sum_{i=1}^{j} q^B(a_i|0), \quad j = 1, \ldots, k. \qquad (12.8)$$

Finally, A dominates B in the Vardeman–Meeden sense (in symbols $A \geq_{VM} B$) if both $A \geq_{VM(d)} B$ and $A \geq_{VM(nd)} B$ obtain.

Table 12.2 provides an example. It extends Table 12.1 to the case where an additional forecaster is involved. It is easily checked that both forecasters are well calibrated and that B dominates A in the Vardeman–Meeden-non-default-ordering. However, A and B cannot be ranked according to $VM(d)$ (it will emerge below that this is no coincidence).

It is also easily seen that B is more refined than A: Putting all of B's forecasts from the 0.25 and 1 brackets into the 0.4 bracket yields forecasts which are identical in distribution to A.

A related criterion which seems to be favoured in the banking community is based on joining the points

$$(0, 0), \quad \left(\sum_{i=0}^{j-1} q(a_{k-i}), \sum_{i=0}^{j-1} q(a_{k-i}|1) \right), \quad j = 1, \ldots, k \qquad (12.9)$$

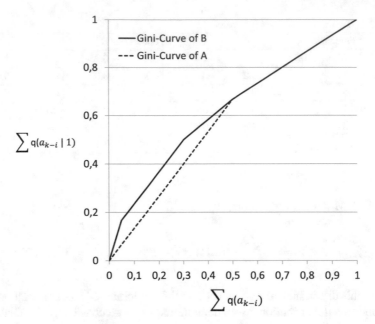

Fig. 12.1 Gini-ordering of forecasters A and B from Table 12.2

by straight lines. The resulting plot is variously called the power curve, the Lorenz curve, the Gini-curve or the cumulative accuracy profile (see, e.g. Engelmann et al. 2003), and a forecaster A is considered better than a forecaster B in this—the Gini-default-sense (formally: $A \geq_G B$)—if A's Gini-curve is nowhere below that of B, so arguing in terms of survivors does not add anything new.

As Fig. 12.1 shows, B is in the above example also Gini-dominating A: Although both curves coincide from 0.5 upwards, A's curve is below that of B to the left of 0.5.

Finally, rather than looking at the Gini-curve, it is common in medical applications to consider the ROC-curve instead, defined by the points

$$(0, 0), \ \left(\sum_{i=0}^{j-1} q(a_{k-i}|0), \sum_{i=0}^{j-1} q(a_{k-i}|1) \right), \quad j = 1, ..., k \ . \tag{12.10}$$

With identical right marginals, this does not imply anything new. Figure 12.2 shows the ROC-curves corresponding to the Gini-curves from Fig. 12.1. It is seen that both Gini-curves shift leftwards, and that B keeps dominating A. This is no coincidence, as it can be shown that, for well-calibrated forecasters, Gini-curves intersect if and only if ROC-curves intersect.

Both the Gini-curve and the ROC-curves do not require predicted default probabilities - sorting the creditors into classes with increasing default probability suffices. They are also both convex if forecasters are semi-calibrated, i.e. if $p(1 \mid a_j)$ is a nondecreasing function of a_j. This is a minimum requirement we will stick to in what follows.

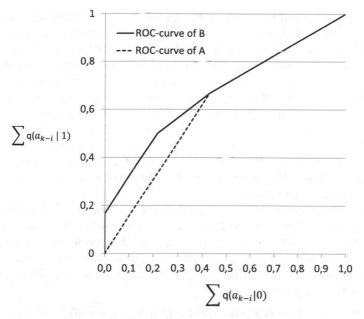

Fig. 12.2 ROC-curves of forecasters A and B from Table 12.2

As both the Gini and the ROC-curve are invariant under monotone transformations of the predicted default probabilities a_j, the ordering implied by them is no longer a partial ordering: From $A \geq B$ and $B \geq A$, one can no longer infer that $A = B$ ('antisymmetry'). But transitivity persists. Orderings of this type are called pre-orderings, and it will be seen in the next section that the VM-ordering likewise violates the antisymmetry condition if the restriction of identical right marginals is relaxed.

12.3   Generalizations and Relationships Among the Orderings

The calibration requirement severely restricts both the entries in the $r(\theta_i, a_j)$-matrix and the chances that two probability forecasters can be compared in the first place. In particular, Krämer (2005) shows that for well-calibrated forecasters A and B, if $q^A(0) = q^B(0) = 0$, then A and B cannot be strictly ordered according to $\geq_{VM(d)}$. And if $q^A(1) = q^B(1) = 0$, then A and B cannot be strictly ordered according to $\geq_{VM(nd)}$. The example in Table 12.2 where $\geq_{VM(nd)}$ obtains is therefore an artefact of $q^B(1) > 0$.

Comparability is much easier if the calibration requirement is abandoned. Even for identical right marginals (Krämer 2005) shows that the unrestricted VM-ordering might then obtain, in which case it implies the Gini-ordering. Given semi-calibration (i.e. $p(1|a_j)$ is non decreasing in a_j), the Gini-ordering is also implied by sufficiency. The Gini-ordering is thus the least demanding of the bunch in the case of identical right marginals. Still, in practice, most forecasters do not seem to be comparable at all.

Therefore, we now also abandon the restriction $p^A(1) = p^B(1)$. In practice, this means that we can now consider non-calibrated forecasters for different populations with different overall default probabilities. In particular, it can now be shown via simple examples that the VM-ordering might obtain even in case of calibration.

For the Gini-ordering, it can now happen that B's Gini-curve is better than A's for defaults and worse for non-defaults, and we say that B dominates A in the Gini-sense if it does so both for defaults and non-defaults. In that case, it is easily seen that $\geq_G$ keeps on defining a partial ordering.

As to the relationship between $\geq_{VM}$ and $\geq_G$, neither implies the other under these more general circumstances, as can be shown by simple counterexamples. Also, unlike in the case of identical right marginals, the Gini and ROC-orderings are no longer identical when right marginals are different. It is easy to find examples where $A \geq_{ROC} B$ but A and B cannot be ordered according to $\geq_G$. Rather, we now have the following result:

Theorem 1 *For arbitrary bivariate probability functions $r^A(\theta_i, a_j)$, $r^B(\theta_i, a_j)$ and semi-calibrated forecasters A and B, we have*

$$A \geq_G B \quad \Longrightarrow \quad A \geq_{ROC} B.$$

The converse does not hold.

Proof Let $A_i := \sum_{j=0}^{i-1} q^A(a_{k-j})$, $A1_i := \sum_{j=0}^{i-1} q^A(a_{k-j}|1)$ and $A0_i := \sum_{j=0}^{i-1} q^A(a_{k-j}|0)$, similarly for B. With the equivalence of the Gini- and ROC-orderings, $A \geq_G B$ is equivalent to

$$\frac{A1_i}{A_i} \geq \frac{B1_i}{B_i} \tag{12.19}$$

(i.e. A's Gini-curve for defaults is above that of B's) and

$$\frac{A0_i}{A_i} \leq \frac{B0_i}{B_i} \tag{12.20}$$

(i.e. A's Gini-curve for non-defaults is below that of B's). However, from (12.20) we have

$$\frac{A_i}{A0_i} \geq \frac{B_i}{B0_i} \tag{12.21}$$

and multiplying the left and right side of (12.19) with the left and right side of (12.21) yields

$$\frac{A1_i}{A0_i} \geq \frac{B1_i}{B0_i}, \tag{12.22}$$

which by the definition means $A \geq_{ROC} B$. That the converse is false can be shown by simple counterexamples. $\square$

From Theorem 1 it is clear that the ROC-ordering is the least demanding in practice, as will also be verified by our empirical example below.

12.4   Application

Our data are from the Moody's and S&P websites, respectively, from where we obtained the rating history of 5333 (Moody's) and 6505 private companies (S&P), covering the periods 1971–2014 (Moody's) and 1981–2014 (S&P) (see Vazza 2015 and Moody's 2015). For each company, we recorded its first rating and its default state 10 years after. Table 12.3 shows the results. PD is the percentage of defaults, and $q(a_i)$ denotes the relative frequency of rating class a_i, as defined in Sect. 12.1. For instance, 24.26% of Moody's customers and 22.94% of S&P's customers were initially rated A (among which 2.09% defaulted within 10 years in case of Moody's and among which 1.71% defaulted in the case of S&P). Overall, we recorded 2301 defaults among debtors rated by S&P and 1938 defaults rated by Moody's, corresponding to $P^M(1) = 15.43\%$ and $P^{S\&P}(1) = 12.74\%$.

Table 12.4 gives the resulting bivariate probability distribution if we view the empirical default rates in the various rating classes as predicted default probabilities (i.e. if we assume that both agencies are well calibrated), after incorporating the respective distribution of creditors across rating grades.

From Table 12.4, one readily obtains the conditional probabilities $q^{Moody's}(a_i|1)$, $q^{S\&P}(a_i|1)$, $q^{S\&P}(a_i|0)$, $q^{S\&P}(a_i|1)$, and it emerges that none of the relationships (12.7) or (12.8) obtains (see Table 12.5). Thus, Moody's and S&P cannot be ranked in either the VM-default nor non-default sense. However, as Figs. 12.3 and 12.4 show, the Gini-curve for defaults of S&P is below and the Gini-curve for survivors is above that of Moody's, so Gini-domination obtains.

From Theorem 1 it is, therefore, no surprise that Moody's dominates S&P also with respect to the ROC-criterion as shown in Fig. 12.5.

Table 12.3 Empirical 10-year default rates (PD) and distribution of debtors among rating classes

Rating class	Moody's		S&P	
	PD	$q(a_j)$	PD	$q(a_j)$
AAA/Aaa	0.49	3.41	0.71	1.07
AA/ Aa	0.89	11.50	0.78	7.13
A	2.09	24.26	1.71	22.94
BBB/Baa	4.95	23.18	4.98	26.15
BB/Ba	19.79	14.32	16.38	17.37
B	10.25	17.86	29.97	22.77
CCC/Caa-C	65.97	5.54	51.35	2.56

Table 12.4 Bivariate probability distribution across rating classes and default states

a_j	Moody's			S&P		
	$q(a_j)$	$r(0, a_j)$	$r(1, a_j)$	$q(a_j)$	$r(0, a_j)$	$r(1, a_j)$
0.49	3.41	3.39	0.02	0.00	0.00	0.00
0.71	0.00	0.00	0.00	1.07	1.06	0.01
0.78	0.00	0.00	0.00	7.13	1.07	0.06
0.89	11.50	11.40	0.10	0.00	0.00	0.00
1.71	0.00	0.00	0.00	22.94	22.55	0.39
2.09	24.26	23.75	0.51	0.00	0.00	0.00
4.95	23.18	22.03	1.15	0.00	0.00	0.00
4.98	0.00	0.00	0.00	26.15	24.85	1.30
16.38	0.00	0.00	0.00	17.37	14.52	2.85
19.79	14.23	11.41	2.82	0.00	0.00	0.00
29.97	0.00	0.00	0.00	22.77	15.95	6.82
40.25	17.86	10.67	7.19	0.00	0.00	0.00
51.35	0.00	0.00	0.00	2.56	1.25	1.31
65.97	5.54	1.89	3.65	0.00	0.00	0.00

Table 12.5 Cumulated sums of defaults and non-defaults

a_j	Moody's		S&P	
	$\sum_{i=1}^{j} q(a_i\|0)$	$\sum_{i=1}^{j} q(a_i\|1)$	$\sum_{i=1}^{j} q(a_i\|0)$	$\sum_{i=1}^{j} q(a_i\|1)$
0.49	4.01	0.11	0.00	0.00
0.71	4.01	0.11	1.22	0.06
0.78	4.01	0.11	9.33	0.50
0.89	17.49	0.77	9.33	0.5
1.71	17.49	0.77	35.17	3.57
2.09	45.59	4.06	35.17	3.57
4.95	71.65	11.49	35.17	3.57
4.98	71.65	11.49	63.65	13.80
16.38	71.65	11.49	80.30	36.13
19.79	85.18	29.74	80.30	36.13
29.97	85.15	29.74	98.57	89.68
40.25	97.77	76.32	98.57	89.68
51.35	97.77	76.32	100.00	100.00
65.97	100.00	100.00	100.00	100.00

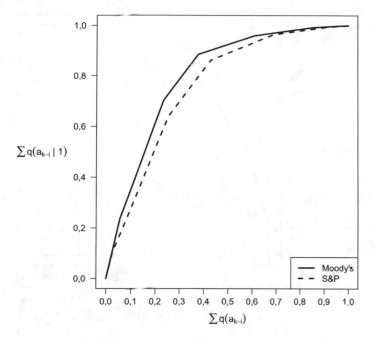

Fig. 12.3 Gini-ordering (default)

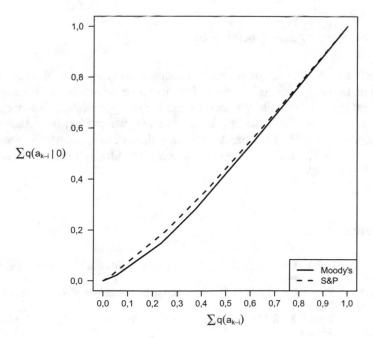

Fig. 12.4 Gini-ordering (non-default)

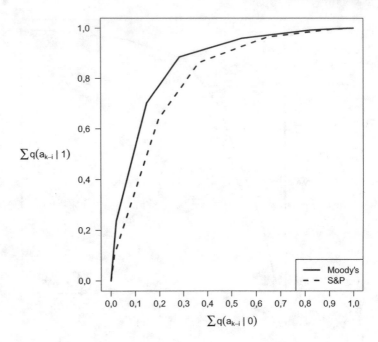

Fig. 12.5 ROC-curves

12.5 Summary and Conclusion

We illustrated that the requirements of calibrated forecasts and identical sets of
debtors make the comparison of two probability forecasters difficult. Therefore, we
abandoned these restrictions. We showed that the ROC-ordering is less demanding
in practice than the Gini-ordering. In an example, we showed for a database of
5333 (Moody's) and 6505 10-year default predictions (S&P), that Moody's and S&P
cannot be ordered according to the Gini-ordering, but Moody's dominate S&P with
respect to the ROC-criterion.

References

Bischl, B., Schiffner, J., & Weihs, C. (2013). Benchmarking local classification methods. *Computational Statistics, 28*, 2599–2619.
DeGroot, M., & Fienberg, S. E. (1983). The comparison and evaluation of forecasters. *The Statistician, 32*, 12–22.
Engelmann, B., Hayden, E., & Tasche, D. (2003). Testing rating accuracy. *Discussion paper 2003*, Deutsche Bundesbank.
Krämer, W. (2005). On the ordering of probability forecasts. *Sankhya: The Indian Journal of Statistics, 67*, 662–669.

Krämer, W. (2006). Evaluating probability forecasts in terms of refinement and strictly proper scoring rules. *Journal of Forecasting.*, *25*, 223–226.

Krämer, W. (2017). On assessing the relative performance of default predictions. *Journal of Forecasting*, *36*, 854–858.

Krämer, W., & Güttler, A. (2008). On comparing the accuracy of default predictions in the rating industry. *Empirical Economics*, *34*, 343–356.

Moody's (2015). Annual Default Study: Corporate Default and Recovery Rates 1920–2014, Moody's Investor Service.

Schervish, M. (1989). A general method for comparing probability assessors. *The Annals of Statistics*, *17*, 1856–1879.

Vardeman, S., & Meeden, G. (1983). Calibration, sufficiency, and domination considerations for bayesian probability assesors. *Journal of American Statistical Assocciation*, *78*, 808–816.

Vazza, D. et al. (2015). Annual global corporate default study and rating transitions, Standard & Poor's ratings direct.

Weihs, C., Ligges, U., Möhrchen, F., & Müllensiefen, D. (2007). Classification in music research. *Advances in Data Analysis and Classification*, *1*, 255–291.

Winkler, R. L. (1996). Scoring rules and the evaluation of prbabilities. *Test*, *5*, 1–60.

Chapter 13
Improving GMM Efficiency in Dynamic Models for Panel Data with Mean Stationarity

Giorgio Calzolari and Laura Magazzini

Abstract Estimation of dynamic panel data models largely relies on the generalized method of moments (GMM), and adopted sets of moment conditions exploit information up to the second moment of the variables. However, in many microeconomic applications, the variables of interest are skewed (typical examples are individual wages, size of the firms, number of employees, etc.); therefore, third moments might provide useful information for the estimation process. In this paper, we propose a moment condition, to be added to the set of conditions customarily exploited in GMM estimation of dynamic panel data models, that exploits third moments. The moment condition we propose is based on the data generating process that, under mean stationarity, characterizes the initial observation y_{i0} and the long-run mean of the dependent variable. In the literature on dynamic panel data models and in the way how Monte Carlo simulations are implemented therein for mean stationary processes, this condition is always fulfilled, but never explicitly exploited for estimation. Monte Carlo experiments show remarkable efficiency improvements when the distribution of individual effects, and thus of y_{i0}, is indeed skewed.

13.1 Introduction

This paper aims at improving efficiency of generalized method of moments (GMM) estimation of dynamic panel data models at only moderate costs in terms of computational effort. The (moderately) larger cost of the computation is essentially driven by the use of one additional nonlinear moment condition. We also evaluate the possible

G. Calzolari
Department of Statistics, Computer Science, Applications, Università di Firenze, Firenze, Italy
e-mail: giorgio.calzolari@unifi.it

L. Magazzini (✉)
Department of Economics, Università di Verona, Verona, Italy
e-mail: laura.magazzini@univr.it

© Springer Nature Switzerland AG 2019   201
N. Bauer et al. (eds.), *Applications in Statistical Computing*,
Studies in Classification, Data Analysis, and Knowledge Organization,
https://doi.org/10.1007/978-3-030-25147-5_13

small sample efficiency benefits derived from the computational intensive continuous updated version of the covariance matrix estimation.

Interestingly, the additional moment condition has an important justification in microeconomic applications, and the additional cost of the computation is well balanced by the statistical efficiency benefits.

Traditional econometric literature has largely relied on maximum likelihood estimation, in which full distributional assumption allows to obtain efficiency at the cost of robustness of the analysis. In more recent times, methods that are robust to distributional misspecification are preferred for empirical analysis, and GMM is widely applied.

We consider the estimation of dynamic linear panel data models on N independent units observed over T time periods ($i = 1, ..., N; \ t = 1, ..., T$):

$$y_{it} = \beta y_{it-1} + \alpha_i + e_{it} = \beta y_{it-1} + \varepsilon_{it} \tag{13.1}$$

where α_i captures individual 'unobserved heterogeneity', constant over time and different across units, and e_{it} is an idiosyncratic component changing both over time and across units. We consider the typical case of micro-panels, where N is 'large' and T is 'small', the error components α_i and e_{it} are independently distributed across individuals and $E(\alpha_i) = 0,$[1] $E(e_{it}) = 0, E(\alpha_i e_{it}) = 0 \ (t = 1, ..., T)$ and $E(e_{it} e_{is}) = 0$ for $t \neq s$. The initial y_{i0} is observed. We further assume $|\beta| < 1$.

Since the seminal work of Arellano and Bond (1991), estimation of dynamic panel data models relies on GMM (Hansen 1982).[2]

In the context of the linear dynamic panel data models, two GMM approaches are customarily considered for estimation: the difference GMM estimator and the system GMM estimator (Arellano and Bond 1991; Arellano and Bover 1995; Blundell and Bond 1998). The latter approach is known to produce efficiency improvements, which are particularly strong when the autoregression coefficient (β) approaches unity. This is obtained at the cost of a more restrictive hypothesis on the initial y_{i0}, known as the 'mean stationarity' assumption. Throughout the paper, we will assume that such initial conditions are satisfied (Blundell and Bond 1998). GMM estimation easily allows for the presence of additional regressors, x_{it}, that can be treated as strictly exogenous, predetermined or simultaneously determined (Blundell and Bond 2000; Blundell et al. 2000).

What we shall show in this paper is that if the initial conditions of mean stationarity are interpreted as they are in 'all' simulation experiments presented in the rich literature on this subject under the assumption of mean stationarity, further remark-

---

[1]The zero mean of the individual effect restriction is imposed without loss of generality. In case $E(\alpha_i) \neq 0$, a model with an intercept can be considered: $y_{it} = \beta_1 + \beta_2 y_{it-1} + \alpha_i + e_{it}$ with $E(\alpha_i) = 0$ (see e.g. Baltagi 2013).

[2]Other approaches are also available for estimation of linear dynamic panel data models. Hsiao et al. (2002) developed a transformed likelihood approach for the estimation of linear dynamic model within a fixed effect framework. Kiviet (1995) proposed a method to correct the small sample bias of the least squares dummy variable estimator. The performance of simulation methods (indirect inference) has also been analysed (Gouriéroux et al. 2010). This paper focuses on GMM.

able efficiency improvements can be obtained from the simple addition of 'one more' moment condition that exploits third moments to the system GMM estimator.

In the next section, we briefly describe the two traditional approaches with their assumptions, and the additional moment condition we propose. A detailed set of Monte Carlo experiments in Sect. 13.3 shows the efficiency improvements over the system GMM estimator. The efficiency gains are quite remarkable when y is skewed, with a variance reduction of the Monte Carlo estimates that is more than 75% in some cases. An empirical application is presented in Sect. 13.4. Section 13.5 concludes.

13.2   GMM Estimation of Dynamic Panel Data Models

In dynamic linear panel data model (13.1), a consistent estimator of β can be obtained by transforming the model in first differences, thus removing the individual effect α_i. Then, under suitable assumptions on the correlation structure of e_{it}, lagged values of y_{it} can be used as instruments in the differenced equation (Anderson and Hsiao 1981, 1982; Arellano and Bond 1991). In particular, as e_{it} is customarily assumed to be uncorrelated over time, lag 2 or more of y_{it} are valid instruments for Δy_{it-1}. Thus, the following $T(T-1)/2$ linear moment conditions can be exploited for estimation (Arellano and Bond 1991):

$$E(y_{it-j}\Delta\varepsilon_{it}) = 0, \tag{13.2}$$

with $j, t \geq 2$. The GMM estimator based on the set of moment conditions in (13.2) is known in the literature as difference GMM estimator.

Simulation studies have shown that the difference GMM estimator may have non-negligible finite sample bias and poor performances when β approaches 1, or when the variance of α_i is large with respect to the variance of e_{it} (Blundell and Bond 1998).[3]

Blundell and Bond (1998) introduce an additional assumption on the initial observation (see also Arellano and Bover 1995):

$$E(\varepsilon_{i2}\Delta y_{i1}) = 0 \tag{13.3}$$

This is known as the 'mean stationarity' assumption.

If this is satisfied, $T-1$ additional moment conditions may be exploited for estimation $(t \geq 2)$[4]:

$$E(\varepsilon_{it}\Delta y_{it-1}) = 0 \tag{13.4}$$

---

[3] Ahn and Schmidt (1995) show that the difference GMM estimator is not fully exploiting all the information in the data: under the standard assumptions, they propose additional moment conditions that allow improving efficiency (see also Schmidt et al. 1992).

[4] Of course, longer lags could be considered, but these conditions are redundant when the moment conditions (13.2) are included for estimation (Arellano and Bover 1995; Blundell and Bond 1998).

These conditions exclude cases in which y_{i0} is exogenous or when its correlation with the individual effect α_i is different from the correlation of α_i with y_{it} at different times.

The GMM estimator based on the moment conditions in (13.2) and (13.4) is known as the system GMM estimator. When the moment conditions (13.4) are added to the moment conditions in (13.2), substantial efficiency gains are achieved with respect to difference GMM estimator (Hahn 1997). Besides gains in (asymptotic) efficiency, the system GMM estimator also performs better in terms of small sample bias (Hayakawa 2007).[5]

Blundell and Bond (1998) state that the key requirement for condition (13.3) to hold is that the deviations of the initial conditions from $\alpha_i/(1 - \beta)$ are uncorrelated with the level of the long-run average $\alpha_i/(1 - \beta)$ itself. Condition (13.3) is therefore satisfied in cases in which

$$y_{i0} = \frac{\alpha_i}{1 - \beta} + e_{i0} \tag{13.5}$$

with e_{i0} uncorrelated with α_i and randomly distributed across agents (see also Roodman 2009). Equation (13.5) has become the hallmark of simulation studies in which the data generating process for y_{i0} obeys the mean stationarity assumption. It is adopted in 'all' Monte Carlo designs of the literature that assume mean stationarity (at least to the best of our knowledge), in which in fact the data generating process in (13.5) is adopted for the first observation (Arellano and Bover 1995; Blundell and Bond 1998; Windmeijer 2005; Roodman 2009). However, none of the above GMM estimators fully exploits such particular structure of y_{i0}, even though settled in the Monte Carlo experiments. Note that for all individuals i, α_i enters the first observation y_{i0} multiplied by the *same* factor for each i (equal to $1/(1 - \beta)$). Furthermore, the stronger assumption of independence between α_i and e_{i0} is considered. It seems therefore quite reasonable to try to further improve the GMM efficiency by explicitly exploiting these conditions.[6]

It is quite unlikely (perhaps impossible) that linear conditions analogous to those used by the methods available in the literature can distinguish between initial values y_{i0} generated as in (13.5) from values generated by more complicated structures (still satisfying mean stationarity). Nonlinear moment conditions should therefore be considered. In principle, many new moment conditions could be introduced. Being nonlinear, it is not difficult to prevent possible redundancies. But, to avoid excess of moment conditions that could be dangerous for moderately small samples, there should only be very few additional moment conditions (Roodman 2009).

---

[5]In addition, moment conditions (13.4) and (13.2) are shown to encompass the moment conditions proposed by Ahn and Schmidt (1995), Blundell and Bond (1998).

[6]However, the data generating process for y_{i0} in (13.5) is more restrictive than condition (13.3). As a result, the condition we propose is more restrictive than assumptions needed for consistency of the system GMM estimator. For instance, condition (13.3) would hold even if $y_{i0} = \gamma_i \alpha_i + e_{i0}$ with a random γ_i independent of α_i and $E(\gamma_i) = 1/(1 - \beta)$. Instead, the data generating process in (13.5) also assumes $V(\gamma_i) = 0$ ($\gamma_i = 1/(1 - \beta)$ for each i). Furthermore, α_i and e_{i0} are generated as independent random variables.

We therefore propose just 'one' nonlinear moment condition, to be added to the system GMM moments (in the case in which mean stationarity can be assumed), leading to efficiency gains (sometimes quite remarkable, see our experiments in Sect. 13.3)[7]:

$$E\left[y_{i0}\alpha_i\left(y_{i0} - \frac{\alpha_i}{1-\beta}\right)\right] = E\left[y_{i0}^2\alpha_i - \frac{1}{1-\beta}y_{i0}\alpha_i^2\right] = 0 \qquad (13.6)$$

or, equivalently

$$E\left[(1-\beta)y_{i0}^2\alpha_i - y_{i0}\alpha_i^2\right] = 0 \qquad (13.7)$$

If y_{i0} is generated as in (13.5), then condition (13.6) holds.

The variance of the GMM estimator based on the moment condition we propose involves third moments of y_{i0}. Monte Carlo experiments in Sect. 13.3 will show that the remarkable efficiency gains can indeed be obtained when the distribution of y_{i0} is asymmetric. However, notice that no explicit assumption on the distribution is ever required.

The additional condition (13.6) can be particularly helpful in applications in which a skewed distribution for the individual effects cannot be ruled out, such as in the case of 'wage of individuals', or 'size of the firms', characterized by strongly skewed distributions (in the Monte Carlo experiments we shall use log-normal or χ^2 distributions).

The proposed set up can be extended to cases in which additional regressors are included in the equation, and we consider the model:

$$y_{it} = \beta y_{it-1} + \delta x_{it} + \alpha_i + e_{it} = \beta y_{it-1} + \delta x_{it} + \varepsilon_{it} \qquad (13.8)$$

Additional moment conditions will be considered, spanning from the assumption on the relationship between x_{it} and the composite error term (Blundell and Bond 2000; Blundell et al. 2000). The moment condition we propose would be changed to

$$E\left[y_{i0}\alpha_i\left(y_{i0} - \frac{\alpha_i + \delta x_{i0}}{1-\beta}\right)\right] = 0 \qquad (13.9)$$

In practice, in the moment conditions (13.6) for a pure dynamic model, or (13.9) when additional regressors are included in the equation, α_i is replaced with $\sum_{t=1}^{T} \varepsilon_{it}/T$, and α_i^2 with $\sum_{t=2}^{T}\sum_{s<t} \varepsilon_{it}\varepsilon_{is}/(T(T-1)/2)$, so that it can be easily written as a function of the parameter(s). Experimentally, we also found that it seems more convenient to replace the initial value y_{i0} and x_{i0} with the demeaned value $\tilde{z}_{i0} = z_{i0} - \bar{z}_0$ where $\bar{z}_0 = \sum_{i=1}^{N} z_{i0}/N$ and $z = [y, x]'$. Monte Carlo experiments are presented in the next section.

---

[7]We have also explored moment conditions that involve the fourth-order moment of α_i, but without obtaining any significant benefit.

13.3   Monte Carlo Experiments

In this section, a set of Monte Carlo experiments is conducted in order to assess the finite sample performance of the GMM estimator that also exploits the proposed moment conditions as compared to the traditional system GMM estimator. The experiments are run both for a pure dynamic model and a model with an additional simultaneously determined regressor.

13.3.1   The Model with No x_{it}

We generate a dynamic panel data model as

$$y_{it} = \beta y_{it-1} + \varepsilon_{it} = \beta y_{it-1} + \alpha_i + e_{it} \tag{13.10}$$

with $|\beta| < 1$, and e_{it} normally distributed with mean zero and unconditional variance equal to 1, independent of α_i. As for the distribution of α_i, we considered the normal distribution, and two asymmetric distributions, i.e. the log-normal distribution and the χ^2 distribution with 1 degree of freedom.[8]

The initial observation y_{i0} is generated as

$$y_{i0} = \frac{\alpha_i}{1 - \beta} + e_{i0} \tag{13.11}$$

and several distributions are also considered for e_{i0}. N is set equal to 300, and we consider different ratios between the unconditional variance of α_i and e_{it}, which is set equal to, respectively, 1 and 4.[9]

As for estimation, we consider both a two-step approach and the continuously updated version of the GMM estimator (Hansen et al. 1996).[10]

The full set of results of the Monte Carlo experiments is reported in Appendix A (Table 13.4). In Table 13.1, we only report the ratio of variances and, in parentheses,

---

[8]In the case of log-normal and χ^2 distributions, which are characterized by a non-zero mean, we first simulated the value of α_i and then considered its demeaned transformation in the data generating process. We also multiplied it by an appropriate constant to get the chosen value for the unconditional variance.

[9]When the ratio of unconditional variances equals 1, the two error components give equally weighted contribution to the unconditional variance of y_{it}. A ratio equal to four mimics the condition of a variance of α_i much larger than the variance of e_{it}, identified by Bun and Windmeijer (2010) as responsible for large bias in small samples.

[10]No closed-form solution is available for the two-step version of the GMM estimator that exploits the additional moment condition, and for the continuously updated version of GMM (also for system GMM estimation). Algorithms for the minimization of the GMM function are therefore required. Computation efficiency might therefore be an issue, and 'efficient computation algorithm could be implemented by mixing the use of the approximations to the Hessian and of the exact Hessian' (Calzolari et al. 1987, p. 299).

Table 13.1 Monte Carlo results: ratio (CM/SYS) of variances and, in parentheses, ratio (CM/SYS) of interquartile ranges; CUE and two-step version of GMM estimation

$T+1$	β	$\alpha_i, e_{i0} \sim N$		$\alpha_i, e_{i0} \sim \log\text{-}N$		$\alpha_i, e_{i0} \sim \chi_1^2$	
		CUE	2-step	CUE	2-step	CUE	2-step
$\sigma_\alpha^2/\sigma_e^2 = 1$							
4	0.5	0.9988	0.9573	0.5002	0.5418	0.5467	0.5844
		(1.0027)	(0.9764)	(0.7062)	(0.7328)	(0.7104)	(0.7458)
8	0.5	1.0079	1.0083	0.8107	0.8225	0.7995	0.8137
		(1.0039)	(1.0001)	(0.8997)	(0.9130)	(0.8831)	(0.9031)
4	0.9	1.0470	1.0101	0.2362	0.5664	0.2739	0.6352
		(1.0736)	(0.9564)	(0.5541)	(0.5826)	(0.5803)	(0.6319)
8	0.9	1.0110	0.9155	0.4446	0.5576	0.4515	0.5572
		(1.0180)	(0.9497)	(0.7152)	(0.7326)	(0.6896)	(0.7015)
$\sigma_\alpha^2/\sigma_e^2 = 4$							
4	0.5	1.0061	0.9806	0.5836	0.6481	0.6360	0.6894
		(1.0050)	(0.9958)	(0.7415)	(0.7854)	(0.7753)	(0.8063)
8	0.5	1.0114	1.0158	0.8651	0.8771	0.8721	0.8817
		(1.0117)	(1.0083)	(0.9408)	(0.9419)	(0.9314)	(0.9417)
4	0.9	0.9737	0.8627	0.3644	0.6680	0.4809	0.7469
		(0.9542)	(0.8680)	(0.6323)	(0.6769)	(0.6685)	(0.7169)
8	0.9	0.9642	0.9033	0.4987	0.6521	0.5510	0.6875
		(0.9749)	(0.9366)	(0.7295)	(0.7547)	(0.7361)	(0.7675)

interquartile ranges of the two estimators, both when considering a two-step version of the estimator and the continuously updated estimator (CUE). The mean and the median values of the Monte Carlo estimates show lack of bias of both the traditional system GMM estimator (SYS) and the estimator we propose (CM) that additionally exploits the moment condition (13.6).[11] When computing the two-step version of the proposed estimator based on the additional moment condition, the first step system GMM estimator is used to estimate the optimal GMM weighting matrix.[12]

As expected, when the distribution of α_i and e_{i0} is symmetric (as it is the case with the normal distribution), no gain in performance is detected and the ratio of the

---

[11]The CUE is known to have a problem of possible multiple minima. In order to control for this issue, we considered two different initial values (i.e. OLS estimator and the two-step system GMM estimator) and disregarded those simulation runs in which we found multiple minima. In most experiments, less than 1% of estimated values is disregarded with the exception of the simulation runs with $T + 1 = 4$ and $\rho = 0.9$ in which we disregarded about 2.5% of cases. The issue of multiple minima in panel data model estimation also arises within a static framework, when an autoregressive process characterizes the idiosyncratic component (Calzolari and Magazzini 2012). The model is closely related to the dynamic specification considered in this paper.

[12]Asymptotically, any consistent estimator of the weighting matrix should provide the same final (second step) result. However, experimentally, on small- or medium-sized samples, this was the best choice.

variance of the CM approach to the SYS estimator is about 1. On the contrary, the results of the Monte Carlo experiments show remarkable gains in performance when the distribution of α_i and e_{i0} (thus, of y_{i0}) is asymmetric (log-normal or chi-squared distribution).[13] Larger gains in performance are observed when T is small and β gets closer to 1.

13.3.2 The Case of a Simultaneously Determined x_{it}

In this section, a simultaneously determined (endogenous) regressor is added to the data generating process. In building this simulation experiment, we rely on Blundell et al. (2000). The model is generated as

$$y_{it} = \beta y_{it-1} + \delta x_{it} + \alpha_i + e_{it} = \beta y_{it-1} + \delta x_{it} + \varepsilon_{it} \qquad (13.12)$$

with $x_{it} = \rho x_{it-1} + \tau \alpha_i + \theta e_{it} + w_{it}$. The error components are simulated as in the pure dynamic case (with both variances equal to 1), and $w_{it} \sim N(0, \sigma_w^2 = 0.16)$. The value of the autoregressive parameters are $\beta, \rho = 0.5, 0.9$, and $\delta = 1, \tau = 0.25$ and $\theta = -0.1$. As a result, x_{it} is correlated both with the individual effect α_i and the contemporaneous idiosyncratic components e_{it}. Simulations are run with $N = 500$ and $T = 3, 7$.

The moment conditions (13.2) and (13.4) remain valid within this setup. The following moment conditions also hold true:

$$E(\Delta \varepsilon_{it} x_{is}) = 0 \quad \text{with } t = 2, ..., T \text{ and } s = 0, ..., t - 2$$
$$E(\varepsilon_{it} \Delta x_{it-1}) = 0 \quad \text{with } t = 2, ..., T$$

The additional moment condition we propose is modified as in (13.9).

Comparison of variance and interquartile range of the estimated coefficients is reported in Table 13.2. We report the results based on the log-normal distribution of the error terms (and variables).[14]

The additional moment condition we propose generates gains in performance also in models that include an endogenous regressor. Larger gains are observed with a smaller T. However, increases in the autoregressive parameters of y and x allow for more efficient estimation of δ, whereas larger gains in efficiency for the estimation of the autoregressive parameter β are observed with less persistent variables. Looking at the full set of results, the two-step estimator performs better in this setup, so that it will be considered in the empirical analysis in the next section.

---

[13]Better performance of the estimator we propose also arises with normally distributed y_{i0} and $\beta = 0.9$. However, in unreported Monte Carlo experiments, this result vanishes with a larger N, and therefore we do not place too much emphasis on this result.

[14]The full set of results, comprising also normally and chi-squared distributed error terms, is reported in Appendix A.

Table 13.2 Monte Carlo results with an endogenous regressor: ratio (CM/SYS) of variances and, in parentheses, ratio (CM/SYS) of interquartile ranges; log-normally distributed error terms α_i and e_{i0}

$T+1$	β	ρ	$\hat{\beta}$		$\hat{\delta}$	
			CUE	2-step	CUE	2-step
4	0.5	0.5	0.7030	0.7298	1.0303	1.0189
			(0.8445)	(0.8593)	(1.0037)	(1.0081)
8	0.5	0.5	0.8727	0.8769	1.0086	1.0016
			(0.9123)	(0.9273)	(1.0004)	(1.0073)
4	0.9	0.9	0.8911	0.8063	0.9176	0.8574
			(0.9399)	(0.8989)	(0.9401)	(0.9179)
8	0.9	0.9	1.0225	0.9448	1.0084	0.9541
			(1.0259)	(0.9661)	(1.0050)	(0.9737)

13.4   Empirical Application

The estimation method proposed in this paper is applied to the data used by Blundell and Bond (2000), which include information on a balanced sample of R&D performing US manufacturing firms ($N = 509$) observed over the time period 1982–1989. The following dynamic Cobb-Douglas production function is considered:

$$y_{it} = \beta_1 y_{it-1} + \beta_2 n_{it} + \beta_3 n_{it-1} + \beta_4 k_{it} + \beta_5 k_{it-1} + \tau_t + \alpha_i + \varepsilon_{it} \qquad (13.13)$$

with y_{it} the log sales of firm i at time t, n_{it} the log employment, k_{it} the log capital and τ_t year-specific dummy variables.

Blundell and Bond (2000) finds that, consistent with the presence of measurement errors, the validity of standard instruments in the difference GMM estimator is not satisfied, so that lagged levels of the variables dated $t - 2$ and earlier are not valid instruments in this setting. Accordingly, the system GMM estimator will rely on instruments dated $t - 3$ (and earlier) for the first difference equation and on lagged first differences dated $t - 2$ as instruments in the level equations. Two-step results are considered, and the standard errors will be computed using the formula provided by Windmeijer (2005). The one-step system GMM estimator is also reported for comparison. Results of the system GMM (SYS) estimator and of the estimator that exploits the additional moment condition (CM) are reported in Table 13.3. Standard diagnostic statistics (p-value of m_1, m_2 and the Hansen test of overidentifying restrictions) are reported at the bottom of the table. All tests support the validity of GMM estimation, at least at the one percent level of significance. We also report the difference-in-Hansen test that verifies the validity of the additional moment condition we propose. The null hypothesis is not rejected, at least at the 1% level of significance.

Table 13.3 Estimation results

	1982–1989			1984–1989			1982-1985		
	SYS-1	SYS-2	CM-2	SYS-1	SYS-2	CM-2	SYS-1	SYS-2	CM-2
y_{it-1}	0.612***	0.649***	0.655***	0.725***	0.769***	0.786***	0.789*	0.794	0.801
	(0.100)	(0.090)	(0.096)	(0.105)	(0.108)	(0.104)	(0.406)	(0.584)	(0.521)
n_{it}	0.470***	0.516***	0.502***	0.446***	0.379**	0.364**	0.739	1.17	1.04
	(0.113)	(0.094)	(0.099)	(0.166)	(0.156)	(0.153)	(1.92)	(2.57)	(0.702)
n_{it-1}	−.283**	−0.305***	−0.322***	−0.370**	−0.245**	−0.261**	−0.644	−1.08	−0.935
	(0.121)	(0.089)	(0.098)	(0.145)	(0.125)	(0.131)	(2.17)	(3.01)	(0.918)
k_{it}	0.400***	0.428***	0.476***	0.339*	0.378*	0.422**	0.087	−0.043	−0.047
	(0.153)	(0.142)	(0.151)	(0.201)	(0.195)	(0.185)	(0.980)	(1.33)	(.988)
k_{it-1}	−0.215*	−0.289**	−0.312**	−0.174	−0.281*	−0.313**	0.012	0.132	0.119
	(0.120)	(0.114)	(0.129)	(0.163)	(0.154)	(0.155)	(0.824)	(1.14)	(0.725)
τ_t	yes	yes	yes	yes	yes	yes	yes	yes	yes
p-value of:									
m_1	0.000	0.000		0.000	0.000		0.306	0.537	
m_2	0.672	0.763		0.814	0.909				
Hansen test	–	0.033	0.019	–	0.038	0.041		0.246	0.506
Diff.H. test	–	–	0.041	–	–	0.336			0.905
LRe-n	0.481***	0.601***	0.516***	0.275	0.581*	0.481*	0.452	0.431	0.515
	(0.144)	(0.150)	(0.163)	(0.280)	(0.303)	(0.264)	(0.900)	(1.27)	(0.492)
LRe-k	0.477	0.396**	0.475***	0.603***	0.420*	0.501**	0.470	0.430	0.362
	(0.110)	(0.127)	(0.134)	(0.188)	(0.243)	(0.202)	(0.626)	(0.688)	(0.597)
CRS: t-test	−0.791	−0.042	−0.173	−1.07	0.016	−0.136	−0.160	−0.168	−0.239

*** 1%; ** 5%; * 10%

In terms of point estimates of the coefficients, CM estimation results are very similar to the results obtained by SYS. Coherently with the Monte Carlo results, no substantial efficiency gain is obtained when the full-time period is considered for estimation ($T + 1 = 8$). In order to highlight the advantages provided by the additional moment condition, we also estimate the model using a selection of observation years, in particular, we report the estimates over the period 1984–1989 ($T + 1 = 6$) and 1982–1985 ($T + 1 = 4$). In cases in which the observation period is shortened, efficiency gains are detected and these are particularly remarkable with $T + 1 = 4$. However, in this case, estimated coefficients lack statistical significance.

The table also reports the long-run elasticity of labour (LRe-n) and of capital (LRe-k),[15] and the value of the t-test for the constant returns to scale hypothesis, which is never rejected.

---

[15]The hypothesis of common factor restriction is never rejected at the 5% level of significance. See Blundell and Bond (2000) for details.

13.5   Discussion and Conclusions

In the context of dynamic panel data models under the assumption of mean station-arity, we propose an additional moment condition that can enhance performance of GMM estimator customarily employed in applications. GMM methods employed in empirical applications (i.e. the difference GMM estimator and the system GMM esti-mator) exploit information up to the second moment of the variables. The moment condition we propose exploits third moments, and we propose to add it to the set of moment conditions that characterize the system GMM estimator. Remarkable efficiency gains can be achieved, especially when T is small, the autoregression coefficient β approaches 1, and the individual effects are skewed (as it is indeed the case in many microeconomic applications).

Acknowledgements   We gratefully acknowledge comments and suggestions from two anonymous Reviewers, Francesca Mantese, and conference participants at the Fifth and Sixth Italian Congress of Econometrics and Empirical Economics.

Appendix A: Detailed Results of Monte Carlo Experiments

In this Appendix, detailed tables reporting the results of the Monte Carlo experiments are presented. Table 13.4 reports the full set of results for the Monte Carlo experiments related to the pure dynamic model (Sect. 13.3.1; Table 13.1), whereas Table 13.5 reports results for the experiments including a simultaneously determined regressor (Sect. 13.3.2; Table 13.2). The tables show the mean, variance, median and interquartile range (IQR) of Monte Carlo estimates.

Appendix B: Variance Comparison

In this Appendix, we derive the variance of the estimator we propose, based on moment conditions (13.2), (13.4) and (13.6), and we compare it with the variance of the traditional system GMM estimator.

We denote the set of R moment conditions in the population as $E(\mathbf{f}(\beta)) = 0$. In order to simplify computations, we take into account the case of model (13.1), so that β (and the variance of its estimator) is a scalar.

From general GMM theory (Hansen 1982), the asymptotic variance of a GMM estimator is given by

$$V(\hat{\beta}) = (B'\Lambda^{-1}B)^{-1}$$

with $B = \partial \mathbf{f}(\beta)/\partial \beta$ and $\Lambda = E(\mathbf{f}(\beta)\mathbf{f}(\beta)')$ the efficient weighting matrix of the GMM procedure.

Table 13.4 Detailed results of Monte Carlo experiments within a pure dynamic framework (Sect. 13.3.1; Table 13.1)

$T+1$	β	Dist.	GMM-SYS (CUE)				GMM-SYS (2-st.)				GMM-CM (CUE)				GMM-CM (2-st.)			
			Mean	Var	Med.	IQR	Mean	Var	Med.	IQR	Mean	Var	Med.	IQR	Mean	Var	Med.	IQR
$\sigma_\alpha^2/\sigma_e^2 = 1$																		
4	0.5	N	0.5002	0.0053	0.5018	0.1007	0.4979	0.0050	0.4987	0.0966	0.4988	0.0053	0.4995	0.1009	0.4962	0.0047	0.4968	0.0943
8	0.5	N	0.5013	0.0010	0.5016	0.0430	0.4982	0.0010	0.4987	0.0424	0.5015	0.0011	0.5017	0.0432	0.4990	0.0010	0.4994	0.0424
4	0.9	N	0.8951	0.0080	0.9061	0.0991	0.8873	0.0068	0.8946	0.1018	0.8904	0.0083	0.9019	0.1064	0.8762	0.0069	0.8793	0.0974
8	0.9	N	0.9000	0.0012	0.9032	0.0433	0.8892	0.0012	0.8923	0.0461	0.8981	0.0012	0.9008	0.0441	0.8862	0.0011	0.8888	0.0437
4	0.5	$\log$-N	0.5005	0.0053	0.5025	0.0991	0.4979	0.0048	0.4992	0.0938	0.4935	0.0027	0.4937	0.0700	0.4922	0.0026	0.4927	0.0688
8	0.5	$\log$-N	0.5009	0.0010	0.5011	0.0430	0.4969	0.0010	0.4976	0.0421	0.4944	0.0008	0.4947	0.0387	0.4912	0.0008	0.4920	0.0384
4	0.9	$\log$-N	0.8926	0.0083	0.9043	0.0958	0.8875	0.0065	0.8953	0.0980	0.9004	0.0020	0.9011	0.0531	0.9077	0.0037	0.9015	0.0571
8	0.9	$\log$-N	0.8998	0.0010	0.9027	0.0386	0.8901	0.0011	0.8933	0.0428	0.8994	0.0004	0.9002	0.0276	0.8899	0.0006	0.8915	0.0314
4	0.5	χ_1^2	0.5006	0.0053	0.5025	0.0986	0.4981	0.0048	0.4988	0.0934	0.4921	0.0029	0.4930	0.0700	0.4917	0.0028	0.4925	0.0697
8	0.5	χ_1^2	0.5011	0.0011	0.5014	0.0444	0.4975	0.0010	0.4982	0.0428	0.4938	0.0009	0.4944	0.0392	0.4917	0.0008	0.4923	0.0387
4	0.9	χ_1^2	0.8942	0.0083	0.9044	0.0974	0.8871	0.0067	0.8943	0.0982	0.9001	0.0023	0.9009	0.0565	0.9059	0.0043	0.9006	0.0621
8	0.9	χ_1^2	0.9001	0.0011	0.9025	0.0407	0.8897	0.0012	0.8929	0.0448	0.8988	0.0005	0.8998	0.0281	0.8903	0.0007	0.8923	0.0314
$\sigma_\alpha^2/\sigma_e^2 = 4$																		
4	0.5	N	0.5011	0.0076	0.5006	0.1222	0.5071	0.0068	0.5081	0.1135	0.5040	0.0076	0.5038	0.1228	0.5132	0.0067	0.5150	0.1130
8	0.5	N	0.5017	0.0011	0.5019	0.0452	0.5010	0.0012	0.5011	0.0455	0.5022	0.0012	0.5022	0.0457	0.5026	0.0012	0.5026	0.0459
4	0.9	N	0.8823	0.0119	0.9038	0.1193	0.8837	0.0093	0.8982	0.1146	0.8892	0.0104	0.9056	0.1138	0.8845	0.0080	0.8919	0.0995
8	0.9	N	0.8972	0.0017	0.9032	0.0537	0.8902	0.0016	0.8948	0.0546	0.8977	0.0016	0.9027	0.0523	0.8908	0.0015	0.8955	0.0511
4	0.5	$\log$-N	0.5022	0.0073	0.5035	0.1201	0.5059	0.0065	0.5074	0.1113	0.4891	0.0043	0.4905	0.0890	0.4926	0.0042	0.4945	0.0874
8	0.5	$\log$-N	0.5015	0.0012	0.5018	0.0456	0.5001	0.0012	0.5005	0.0459	0.4950	0.0010	0.4954	0.0429	0.4957	0.0010	0.4962	0.0432
4	0.9	$\log$-N	0.8823	0.0119	0.9022	0.1146	0.8849	0.0089	0.8983	0.1108	0.8960	0.0043	0.9019	0.0725	0.9041	0.0059	0.9048	0.0750
8	0.9	$\log$-N	0.8971	0.0015	0.9025	0.0471	0.8903	0.0015	0.8946	0.0515	0.8972	0.0007	0.8995	0.0343	0.8914	0.0010	0.8947	0.0388
4	0.5	χ_1^2	0.5021	0.0073	0.5028	0.1179	0.5065	0.0066	0.5074	0.1117	0.4881	0.0046	0.4901	0.0914	0.4935	0.0046	0.4956	0.0900
8	0.5	χ_1^2	0.5015	0.0012	0.5016	0.0467	0.5004	0.0012	0.5009	0.0466	0.4957	0.0010	0.4962	0.0434	0.4969	0.0011	0.4972	0.0438
4	0.9	χ_1^2	0.8844	0.0118	0.9031	0.1143	0.8851	0.0095	0.8994	0.1100	0.8929	0.0057	0.9013	0.0764	0.9006	0.0071	0.9030	0.0789
8	0.9	χ_1^2	0.8975	0.0016	0.9020	0.0496	0.8906	0.0016	0.8952	0.0534	0.8962	0.0009	0.8993	0.0365	0.8913	0.0011	0.8949	0.0410

Table 13.5 Detailed results of Monte Carlo experiments with an endogenous regressor (Sect. 13.3.2; Table 13.2)

T+1	β	ρ	dist.		GMM-SYS (CUE)				GMM-SYS (2-st.)				GMM-CM (CUE)				GMM-CM (2-st.)			
					Mean	Var	Med.	IQR	Mean	Var	Med.	IQR	Mean	Var	Med.	IQR	Mean	Var	Med.	IQR
4	0.5	0.5	N	$\hat{\beta}$	0.5006	3.195E-03	0.5017	0.0761	0.5034	2.910E-03	0.5044	0.0733	0.4994	3.186E-03	0.5003	0.0763	0.5032	2.816E-03	0.5042	0.0722
				$\hat{\delta}$	1.005	5.166E-02	0.9992	0.2987	0.9957	4.377E-02	0.9915	0.2765	1.007	5.338E-02	1.003	0.3007	1.001	4.308E-02	0.9973	0.2736
8	0.5	0.5	N	$\hat{\beta}$	0.5023	6.188E-04	0.5023	0.0335	0.5053	5.787E-04	0.5051	0.0326	0.5023	6.191E-04	0.5022	0.0337	0.5059	5.803E-04	0.5057	0.0327
				$\hat{\delta}$	1.000	1.304E-02	0.9994	0.1541	0.9801	1.001E-02	0.9786	0.1354	1.001	1.321E-02	1.000	.1550	0.9816	1.002E-02	0.9804	0.1355
4	0.9	0.9	N	$\hat{\beta}$	0.8989	3.004E-04	0.9000	0.0223	0.8997	1.838E-04	0.9033	0.0165	0.8989	2.536E-04	0.8998	0.0206	0.8997	1.436E-04	0.9002	0.0145
				$\hat{\delta}$	0.9949	1.716E-02	0.9965	0.1866	0.9973	1.108E-02	0.9976	0.1388	0.9942	1.521E-02	0.9953	0.1724	0.9984	9.259E-03	0.9985	0.1243
8	0.9	0.9	N	$\hat{\beta}$	0.9002	4.875E-05	0.9004	0.0096	0.9002	2.729E-05	0.9004	0.0068	0.9002	4.940E-05	0.9003	0.0098	0.9002	2.524E-05	0.9003	0.0066
				$\hat{\delta}$	1.002	1.944E-03	1.002	0.0634	0.9997	1.162E-03	1.000	0.0458	1.002	1.951E-03	1.000	0.0643	1.000	1.093E-03	1.001	0.0443
4	0.5	0.5	$\log$-N	$\hat{\beta}$	0.5008	3.056E-03	0.5024	0.0741	0.5029	2.747E-03	0.5045	0.0701	0.4980	2.148E-03	0.4987	0.0626	0.4996	2.004E-03	0.5003	0.0602
				$\hat{\delta}$	1.003	5.108E-02	0.9992	0.3021	0.9928	4.356E-02	0.9898	0.2776	1.007	5.263E-02	1.004	0.3032	0.9996	4.439E-02	0.9981	0.2799
8	0.5	0.5	$\log$-N	$\hat{\beta}$	0.5020	6.040E-04	0.5020	0.0338	0.5044	5.601E-04	0.5045	0.0321	0.4988	5.271E-04	0.4987	0.0308	0.5015	4.912E-04	0.5015	0.0297
				$\hat{\delta}$	0.9998	1.272E-02	0.9992	0.1528	0.9776	9.776E-03	0.9774	0.1341	1.008	1.283E-02	1.008	0.1528	0.9843	9.791E-03	0.9839	0.1351
4	0.9	0.9	$\log$-N	$\hat{\beta}$	0.8989	2.945E-04	0.9000	0.0224	0.8997	1.834E-04	0.9003	.165	0.8987	2.624E-04	0.8997	0.0211	0.8998	1.479E-04	0.9003	0.0148
				$\hat{\delta}$	0.9948	1.704E-02	0.9974	0.1879	0.9971	1.105E-02	0.9968	0.1374	0.9932	1.564E-02	0.9954	0.1767	0.9986	9.473E-03	0.9999	0.1261
8	0.9	0.9	$\log$-N	$\hat{\beta}$	0.9002	4.780E-05	0.9004	0.0096	0.9001	2.701E-05	0.9003	0.0068	0.9001	4.888E-05	0.9003	0.0098	0.9002	2.552E-05	0.9003	0.0066
				$\hat{\delta}$	1.002	1.932E-03	1.002	0.0626	0.9991	1.152E-03	0.9996	0.0453	1.002	1.948E-03	1.002	0.0629	0.9996	1.099E-03	1.000	0.0446
4	0.5	0.5	χ_1^2	$\hat{\beta}$	0.5014	3.084E-03	0.5023	0.0737	0.5037	2.809E-03	0.5047	0.0707	0.4982	2.310E-03	0.4990	0.0648	0.5007	2.163E-03	0.5014	0.0624
				$\hat{\delta}$	1.003	5.123E-02	1.001	0.2998	0.9935	4.359E-02	0.9938	0.2762	1.008	5.313E-02	1.008	0.3018	1.002	4.410E-02	1.003	0.2780
8	0.5	0.5	χ_1^2	$\hat{\beta}$	0.5021	6.070E-04	0.5022	0.0332	0.5049	5.678E-04	0.5050	0.0319	0.4990	5.357E-04	0.4992	0.0311	0.5023	5.028E-04	0.5025	0.0304
				$\hat{\delta}$	0.9995	1.313E-02	0.9989	0.1540	0.9785	1.007E-02	0.9787	0.1346	1.007	1.325E-02	1.007	0.1544	0.9852	1.003E-02	0.9853	0.1349
4	0.9	0.9	χ_1^2	$\hat{\beta}$	0.8990	2.970E-04	0.9000	0.0226	0.8997	1.853E-04	0.9002	0.0165	0.8989	2.561E-04	0.8998	0.0206	0.8998	1.473E-04	0.9002	0.0148
				$\hat{\delta}$	0.9949	1.703E-02	0.9979	0.1893	0.9974	1.091E-02	0.9975	0.1365	0.9942	1.517E-02	0.9959	0.1717	0.9984	9.248E-03	0.9988	0.1228
8	0.9	0.9	χ_1^2	$\hat{\beta}$	0.9002	4.819E-05	0.9003	0.0097	0.9002	2.688E-05	0.9003	0.0068	0.9001	4.866E-05	0.9002	0.0097	0.9002	2.496E-05	0.9003	0.0066
				$\hat{\delta}$	1.001	1.941E-03	1.002	0.0628	0.9992	1.147E-03	1.000	0.0453	1.001	1.944E-03	1.001	0.0636	0.9997	1.084E-03	1.000	0.0443

In this case, we can partition B into two components (and Λ accordingly): (i) the elements corresponding to the moment conditions that characterize the traditional system GMM estimator, and (ii) the one additional moment condition we propose in this paper. Assuming that the moment condition we propose is the last moment condition in the full set composed of (13.2), (13.4) and (13.6), we can write

$$
B = \begin{pmatrix} B_1 \\ B_2 \end{pmatrix} \quad \text{and} \quad \Lambda = \begin{pmatrix} \Lambda_1 & \Lambda_2 \\ \Lambda_3 & \Lambda_4 \end{pmatrix}
$$

with B_1 related to the moment conditions in (13.2) and (13.4), and B_2 related to the moment condition in (13.6). Λ is partitioned accordingly (with $\Lambda_3 = \Lambda_2'$). Therefore, we have

$$
V(\hat{\beta}) = (B'\Lambda^{-1}B)^{-1} = \left\{ \begin{pmatrix} B_1' & B_2' \end{pmatrix} \begin{pmatrix} \Lambda_1 & \Lambda_2 \\ \Lambda_3 & \Lambda_4 \end{pmatrix}^{-1} \begin{pmatrix} B_1 \\ B_2 \end{pmatrix} \right\}^{-1} \tag{13.14}
$$

By using the formula of the inverse of a partitioned matrix, we get

$$
V(\hat{\beta}) = (B_1'\Lambda_1^{-1}B_1 + (\Lambda_3\Lambda_1^{-1}B_1 - B_2)'A(\Lambda_3\Lambda_1^{-1}B_1 - B_2))^{-1} \tag{13.15}
$$

with $A = (\Lambda_4 - \Lambda_3\Lambda_1^{-1}\Lambda_2)^{-1} = (\Lambda_4 - \Lambda_2'\Lambda_1^{-1}\Lambda_2)^{-1}$ (a scalar). Note that the first element in (13.15) corresponds to the inverse of the variance of the system GMM estimator of β. Therefore, the variance of the GMM estimator based on (13.2), (13.4), and the additional moment condition (13.6) will be smaller than the variance of the system GMM estimator if

$$
(\Lambda_3\Lambda_1^{-1}B_1 - B_2)'A(\Lambda_3\Lambda_1^{-1}B_1 - B_2) > 0
$$

As the scalar $A = (\det(\Lambda)/\det(\Lambda_1))^{-1} > 0,$[16] this condition will be always satisfied unless

$$
\Lambda_3\Lambda_1^{-1}B_1 - B_2 = 0 \tag{13.16}
$$

No efficiency gain will be obtained adding the moment condition we propose when condition (13.16) is satisfied. On the contrary, if condition (13.16) is not satisfied ($\Lambda_3\Lambda_1^{-1}B_1 - B_2 \neq 0$), efficiency gains are obtained by adding the moment condition we propose to the moment conditions that characterize the system GMM estimator.

By relying on further simulations (not reported in the main text of the paper) with $T + 1 = 3$, it is possible to show that $\Lambda_3\Lambda_1^{-1}B_1 - B_2 \neq 0$ if α_i or e_{i0} have an asymmetric distribution (the third moment is not zero). Coherently, the Monte Carlo experiments reported in Sect. 13.3 show gains in performance in the cases in which

---

[16]The equality can be obtained by relying on the fact that $\Lambda_1^{-1} = \Lambda_1^*/\det(\Lambda_1)$ where Λ_1^* is adjugate of Λ_1. Furthermore, for the way in which they are computed, both Λ and Λ_1 are positive definite matrices, so their determinant will be positive.

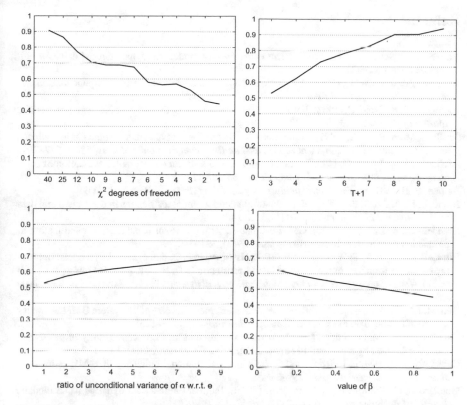

Fig. 13.1 Ratio of the asymptotic variance of the estimator we propose with respect to the asymptotic variance of the system GMM estimator, for different (i) levels of asymmetry in the distribution of α_i, e_{i0}; (ii) time periods; (iii) ratios of unconditional variance of α with respect to e; (iv) values of β

the error components are assumed to have an asymmetric (log-normal or chi-squared) distribution.

Figure 13.1 shows the ratio of the asymptotic variance of the estimator we propose with respect to the asymptotic variance of the system GMM estimator. The baseline experiment consider $\beta = 0.5$, $T + 1 = 3$, a χ_3^2 distribution for e_{i0} and α_i, and equal unconditional variances of α_i and e_{it} (both equal to 1). From the baseline experiment, we changed: (i) the level of asymmetry in the distribution of α_i, e_{i0}, accomplished by changing the number of degrees of freedom of the χ^2 distribution; (ii) the number of time periods $T + 1$ from 3 to 10; (iii) the ratio of unconditional variance of α with respect to e (from 1 to 9; we changed the value of the unconditional variance of α_i, leaving the unconditional variance of e fixed at 1); (iv) the values of β from 0.1 to 0.9. Coherently with the results in the Monte Carlo experiments in Sect. 13.3, the gains in performance led by the additional moment condition we propose are stronger for larger asymmetry in the distribution of α_i and e_{i0} (thus, of y_{i0}), shorter observation in time, and values of β that approaches 1. Also, the gains in performance decrease when the unconditional variance of α_i increases with respect to the unconditional variance of e_{it}.

References

Ahn, S. C., & Schmidt, P. (1995). Efficient Estimation of models for dynamic panel data. *Journal of Econometrics, 68*(1), 5–27.

Anderson, T., & Hsiao, C. (1981). Estimation of dynamic models with error components. *Journal of the American Statistical Association, 76*, 598–606.

Anderson, T., & Hsiao, C. (1982). Formulation and estimation of dynamic models using panel data. *Journal of Econometrics, 18*, 47–82.

Arellano, M., & Bond, S. (1991). Some tests of specification for panel data: Monte Carlo evidence and an application to employment equations. *The Review of Economic Studies, 58*(2), 277–297.

Arellano, M., & Bover, O. (1995). Another look at the instrumental variables estimation of error components models. *Journal of Econometrics, 68*, 29–51.

Baltagi, B. H. (2013). *Econometric analysis of panel data* (5th ed.). New York: Wiley.

Blundell, R., & Bond, S. (1998). Initial conditions and moment restrictions in dynamic panel data models. *Journal of Econometrics, 87*(1), 115–143.

Blundell, R., & Bond, S. (2000). GMM estimation with persistent panel data: An application to production functions. *Econometric Reviews, 19*(3), 321–340.

Blundell, R., Bond, S., & Windmeijer, F. (2000). Estimation in dynamic panel data models: Improving on the performance of the standard GMM estimator. In B. H. Baltagi, T. B Fomby, & C. Hill (Eds.), *Nonstationary panels, panel cointegration, and dynamic panels*. Advances in econometrics, *15*.

Bun, M. J. C., & Windmeijer, F. (2010). The weak instrument problem of the system GMM estimator in dynamic panel data models. *Econometrics Journal, 13*(1), 95–126.

Calzolari, G., & Magazzini, L. (2012). Autocorrelation and masked heterogeneity in panel data models estimated by maximum likelihood. *Empirical Economics, 43*(1), 145–152.

Calzolari, G., Panattoni, L., & Weihs, C. (1987). Computational efficiency of FIML estimation. *Journal of Econometrics, 36*(3), 299–310.

Gouriéroux, C., Phillips, P. C. B., & Yu, J. (2010). Indirect inference for dynamic panel models. *Journal of Econometrics, 157*(1), 68–77.

Hahn, J. (1999). How informative is the initial condition in the dynamic panel model with fixed effects? *Journal of Econometrics, 93*, 309–326.

Hansen, L. P. (1982). Large sample properties of generalized method of moments estimators. *Econometrica, 50*, 1029–1054.

Hansen, L. P., Heaton, J., & Yaron, A. (1996). Finite-sample properties of some alternative GMM estimators. *Journal of Business and Economic Statistics, 14*(3), 262–280.

Hayakawa, K. (2007). Small sample bias properties of the system GMM estimator in dynamic panel data models. *Economics Letters, 95*, 32–38.

Hsiao, C., Pesaran, M. H., & Tahmiscioglu, A. K. (2002). Maximum likelihood estimation of fixed effects dynamic panel data models covering short time periods. *Journal of Econometrics, 109*(1), 107–150.

Kiviet, J. F. (1995). On bias, inconsistency, and efficiency of various estimators in dynamic panel data models. *Journal of Econometrics, 68*(1), 53–78.

Roodman, D. (2009). A note on the theme of too many instruments. *Oxford Bulletin of Economics and Statistics, 71*(1), 135–158.

Schmidt, P., Ahn, S. C., & Wyhowski, D. (1992). Comment. *Journal of Business and Economic Statistics, 10*, 10–14.

Windmeijer, F. (2005). A finite sample correction for the variance of linear efficient two-step GMM estimators. *Journal of Econometrics, 126*(1), 25–51.

Part V
Statistics in Industrial Applications

Chapter 14
Economically Designed Bayesian np Control Charts Using Dual Sample Sizes for Long-Run Processes

Imen Kooli and Mohamed Limam

Abstract The implementation of a control chart requires the determination of three design parameters: the sample size, the sampling interval, and the control limits under which the production process will be stopped for potential repair. For a static control chart, the design parameters are maintained at the same level from an inspection epoch to another. Several research papers showed that adopting dynamic control charts in which one or more of the design parameters are allowed to vary from an inspection epoch to another leads to substantial cost savings compared to the classical ones. In this paper, we develop the expected long-run costs of two Bayesian np schemes, namely, the basic Bayesian and the Bayes-n charts for processes operating over an infinite horizon length. Optimal solutions leading to least-cost plans are searched for different sets of process and cost parameters. Experimental results show that moving from classical np control charts to Bayesian ones results in significant economic savings.

14.1 Introduction

The continuing endeavor to improve the efficiency of static control charts with respect to statistical or economic considerations has led to the development of a new class of control charts called dynamic control charts. In these charts, one or more of the design parameters are allowed to vary from an inspection epoch to another based on the past data information. Dynamic control charts are divided into two categories: adaptive control charts and Bayesian control charts (Tagaras 1998). In adaptive control charts, the choice of the dynamic design parameter for the next sampling instance

I. Kooli (✉)
Department of Quantitative Methods and Computer Sciences,
Higher Institute of Management of Sousse (ISG), University of Sousse, Sousse, Tunisia
e-mail: imene.kooli@gmail.com

M. Limam
Dhofar University, Box 2509, Salalah 211, Oman
e-mail: Limam@du.edu.om

© Springer Nature Switzerland AG 2019
N. Bauer et al. (eds.), *Applications in Statistical Computing*,
Studies in Classification, Data Analysis, and Knowledge Organization,
https://doi.org/10.1007/978-3-030-25147-5_14

is expressed as a function of the observed current value of the monitored quality characteristic. In Bayesian control charts, this choice incorporates all the previous information about the state of the process.

Concerning adaptive control charts, (Reynolds et al. 1988) were the first authors to show that the application of a variable sampling interval (VSI) scheme leads to significant improvement of the statistical performance of $\bar{X}$ control charts. In Bai and Lee (1998), it was shown that the proposed sampling scheme of the aforementioned authors also leads to least-cost plans compared to the static one. $\bar{X}$ control charts operating with a variable sample size (VSS) by switching the size of the subsequent sample between two possible values depending on the current recorded sample mean were independently investigated by Prabhu et al. (1993), Costa (1994). Other adaptive sampling schemes have been proposed by Prabhu et al. (1994), Park and Reynolds (1999), Lee (2013). Recent research papers dealing with adaptive control charts for measurable data include those by Aly et al. (2015), Mahadik (2017), Khaw et al. (2017), Wang et al. (2018). Few researchers have focused on adaptive control charts for attribute data. The statistical performance of adaptive attribute control charts was evaluated by Luo and Wu (2002), Epprecht et al. (2003), Epprecht et al. (2010), and the economic effectiveness of np control charts using adaptive sampling strategies was investigated by Kooli and Limam (2011), Kooli and Limam (2015).

The pioneering works dealing with Bayesian control charts focused on production processes operating over a finite period of time. A partially observable Markov decision process was applied by Calabrese (1995) to derive the expected cost of Bayesian np control charts. It was shown that even if the sample size and sampling interval are kept fixed during the process monitoring, the proposed Bayesian chart is equivalent to an np control chart with an adaptive control limit. The Bayesian sampling scheme was considered to be a basic Bayesian one by Tagaras and Nikolaidis (2002) and was extended to the case where one or two of the remaining chart parameters are allowed to change during the monitoring of a process operating for a finite time period. The Bayes-n chart, the Bayes-h chart, and Bayes-n-h chart were taken as shorthand notations for a Bayesian sampling strategy using a variable sample size, a variable sampling interval, and both variable sample size and sampling interval, respectively. Then, the least-cost sampling plans of the static and the four types of the proposed dynamic charts were evaluated for different sets of processes and cost parameters. With respect to the static one-sided $\bar{X}$ chart, the cost savings ranged from very low to pronounced depending on the type of the Bayesian chart. The economic model was extended by Nenes and Tagaras (2007) to treat a fully adaptive Bayesian two-sided $\bar{X}$ chart for process operating for a short period of time. The economic design of Bayes-n np control charts was studied by Kooli and Limam (2009).

Economically designed Bayesian $\bar{X}$ control charts for processes operating indefinitely were first treated by Nenes (2013). In a subsequent work, the economic design of a control chart used to monitor a production process where two assignable causes may occur, one affecting the mean level and the other the variance, was developed (Nenes and Panagiotidou 2013). Bayesian attribute control charts for processes operating over an infinite horizon length were not considered previously in the literature. In this paper, we propose economic models of two Bayesian np schemes, namely, the basic Bayesian and the Bayes-n charts for processes operating indefinitely. Section 14.2 presents model assumptions and the different components of the set of the input parameters. Thereafter, the expected long-run cost per hour of basic Bayesian *np* control charts is developed. Section 14.4 considers the economic design of Bayesian *np* control charts operating with dual sample sizes. Optimal solutions and economic comparison between the different proposed sampling schemes are provided in Sect. 14.5. Main conclusions complete the paper.

14.2 Model Assumptions and Set of the Input Parameters

Throughout this paper, it is assumed that an attribute *np* control chart is used to maintain the statistical control of a manufacturing process operating indefinitely. The quality characteristic of interest is the number of nonconforming units, D_t, observed at the tth sampling instance. Given that the primary concern of an attribute *np* is the detection of quality deterioration, only the upper control limit (UCL) is considered to be active during the process monitoring.

The process is described by two states: State 0 represents the in-control state, while State 1 represents the out-of-control state. The process is considered to be in-control if the number of defective items falls below the UCL; otherwise, it is considered to be out-of-control. The occurrence of the assignable cause shifts the proportion of nonconforming units, p_{ncu}, from the target level p_0 to the undesirable level p_1, such that $p_1 > p_0$. A common assumption widely adopted for the development of economic models is that the time of occurrence of the assignable cause is an exponentially distributed random variable with mean $1/\lambda$. Therefore, the probability of the occurrence of a process shift in an interval of length h hours is $\gamma_h = 1 - \exp(-\lambda h)$.

In order to search for optimal solutions of the different economic models, the cost parameters associated with the operation of the chart and the process parameters related to the behavior of the production process need to be evaluated. The sampling cost is composed of both fixed and variable components denoted by b and c, respectively. A false alarm cost, L_0, is incurred if the process is stopped, whereas no assignable cause is really present. The cost of removing the root cause and repairing the process is L_1. The operation of the process in the out-of-control state incurs an hourly cost, M, due to the production of an increased number of defective items.

For static *np* control charts, a sample of size n is picked every h hours and the computed number of defective items is plotted against the sampling time. A signal is issued by the attribute *np* control chart if a single sample point exceeds the UCL and the process is stopped for potential repair. As defined by Kooli and Limam (2015), the expected long-run cost per hour, E(TCU), of the static *np* control chart is computed by

$$E(TCU) = \frac{b+cn}{h} + \frac{1-\beta}{h\,(\exp(\lambda h) - \beta)}\left\{\alpha\,L_0 + \frac{M(1-\exp(\lambda h))}{\lambda} + L_1(\exp(\lambda h) - 1)\right\} + M,$$

$$(14.1)$$

where $\alpha = P(D_t > UCL | p_{ncu_t} = p_0) = \sum_{j=UCL+1}^{n} b_{0j}$ is the probability of getting a false

alarm, $\beta = \sum_{j=0}^{UCL} b_{1j}$ is the probability that the np control chart fails to detect a signal

while a process shift has really occurred and $b_{ij} = \dfrac{n!}{j!(n-j)!} p_i^j (1-p_i)^{n-j}$, for $i = 0, 1$, is the binomial distribution with parameters n and p_i at the occurrence j.

14.3 Basic Bayesian Scheme

For classical np control charts, the decision a_t, whether or not to stop the process for potential investigation at the sampling time t, is made by making a direct comparison of the observed number of defective units D_t to the UCL of the chart. For Bayesian np control charts, this decision is based on the probability that the process is operating out-of-control, p_t, which is updated every time a new sample is collected and the number of defective units is observed using Bayes theorem. When a basic Bayesian chart has to be designed for long-run processes, the decision variables are the size of the sample, n, the time separating two successive samples, h, and the control limit p_B. If at a given time t, the probability that the process is operating under the occurrence of a process shift exceeds the value of p_B, then the process is stopped for investigation ($a_t = 1$); otherwise, no intervention is required ($a_t = 0$).

Let p denotes the probability that the process is operating off-target at the previous sampling instance. After observing the number of defective items d of the current sample of size n and given the action a chosen previously, the transformation of the value of p is computed as

$$T(p|d, n, a = 0) = \frac{[(1 - \gamma_h)p + \gamma_h] b_{1d}}{(1 - \gamma_h)(b_{1d} - b_{0d})p + (1 - \gamma_h)b_{0d} + \gamma_h b_{1d}},$$

and

$$T(p|d, n, a = 1) = \frac{\gamma_h b_{1d}}{(1 - \gamma_h)b_{0d} + \gamma_h b_{1d}}.$$

Let X_t be the state of the production process at the tth sampling epoch which can be either in-control (State 0) or out-of-control (State 1), and $C(i, a)$ denote the expected cost incurred during a period of length h hours if the process is in State i at its beginning and action a is chosen. Then, $C(0, 0) = \gamma_h M(h - \tau) + b + cn = M\left(h - \dfrac{\gamma_h}{\lambda}\right) + b + cn$ incorporates the expected cost incurred if the shift occurs during the next h hours and the sampling cost of the following sample. If the process has been stopped while it operates properly, then the cost of a false alarm is added and $C(0, 1) = L_0 + M\left(h - \frac{\gamma_h}{\lambda}\right) + b + cn$. The term $C(1, 0) = Mh + b + cn$ is the cost of letting the process as it is while it is off-target. Finally, $C(1, 1) = L_1 + M\left(h - \frac{\gamma_h}{\lambda}\right) + b + cn$ is the expected cost incurred by a newly repaired system.

To develop the expected long-run cost per hour of a basic Bayesian *np* control chart, we consider a two-dimensional Markov chain (X_t, p_t) as done by Nenes (2013). Let P be the one-step transition matrix of the Markov chain (X, p) and $P_{pp'}^{ij}$ the sub-matrices of P where the elements $p_{pp'}^{ij}$ are computed by

$$
\begin{aligned}
p_{pp'}^{ij} &= P(X_{t+1} = j, p_{t+1} = p' | X_t = i, p_t = p) \\
&= P(X_{t+1} = j | X_t = i) P(p_{t+1} = p' | p_t = p, X_{t+1} = j), \\
&= P(X_{t+1} = j | X_t = i)\, P_j(p'|p).
\end{aligned}
$$

By recalling that the probability that a shift occurs between two successive samples separated by h hours is $\gamma_h = 1 - e^{-\lambda h}$ and given that the probability of moving from the value p to p' depends on the value of the sample size, from here on the probabilities $P_j(p'|p)$ will be denoted by $P_j(p'|p, n)$. The four sub-matrices $P_{pp'}^{ij}$ for $i, j = 0, 1$ are determined by the following relationships:

$$
p_{pp'}^{00} = \begin{cases} e^{-\lambda h} P_0(p'|p, n) & \text{for } p \le p_B \ (a = 0) \\ e^{-\lambda h} P_0(p'|p = 0, n) & \text{for } p > p_B \ (a = 1) \end{cases}, \quad p_{pp'}^{01} = \begin{cases} (1 - e^{-\lambda h}) P_1(p'|p, n) & \text{for } p \le p_B \\ (1 - e^{-\lambda h}) P_1(p'|p = 0, n) & \text{for } p > p_B \end{cases},
$$

$$
p_{pp'}^{10} = \begin{cases} 0 & \text{for } p \le p_B \\ e^{-\lambda h} P_0(p'|p = 0, n) & \text{for } p > p_B \end{cases} \quad \text{and} \quad p_{pp'}^{11} = \begin{cases} P_1(p'|p, n) & \text{for } p \le p_B \\ (1 - e^{-\lambda h}) P_1(p'|p = 0, n) & \text{for } p > p_B \end{cases}.
$$

When the value of p_t exceeds the limit p_B, then the process is stopped to investigate whether the signal issued by the Bayesian control chart is due to a false alarm or a real process shift. In the two cases, after this intervention, the decision-maker is certain that the process resumes its operation in the in-control state. In such a case, the values of $P_j(p'|p, n)$ are equal to $P_j(p'|p = 0, n)$ for $j = 0, 1$.

The value of the out-of-control probability, p, was discretized into 101 possible values by Nenes (2013). In this paper, we set $p \in [0, 1]$ with a 0.001 step. Given that p can take 1001 possible values and X is equal to either 0 or 1, the total number of states of the two-dimensional Markov chain is 2002. It follows that the matrix

$$
P = \begin{bmatrix} P_{pp'}^{00} & P_{pp'}^{01} \\ P_{pp'}^{10} & P_{pp'}^{11} \end{bmatrix}
$$

is of size 2002×2002.

To illustrate the computation of the sub-matrices $P_{pp'}^{ij}$, we consider $i = j = 0$, $n = 2, h = 1$, the fraction of nonconforming units when the process is operating in-

Table 14.1 The updated values of p_t given the new value of D_{t+1} if $a_t = 0$

p_t	$T(p_t\|D_{t+1} = 0, a_t = 0)$	$T(p_t\|D_{t+1} = 1, a_t = 0)$	$T(p_t\|D_{t+1} = 2, a_t = 0)$
0	0.0008	0.0035	0.0146
0.1	0.0854	0.2823	0.6237
0.2	0.1729	0.4683	0.7877
0.3	0.2635	0.6012	0.864
0.4	0.3573	0.7008	0.9080
0.5	0.4546	0.7784	0.9367
0.6	0.5555	0.8404	0.9569
0.7	0.6603	0.8912	0.9718
0.8	0.7691	0.9335	0.9834
0.9	0.8823	0.9693	0.9925
1	1	1	1

control is $p_0 = 3\%$ and it increases to 11.53% under the occurrence of the assignable cause, $p_B = 0.6$, and $p \in [0, 1]$ with a 0.1 step for ease of presentation. Then, $b_{0d} =$

$$\begin{bmatrix} b_{00} \\ b_{01} \\ b_{02} \end{bmatrix} = \begin{bmatrix} 0.9409 \\ 0.0582 \\ 0.0009 \end{bmatrix} \text{ and } b_{1d} = \begin{bmatrix} b_{10} \\ b_{11} \\ b_{12} \end{bmatrix} = \begin{bmatrix} 0.7827 \\ 0.204 \\ 0.0133 \end{bmatrix}.$$

The updated values of the out-of-control probabilities are given in Table 14.1. By rounding down these values to one decimal, the sub-matrices $P_{pp'}^{00}$ is computed as

$$P_{pp'}^{00} = \begin{bmatrix}
0.999^a & 0 & 0 & 0 & 0 & 0 & 0 & 0 & 0 & 0 & 0 \\
0 & 0.94^b & 0 & 0.0581^c & 0 & 0 & 0.0009^d & 0 & 0 & 0 & 0 \\
0 & 0 & 0.94 & 0 & 0 & 0.0581 & 0 & 0 & 0.0009 & 0 & 0 \\
0 & 0 & 0 & 0.94 & 0 & 0 & 0.0581 & 0 & 0 & 0.0009 & 0 \\
0 & 0 & 0 & 0 & 0.94 & 0 & 0 & 0.0581 & 0 & 0.0009 & 0 \\
0 & 0 & 0 & 0 & 0 & 0.94 & 0 & 0 & 0.0581 & 0.0009 & 0 \\
0 & 0 & 0 & 0 & 0 & 0 & 0.94 & 0 & 0.0581 & 0 & 0.0009 \\
0.999 & 0 & 0 & 0 & 0 & 0 & 0 & 0 & 0 & 0 & 0 \\
0.999 & 0 & 0 & 0 & 0 & 0 & 0 & 0 & 0 & 0 & 0 \\
0.999 & 0 & 0 & 0 & 0 & 0 & 0 & 0 & 0 & 0 & 0 \\
0.999 & 0 & 0 & 0 & 0 & 0 & 0 & 0 & 0 & 0 & 0
\end{bmatrix}$$

where $a = \exp(-0.001)(b_{00} + b_{01} + b_{02})$, $b = \exp(-0.001)b_{00}$, $c = \exp(-0.001)b_{01}$ and $d = \exp(-0.001)b_{02}$.

Table 14.2 provides the expected costs per one-step transition for all the possible states of the two-dimensional Markov chain. Let $\pi = [\pi_{ip}]$ be the vector containing the steady-state probabilities π_{ip}, which represent the long-run probabilities that the process is being in any of the 2002 two-dimensional states. These probabilities are

Table 14.2 Expected cost of the basic *np* control chart for long-run processes

State	Steady-state probability	Value of p	Expected cost
$(X = 0, p)$	π_{0p}	$p \le p_B$	$M(h - \frac{\gamma_h}{\lambda}) + b + cn$
		$p > p_B$	$L_0 + M(h - \frac{\gamma_h}{\lambda}) + b + cn$
$(X = 1, p)$	π_{1p}	$p \le p_B$	$Mh + b + cn$
		$p > p_B$	$L_1 + M(h - \frac{\gamma_h}{\lambda}) + b + cn$

obtained by solving the system of linear equations $\pi P = \pi$ and $\displaystyle\sum_{i=0,1}\sum_{p\in[0,1]} \pi_{ip} = 1$.

The rows (1–4) of the last column of Table 14.2 represent the expected costs per period and they correspond to the terms $C(0, 0)$, $C(0, 1)$, $C(1, 0)$, and $C(1, 1)$, respectively.

The expected long-run cost of the basic Bayesian scheme for a period of length h hours is given by

$$E(CP) = b + cn + \sum_{p=0}^{p_B} \pi_{0p} M\left(h - \frac{\gamma}{\lambda}\right) + \sum_{p=p_B+0.001}^{1} \pi_{0p}\left[L_0 + M\left(h - \frac{\gamma}{\lambda}\right)\right] + \sum_{p=0}^{p_B} \pi_{1p} Mh$$

$$+ \sum_{p=p_B+0.001}^{1} \pi_{1p}\left[L_1 + M\left(h - \frac{\gamma}{\lambda}\right)\right],$$

simplifying to

$$E(CP) = b + cn + Mh - \frac{\gamma M}{\lambda}\left[\sum_{p=0}^{1} \pi_{0p} + \sum_{p=p_B+0.001}^{1} \pi_{1p}\right] + L_0 \sum_{p=p_B+0.001}^{1} \pi_{0p} + L_1 \sum_{p=p_B+0.001}^{1} \pi_{1p}.$$

The expected cost per hour, $E(TCU)$, of the basic Bayesian *np* control chart when the production process operates indefinitely is then computed using the relationship

$$E(TCU) = \frac{E(CP)}{h}. \tag{14.2}$$

14.4   Bayesian Scheme Using an Adaptive Sample Size

When the sample size of the Bayesian chart is allowed to vary from an inspection time to another for long-run processes, the size of the sample to pick at the next sampling time has to be chosen from two possible values n_1 and n_2 depending on the value of the out-of-control probability p_t, as proposed by Nenes (2013) for the fully Bayesian $\bar{X}$ charts. Moreover, to the action limit p_s a threshold limit p_R is added to the control chart in order to switch between the two possible values of the sample size. If at the tth sampling epoch, $p_t \leq p_R$ then it is reasonable to adopt the smaller sample size, n_1, at the next decision time. However, if $p_R < p_t \leq p_s$ then larger units should be inspected at the following sampling instance to cope with the large values of the probability of operating off-target. If p_t exceeds the action limit p_S, then a signal is issued by the Bayesian control chart and the action $a_t = 1$ is chosen. Unlike (Nenes 2013), in this case the next sample size will be the larger one to guard against misadjustments after repair as advised by Park and Reynolds (1994), Park and Reynolds (1999), Kooli and Limam (2011).

The decision variables of the Bayes-n np chart are the large sample size n_2, the small sample size n_1, the time separating two successive samples h, the threshold limit p_R, and the action limit p_s. The sub-matrices $P_{pp'}^{ij}$ of the Bayes-n np chart are computed using the following relationships:

$$
P_{pp'}^{00} = \begin{cases} e^{-\lambda h} \, P_0(p'|p, n_1, h) & \text{for } p \leq p_R (a = 0) \\ e^{-\lambda h} \, P_0(p'|p, n_2, h) & \text{for } p_R < p \leq p_S (a = 0) \ , \\ e^{-\lambda h} \, P_0(p'|p = 0, n_2, h) & \text{for } p > p_s (a = 1) \end{cases}
$$

$$
P_{pp'}^{01} = \begin{cases} (1 - e^{-\lambda h}) \, P_1(p'|p, n_1, h) & \text{for } p \leq p_R \\ (1 - e^{-\lambda h}) \, P_1(p'|p, n_2, h) & \text{for } p_R < p \leq p_S \ , \\ (1 - e^{-\lambda h}) \, P_1(p'|p = 0, n_2, h) & \text{for } p > p_s \end{cases}
$$

$$
P_{pp'}^{10} = \begin{cases} 0 & \text{for } p \leq p_s \\ e^{-\lambda h} \, P_0(p'|p = 0, n_2, h) & \text{for } p > p_s \end{cases} ,
$$

and

$$
P_{pp'}^{11} = \begin{cases} P_1(p'|p, n_1, h) & \text{for } p \leq p_R \\ P_1(p'|p, n_2, h) & \text{for } p_R < p \leq p_S \ . \\ (1 - e^{-\lambda h}) P_1(p'|p = 0, n_2, h) & \text{for } p > p_s \end{cases}
$$

The expected costs incurred during a period of length h hours when a Bayesian np control chart using dual sample sizes are used to monitor the quality of the products for processes operating indefinitely are provided in Table 14.3 for the different possible states of the Markov chain (X, p).

Thus, the expected long-run cost of the Bayes-n np chart for a period of length h hours is expressed as

$$
E(CP) = \sum_{p=0}^{p_R} \pi_{0p} \left[M \left(h - \frac{\gamma}{\lambda} \right) + b + cn_1 \right] + \sum_{p=p_R+0.001}^{p_S} \pi_{0p} \left[M \left(h - \frac{\gamma}{\lambda} \right) + b + cn_2 \right]
$$

Table 14.3 Expected cost of the Bayes-n *np* control chart for long-run processes

State	Steady-state probability	Value of p	Expected cost
$(X = 0, p)$	π_{0p}	$p \leq p_R$	$M\left(h - \frac{\gamma_h}{\lambda}\right) + b + cn_1$
		$p_R < p \leq p_S$	$M\left(h - \frac{\gamma_h}{\lambda}\right) + b + cn_2$
		$p > p_S$	$L_0 + M\left(h - \frac{\gamma_h}{\lambda}\right) + b + cn_2$
$(X = 1, p)$	π_{1p}	$p \leq p_R$	$Mh + b + cn_1$
		$p_R < p \leq p_S$	$Mh + b + cn_2$
		$p > p_S$	$L_1 + M\left(h - \frac{\gamma_h}{\lambda}\right) + b + cn_2$

$$+ \sum_{p=p_S+0.001}^{1} \pi_{0p}\left[L_0 + M\left(h - \frac{\gamma}{\lambda}\right) + b + cn_2\right] + \sum_{p=0}^{p_R} \pi_{1p}\left[Mh + b + cn_1\right]$$

$$+ \sum_{p=p_R+0.001}^{p_S} \pi_{1p}\left[Mh + b + cn_2\right] + \sum_{p=p_S+0.001}^{1} \pi_{1p}\left[L_1 + M\left(h - \frac{\gamma}{\lambda}\right) + b + cn_2\right],$$

which reduces to

$$E(CP) = b + Mh - \frac{\gamma M}{\lambda}\left[\sum_{p=0}^{1}\pi_{0p} + \sum_{p=p_S+0.001}^{1}\pi_{1p}\right] + cn_1\left[\sum_{p=0}^{p_R}\pi_{0p} + \sum_{p=0}^{p_R}\pi_{1p}\right]$$

$$+ cn_2\left[\sum_{p=p_R+0.001}^{1}\pi_{0p} + \sum_{p=p_R+0.001}^{1}\pi_{1p}\right] + L_0\sum_{p=p_S+0.001}^{1}\pi_{0p} + L_1\sum_{p=p_S+0.001}^{1}\pi_{1p}.$$

The expected total cost per hour of operation of the Bayesian *np* chart using an adaptive sample size is computed by $E(TCU) = \dfrac{E(CP)}{h}$.

14.5 Numerical Investigation and Comparisons

To find the optimal solutions of the different *np* schemes, minimizing the expected long-run cost per hour, it is necessary to estimate the set $S = \{p_0, \delta, L_0, L_1, \lambda, M, b, c\}$ of the relevant process and cost parameters, where δ represents the size of the shift. For a given value of δ, the corresponding level of p_1 is computed as $p_1 = p_0 + \delta\sqrt{p_0(1 - p_0)}$. In practice, the different elements of the set S are difficult to estimate accurately. A possible solution to this problem involves performing a sensitivity analysis of the input set to detect how parameters' misspecifications

Table 14.4 Level planning
of the input parameters

Model parameter	Level	
	Low	High
$A = p_0$	0.03	0.08
$B = \delta$	0.5	1
$C = L_0$	50	250
$D = L_1$	500	700
$E = \lambda$	0.001	0.008
$F = M$	80	160
$G = b$	0	1
$H = c$	0.1	0.5

affect the optimal results. Table 14.4 provides the high and low levels considered for each component of the set S. These levels were chosen arbitrarily which is a common approach in research dealing with economic design of control charts.

Generally, in the design of experiments, a trial can be repeated or replicated at least twice leading to different observation values for the same combination of factor levels. Thereafter, the estimated mean square of the experimental error serves as a basis for checking the significance of each factor. Since the expected total costs per hour of the different schemes are deterministic and in order to reduce the total number of trials, an unreplicated 2_{IV}^{8-4} fractional factorial design is adopted. In the analysis of variance, three and four factor interactions were pooled to form the error mean (Montgomery 2005). Table 14.5 lists the 16 experimental runs that are used to generate the optimal design parameters for various control charts. The runs are based on the design generators $E = BCD$, $F = ACD$, $G = ABC$ and $H = ABD$.

A direct search procedure was adopted to search for the optimal decision variables $\{n_s^*, UCL_s^*, h_s^*\}$ of the static np control chart minimizing Eq. (14.1). The corresponding minimum expected cost per hour is denoted by $ETCU^*$. The value of the sample size n was selected from the set $\{1, 2, \ldots, 100\}$. For a given value of n, the search for the optimal control limit was conducted over all integer values such that $0 \leq UCL \leq n$. The sampling interval was allowed to vary in 0.01 increments within the range $(0.01, 30)$. The optimal solutions are listed in Table 14.6.

We developed MATLAB programs to search for the optimal design parameters of the Bayesian schemes considered in this paper. Table 14.6 provides the optimal decision variables n^*, h^* and p_B^* minimizing Eq. (14.2) for each of the 16 runs of the sensitivity analysis. The column labeled $\triangle C_{Basic|Fp}$ measures the percentage reduction in cost resulting from the use of a basic Bayesian np chart instead of a static one. The search of the set of the optimal decision variables of the Bayes-n np chart, $\{n_1^*, n_2^*, h^*, p_R^*, p_S^*\}$, is computationally too demanding given that the optimization of the program requires the determination of five design parameters and the size of the transition matrix P is very large. To cope with this inconvenience, we set the values of n_1, n_2, and h equal to the optimal solutions of the VSS np control chart obtained by Kooli and Limam (2011). A search of the optimal limits p_R^* and p_S^* is conducted such that $0.01 \leq p_R \leq 0.2$ and $p_R + 0.01 \leq p_S \leq 0.9$ with a 0.01 step to

Table 14.5 Runs used in the 2_{IV}^{8-4} fractional factorial experiment

Run	p_0	δ	L_0	L_1	λ	M	b	c
1	0.03	0.5	50	500	0.001	80	0	0.1
2	0.08	0.5	50	500	0.001	160	1	0.5
3	0.03	1.0	50	500	0.008	80	1	0.5
4	0.08	1.0	50	500	0.008	160	0	0.1
5	0.03	0.5	250	500	0.008	160	1	0.1
6	0.08	0.5	250	500	0.008	80	0	0.5
7	0.03	1.0	250	500	0.001	160	0	0.5
8	0.08	1.0	250	500	0.001	80	1	0.1
9	0.03	0.5	50	700	0.008	160	0	0.5
10	0.08	0.5	50	700	0.008	80	1	0.1
11	0.03	1.0	50	700	0.001	160	1	0.1
12	0.08	1.0	50	700	0.001	80	0	0.5
13	0.03	0.5	250	700	0.001	80	1	0.5
14	0.08	0.5	250	700	0.001	160	0	0.1
15	0.03	1.0	250	700	0.008	80	0	0.1
16	0.08	1.0	250	700	0.008	160	1	0.5

Table 14.6 Optimal solutions of the static and basic Bayesian *np* control chart for infinite horizon length

Run	Static scheme				Basic Bayesian scheme				
	n_s^*	UCL_s^*	h_s^*	$ETCU^*$	n^*	h^*	p_B^*	$ETCU^*$	$\Delta C_{Basic|Fp}(\%)$
1	41	3	9.04	1.79	36	7.9	0.06	1.77	1.12
2	16	2	10.37	3.50	16	10.3	0.02	3.5	0.00
3	12	1	4.32	8.40	12	4.3	0.11	8.4	0.00
4	17	4	1.48	6.91	16	1.4	0.17	6.88	0.43
5	92	7	3.79	10.28	89	3.7	0.29	10.25	0.29
6	38	6	7.45	11.14	35	6.8	0.3	11.08	0.54
7	17	2	8.94	3.16	25	10.9	0.1	3.1	1.90
8	30	7	9.46	1.45	30	9.5	0.24	1.45	0.00
9	19	1	3.55	14.09	4	0.9	0.09	14.01	0.57
10	47	7	4.22	9.07	46	4.1	0.13	9.06	0.11
11	29	3	6.51	2.06	28	6.4	0.04	2.06	0.00
12	10	2	10.28	2.07	8	8.2	0.06	2.06	0.48
13	42	3	21.59	3.60	40	21	0.11	3.59	0.28
14	74	12	8.95	2.67	64	7.9	0.15	2.65	0.75
15	32	4	2.96	8.19	30	2.7	0.54	8.16	0.37
16	18	4	3.86	12.32	17	3.6	0.28	12.27	0.41

Table 14.7 Optimal solutions of the Bayes-n np control chart for infinite horizon length

Run	n_1	n_2	h	p_R^*	p_S^*	$ETCU^*$	$\Delta C_{Bn|B}(\%)$	$\Delta C_{Bn|Fp}(\%)$	$\Delta C_{Bn|VSS}(\%)$
1	2	24	0.93	0.01	0.21	1.52	14.12	15.08	5.59
2	15	17	10.17	0.01	0.03	3.49	0.29	0.29	0.00
3	10	14	4.07	0.03	0.08	8.36	0.48	0.48	0.00
4	1	6	0.18	0.01	0.48	6.23	9.45	9.84	2.20
5	72	100	3.24	0.02	0.32	10.2	0.49	0.78	0.00
6	1	21	0.70	0.05	0.66	10.28	7.22	7.72	0.48
7	2	14	1.76	0.01	0.35	2.56	17.42	18.99	3.03
8	29	40	9.20	0.01	0.26	1.45	0.00	0.00	0.00
9	5	15	1.09	0.03	0.12	11.83	15.56	16.04	14.89
10	45	57	4.12	0.02	0.13	9.05	0.11	0.22	0.00
11	28	35	6.40	0.01	0.05	2.06	0.00	0.00	0.00
12	1	7	1.68	0.01	0.17	1.83	11.17	11.59	1.61
13	3	45	4.50	0.02	0.31	3.33	7.24	7.50	4.03
14	1	31	0.73	0.01	0.55	2.15	18.87	19.48	4.87
15	1	13	0.17	0.02	0.9	7.31	10.42	10.74	2.40
16	16	24	3.51	0.02	0.31	12.22	0.41	0.81	0.08

reduce the computational time even if the values of p are discretized using a 0.001 step. The optimal values of the threshold limit p_R^* and the action limit p_S^* for each run of the sensitivity analysis are given in Table 14.7. The columns $\Delta C_{Bn|B}$, $\Delta C_{Bn|Fp}$, and $\Delta C_{Bn|VSS}$ represent the percentage cost savings achieved when the dual sample size Bayesian np chart is used instead of the basic Bayesian, static, and VSS np charts, respectively. The economic model of the last scheme was derived by Kooli and Limam (2011).

To study the effects of various components of the input set S on the behavior of the optimal economic solutions of the different sampling schemes, the statistical software Minitab was used to analyze the results of the 2_{IV}^{8-4} experimental design through ANOVA. The results for the optimal static scheme show that the hourly penalty cost of producing in the out-of-control state, M, mainly affects the waiting time between two successive samples and the expected total cost per hour of adopting a static np control chart. Larger values of M imply more frequent sampling during the production cycle leading to an increased cost. It is also observed that the optimal design parameters of the static np control chart are insensitive to the errors in estimating the cost of removing the assignable cause, L_1, and fixed sampling cost, b. However, variation in the sampling cost per item, c, significantly affects all the decision variables of the traditional scheme. Especially, a decrease in the value of c results in an increase in both the optimal sample size and the UCL of the np control chart and leads to monitor the production process more frequently.

Concerning the basic Bayesian sampling procedure, it requires to inspect samples of smaller size more frequently compared to the classical np chart. For processes operating indefinitely, the mean of the cost savings resulting from using the Bayesian scheme instead of the static one is about 0.45% and they rarely exceed 1%. It should be pointed out that better savings are achieved for smaller values of the fixed sampling cost. Regarding the variable sampling cost, it affects significantly all the design parameters of the basic Bayesian *np* chart. It is also observed that when the cost of stopping the process while it is operating properly increases, the optimal basic Bayesian procedure needs more units to be inspected frequently and to plot the updated value of the out-of-control probability using a control chart with a wider control limit.

Results of Table 14.7 show that, compared to classical and basic Bayesian *np* control charts, the cost savings resulting from the adoption of the Bayes-n *np* chart range from low to high with a mean of 7.47 and 7.08%, respectively. Higher cost savings are achieved when the fixed sampling cost, b, is negligible. For the eight runs where the sampling cost contains only a variable component, the average cost savings are equal to 13.69 and 13.02%, respectively. In most of the runs, the optimal threshold limit, p_R, is equal to its minimum allowable value 0.01. In such cases, the large sample size n_2 will be used most of the time during the process monitoring. In the other cases, the threshold limits are still low and they do not exceed the value of 0.05. Generally, the Bayes-n *np* chart exhibits better economic performance than the VSS *np* chart. The highest saving percentage is about 15% and is achieved for the input set of run 9. For the six runs where the two schemes have roughly the same long-run expected cost per hour, certainly the Bayes-n scheme will exhibit more pronounced cost savings if the values of the small and large sample sizes are chosen from larger possible choices at the expense of a high computational time.

14.6 Conclusion

In this paper, we proposed economic models of the basic Bayesian and Bayes-n *np* control charts for long-run processes. A two-state Markov decision process was used to formulate the economic models. The optimal design parameters minimizing the expected total cost per hour of the sampling schemes were searched for the different sets of processes and cost parameters. Results showed that the Bayesian scheme using two sample sizes outperforms significantly the static one and also exhibits better economic performance compared to adaptive *np* charts. The economic effectiveness of other dynamic schemes like the Bayes-h and Bayes-n-h *np* charts is a topic for future research. The proposed economic models were developed for independently distributed measurements. If this assumption is not established, it will be interesting to extend the different models for correlated data. Another interesting issue is to study the economic design of dynamic control charts for processes subject to occurrence of multiple assignable causes.

References

Aly, A. A., Saleh, N. A., Mahmoud, M. A., & Woodall, W. H. (2015). A reevaluation of the adaptive exponentially weighted moving average control chart when parameters are estimated. *Quality and Reliability Engineering International, 31*, 1611–1622.

Bai, D. S., & Lee, K. T. (1998). An economic design of variable sampling interval $\bar{X}$ control charts. *International Journal of Production Economics, 54,* 57–64.

Calabrese, J. M. (1995). Bayesian process control for attributes. *Management Science, 41,* 637–645.

Costa, A. F. B. (1994). $\bar{X}$ charts with variable sample size. *Journal of Quality Technology, 26,* 155–163.

Epprecht, E. K., Costa, A. F. B., & Mendes, F. C. T. (2003). Adaptive control charts for attributes. *IIE Transactions, 35,* 567–582.

Epprecht, E. K., Simoes, B. F. T., & Mendes, F. C. T. (2010). A variable sampling interval EWMA chart for attributes. *The International Journal of Advanced Manufacturing Technology, 49,* 281–292.

Khaw, K. W., Khoo, M. B. C., Yeong, W. C., & Wu, Z. (2017). Monitoring the coefficient of variation using a variable sample size and sampling interval control chart. *Communications in Statistics-Simulation and Computation, 46,* 5772–5794.

Kooli, I., & Limam, M. (2009). Bayesian np control charts with adaptive sample size for finite production runs. *Quality and Reliability Engineering International, 25,* 439–448.

Kooli, I., & Limam, M. (2011). Economic design of an attribute np control chart using a variable sample size. *Sequential Analysis, 30,* 145–159.

Kooli, I., & Limam, M. (2015). Economic design of attribute np control charts using a variable sampling policy. *Applied Stochastic Models in Business and Industry, 31,* 483–494.

Lee, P. H. (2013). Joint statistical design of $\bar{X}$ and s charts with combined double sampling and variable sampling interval. *European Journal of Operational Research, 225,* 285–297.

Luo, H., & Wu, Z. (2002). Optimal np control charts with variable sample sizes or variable sampling intervals. *Economic Quality Control, 17,* 39–61.

Mahadik, S. B. (2017). A unified approach to adaptive Shewhart control charts. *Communications in Statistics-Theory and Methods, 46,* 10272–10293.

Montgomery, D. C. (2005). *Introduction to statistical quality control.* New York: Wiley.

Nenes, G., & Tagaras, G. (2007). The economically designed two-sided Bayesian $\bar{X}$ control chart. *European Journal of Operational Research, 183,* 263–277.

Nenes, G. (2013). Optimization of fully adaptive Bayesian $\bar{X}$ charts for infinite-horizon processes. *International Journal of Systems Science, 44,* 289–305.

Nenes, G., & Panagiotidou, S. (2013). An adaptive Bayesian scheme for joint monitoring of process mean and variance. *Computers and Operations Research, 40,* 2801–2815.

Park, C., & Reynolds, M. R. (1994). Economic design of a variable sample size $\bar{X}$ chart. *Communications in Statistics-Simulation and Computation, 32,* 467–483.

Park, C., & Reynolds, M. R. (1999). Economic design of a variable sampling rate $\bar{X}$ chart. *Journal of Quality Technology, 31,* 427–443.

Prabhu, S. S., Runger, G. C., & Keats, J. B. (1993). $\bar{X}$ chart with adaptive sample sizes. *The International Journal Of Production Research, 31,* 2895–2909.

Prabhu, S. S., Montgomery, D. C., & Runger, G. C. (1994). A combined adaptive sample size and sampling interval $\bar{X}$ control scheme. *Journal of Quality Technology, 26,* 164–176.

Reynolds, M. R., Amin, R. W., Arnold, J. C., & Nachlas, J. A. (1988). $\bar{X}$ charts with variable sampling interval. *Technometrics, 30,* 181–192.

Tagaras, G. (1998). A survey of recent developments in the design of adaptive control charts. *Journal of Quality Technology, 30,* 212–231.

Tagaras, G., & Nikolaidis, Y. (2002). Comparing the effectiveness of various Bayesian $\bar{X}$ control charts. *Operations Research, 50,* 878–888.

Wang, R. F., Fu, X., Yuan, J. C., & Dong, Z. Y. (2018). Economic design of variable-parameter $\bar{X}$ Shewhart control chart used to monitor continuous production. *Quality Technology and Quantitative Management, 15,* 106–124.

Chapter 15
Statistical Analysis of the Lifetime of Diamond-Impregnated Tools for Core Drilling of Concrete

Nadja Malevich, Christine H. Müller, Michael Kansteiner, Dirk Biermann, Manuel Ferreira and Wolfgang Tillmann

Abstract The lifetime of diamond-impregnated tools for core drilling of concrete is studied via the lifetimes of the single diamonds on the tool. Thereby, the number of visible and active diamonds on the tool surface is determined by microscopical inspections of the tool at given points in time. This leads to interval-censored lifetime data if only the diamonds visible at the beginning are considered. If also the lifetimes of diamonds appearing during the drilling process are included, then the lifetimes are doubly interval-censored. We use a well-known maximum likelihood method to analyze the interval-censored data and derive a new extension of it for the analysis of the doubly interval-censored data. The methods are applied to three series of experiments which differ in the size of the diamonds and the type of concrete. It turns out that the lifetimes of small diamonds used for drilling into conventional concrete are much shorter than the lifetimes when using large diamonds or high-strength concrete.

Nadja Malevich and Christine H. Müller—Responsible for statistical methods; Michael Kansteiner, Dirk Biermann, Manuel Ferreira and Wolfgang Tillmann—Responsible for experimental setup.

N. Malevich (✉) · C. H. Müller
Department of Statistics, TU University Dortmund, Dortmund, Germany
e-mail: nadja.malevich@tu-dortmund.de

C. H. Müller
e-mail: cmueller@statistik.tu-dortmund.de

M. Kansteiner · D. Biermann · M. Ferreira · W. Tillmann
Institute of Machining Technology, TU University Dortmund, Dortmund, Germany
e-mail: kansteiner@isf.de

D. Biermann
e-mail: biermann@isf.de

M. Ferreira
e-mail: manuel.ferreira@udo.edu

W. Tillmann
e-mail: wolfgang.tillmann@udo.edu

© Springer Nature Switzerland AG 2019
N. Bauer et al. (eds.), *Applications in Statistical Computing*,
Studies in Classification, Data Analysis, and Knowledge Organization,
https://doi.org/10.1007/978-3-030-25147-5_15

15.1   Introduction

Diamond-impregnated tools for concrete drilling are so-called "self-sharpening tools." This means that these tools need to wear down within the application so that at any time of the process new sharp diamonds are exposed at the tool surface. Several authors already analyzed the wear behavior of diamond-impregnated tools, e.g., Liao and Luo (1992), Di Ilio and Togna (2003), Yu and Xu (2003), Hu et al. (2006), and the main wear mechanisms seem to be identified. Nevertheless, the development of new diamond-impregnated tools requires extensive time- and cost-consuming testing. Hence, scientific work has to be conducted to increase the process knowledge and enable the development of wear models that reduce testing and improve tool performance. Statistical approaches for the analysis of the wear behavior of diamond-impregnated tools were used in Konstanty and Tyrala (2013), Özçelik (2003) and Ersoy et al. (2005). But these articles focused on tools for sawing applications of rock. Only a few authors are dealing with the diamond core drilling process (Carpinteri et al. 2005; Kansteiner et al. 2017).

In particular, the wear of the single diamonds in core drilling tools is not studied well up to now although the wear of the tool depends heavily on it. Having an estimate for the lifetime of the single diamonds, we can estimate, for example, the time until a certain number of diamonds are broken out. This, in turn, gives us insight into the total lifetime of the drilling tool. Hence, this paper deals especially with the statistical analysis of the lifetime of these single diamonds, where the lifetime of a diamond is understood as the time until the diamond is completely broken out.

One of the challenges of the statistical analysis is that the lifetime of the diamonds cannot be observed directly. The drilling process must be interrupted to check how many diamonds on the tool are broken out. This can be done only at predetermined inspection times. Hence, the lifetimes are given as so-called interval-censored data where only intervals are known in which the exact lifetimes are falling.

The statistical analysis of interval-censored data, also called grouped data, is quite an old research area, see, for example, the book of Kulldorff (1961). Nevertheless, it is of high actual interest which is shown by the recent books of Bogaerts et al. (2018) and Sun (2006) and many publications such as Gao et al. (2018), Ismail (2015), Tsai and Lin (2010), Wang et al. (2018), Wu and Huang (2010). See also Malevich and Müller (2019) for determining optimal inspection times for interval-censored data.

Another challenge is that the drilling tool consists of a metal matrix in which the diamonds are embedded. Not all diamonds embedded in this matrix are visible and active in the beginning. Several of them appear only during the drilling process. Hence, the starting point of the lifetimes of these diamonds is not the beginning of the drilling process. Because of the predetermined inspection times, the exact starting points of these lifetimes are not known. They are also interval-censored. Hence, both start and end of the lifetimes are interval-censored so that those data are called doubly interval-censored.

There are not many statistical methods for doubly interval-censored data. Most of them concern nonparametric methods for discrete lifetime distributions with finite support or complicated semiparametric methods for proportional and additive hazard models for the case of covariates, see Bogaerts et al. (2018) and Sun (2006). Since we are mainly interested in the prediction of lifetimes of diamonds and diamond-impregnated tools, we present a rather simple parametric approach for the exponential distribution by extending the maximum likelihood estimator of Kulldorff (1961) and Saleh (1966) for interval-censored failure data to doubly interval-censoring and confidence set estimation.

The description of the experimental setup is given in Sect. 15.2. The statistical methods are presented in Sect. 15.3. Here, we provide the methods for interval-censored lifetime data as well as for doubly interval-censored lifetime data. Section 15.4 contains the results for the experiments. A discussion of these results is given in Sect. 15.5.

15.2   Experimental Setup

15.2.1   Diamond-Impregnated Tools

Diamond-impregnated metal matrix composites are mainly used as cutting and grinding tools for high abrasive mineral materials in the natural stone and construction industry. Due to the high hardness of granite, basalt and high-performance concrete, high demands on the efficiency of the cutting tools exist. To guarantee a high cutting performance, monocrystalline synthetic diamonds with an average grain size of 250–500 μm are used. These abrasive components are embedded into a metal matrix, which basically consisted of pure cobalt, cobalt/copper, or cobalt/bronze until the 2000s. Due to the estimated carcinogenic properties and the rising world market price, in particular, most diamond tool manufacturers tend to substitute cobalt with iron, copper, or bronze. In terms of important materials characteristics like hardness and wear resistance, the newly developed cobalt-reduced and cobalt-free alternative binder systems reached comparable properties as the well-known cobalt-based matrix compositions. Additionally, higher amount of copper or bronze enables the possibility to adjust the hardness of the metal matrix to the requirements of the machined mineral subsoil. Depending on the hardness and abrasiveness of the present mineral materials (concrete, basalt), the metal matrix composition has to be adapted to guarantee a suitable grain protrusion to avoid the negative influence of rounding effects on the diamond grains.

15.2.2   Production of Diamond-Impregnated Tools

The small diamond grinding segments, which are attached to drill bits, saw blades, or wire saws by brazing or laser welding, are fabricated in a powder metallurgical process route. The first step is the premixing of the chosen metal powder components with 5–10 vol-% synthetic monocrystalline diamond grains. Subsequently, the homogenized powder blend is cold pressed and the obtained green bodies are sintered to achieve an almost pore-free structure with an optimal final strength. The sintering procedure is conducted pressureless in a vacuum or inert gas furnace or in a hot-pressing facility which heats and mechanically presses the green bodies simultaneously in a graphite mold. In contrast to vacuum sintering, the hot-pressing process leads to lower porosity with a significant reduction of the sintering time. Besides these conventional sintering methods, the newly developed field-assisted sintering technology (FAST) (Guillon et al. 2014) is similar to the hot-pressing but offers the possibility to increase the heating rate up to max. 1000 K/min. This is realized by a direct current up to 60 kA and a pulse time of a few milliseconds, which leads to small plasma arcs between neighboring powder particles resulting in a very fast partial heating of the powder green body.

15.2.3 The Experiments

The experiments were conducted on a 3-axis machine center, and a special tool holder for a single segment with a diameter of $d = 100$ mm was used, see Fig. 15.1a. The workpieces made of two different types of concrete were clamped on a force dynamometer for force measurements.

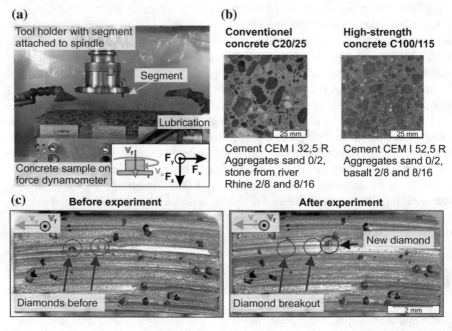

Fig. 15.1 a Experimental setup for single segment tests, **b** Concrete workpieces, **c** Microscopic pictures of diamond-impregnated segments with diamond breakouts (red circles) and a newborn diamond (black circle)

Concrete is made of aggregates such as stones, cement, and water. Hence, concrete has an inhomogeneous material structure. The classification of concrete is based on its compressive strength. Within the tests, two types of concrete were used: a high-strength concrete with a compressive strength of C100/115 and a conventional concrete with compressive strength of C20/25. C100/115 is used, for example, as material for foundations of skyscrapers, whereas C20/25 is used for buildings without special requirements. The main aggregate material of concrete C100/115 is basalt. In C20/25, the main aggregates are stones from the Rhine river, see Fig. 15.1b. Regarding the properties of the two types of concrete, the different material phases must be considered. For simplification, it is possible to differentiate between the aggregate phase and the cement phase. In C20/25, the aggregate phase includes harder and more brittle materials than the aggregate phase of C100/115. But cement phase in C20/25 is of a smaller hardness due to a higher amount of open pores than the cement phase of C100/115. In total, this causes a smaller compressive strength for C20/25.

The concrete workpieces were of a cubic shape, and the length of each side was $b = l = h = 150$ mm. These cubes were mounted on top of the force dynamometer, see Fig. 15.1a. In order to focus on the wear behavior of the segments, the tested parameters for the segments were kept constant. The drilling depth was $s = 4$ mm, the infeed velocity was $v_f = 4$ mm/min, and the circumferential speed was $v_u = 3,225$ m/s. Overall, each segment achieved a cumulated drilling depth of $s_{tot} = 200$ mm. The infeed velocity and the circumferential speed were deduced from force-controlled core drilling tests with eight segments and the same segment specifications. In order to remove slurry from the process, water with an additive of Bechem Avantin 361 (concentration $p = 7\%$) was used.

The tested segments were of laboratory dimensions (height $h_{seg} = 10$ mm, width $w_{seg} = 5$ mm and length $l_{seg} = 10$ mm). The metal matrix was Diabase V21 whose main components are iron, copper, cobalt, and tin. The diamonds were of quality Syngrit SDB 1055 by element six. The grit size of the diamonds was $d_k = 40/50$ mesh and $d_k = 20/30$ mesh. The diamond concentration $c = C20$ was used.

Throughout the article, the segments with the diamond grit size $d_k = 20/30$ mesh are named B18 and B19, and the segments with the diamond grit size $d_k = 40/50$ mesh are referred to as B28 and B29. The assignment of the segments to the two types of concrete is given in Table 15.1.

This provided four sequences of experiments where the experiments of a sequence were done under the same conditions. Each experiment of these sequences is given by the drilling of depth $s = 4$ mm which means a drilling time of 1 min. Since the cumulated drilling depth of each sequence is $s_{tot} = 200$ mm, each sequence consists of 50 experiments and a cumulated drilling time of 50 min.

Table 15.1 Assignment of segment: grit size and type of concrete

Type of concrete	Grit size d_k	
	Large, 20/30 mesh	Small, 40/50 mesh
High strength, C100/115	B18	B29
Conventional, C20/25	B19	B28

After each experiment, the tested segment was microscopically inspected using a digital microscope (DigiMicro Profi by DNT). Hence, the inspection times are given by the drilling depth of $s = 4$ mm and correspond to one minute of drilling. The microscopical analysis comprised the counting of the number of exposed diamonds on the segment surface. Thereby, each exposed diamond got a label so that the visibility of the single diamonds could be followed over the experiments. For simplification, it was only distinguished between visible "1" and non-visible diamonds "0". There was no differentiation between flat/worn diamonds or diamonds with partial outbreak, respectively. Only when a complete diamond breakout occurred, this diamond was classified as "0". Diamonds which are not visible at the beginning are also classified as "0" until they become visible. Figure 15.1c shows the surface of a segment before and after the experiment where two diamonds broke out (marked by red circles) and one new diamond occurred (marked by a black circle).

15.3   Statistical Methods for Interval-Censored Lifetime Data

15.3.1   Analysis of Lifetimes with the Same Starting Times

Let $T_1, \ldots, T_N$ be independent and identically distributed nonnegative random variables (lifetime variables of N objects) with the cumulative distribution function F_θ, where $\theta \in \mathbb{R}^d$ is an unknown parameter. However, the realizations $t_1, \ldots, t_N$ of $T_1, \ldots, T_N$ are not observed. The objects are only observed at given fixed times $0 = \tau_0 < \tau_1 < \ldots < \tau_I < \tau_{I+1} = \infty$, which are called inspection times. This means that only realizations z_n of Z_n with

$$Z_n = i, \quad \text{if} \quad T_n \in (\tau_{i-1}, \tau_i], \quad i = 1, \ldots, I + 1,$$

are observed for $n = 1, \ldots, N$. Such data are called **interval-censored data**. The goal is to estimate the unknown parameter θ and to construct corresponding confidence intervals.

The independence of $T_1, \ldots, T_N$ and the definition of $Z_1, \ldots, Z_N$ yield the following likelihood function for interval-censored lifetimes $z_1, \ldots, z_N$ (see, e.g., Bogaerts et al. 2018, p. 15 ff., Sun 2006, p. 28):

$$l(\theta) := l(\theta; z_1, \ldots, z_N) = \prod_{n=1}^{N} P_\theta(Z_n = z_n) = \prod_{n=1}^{N} \prod_{i=1}^{I+1} P_\theta(Z_n = i)^{\mathbb{1}_{\{z_n = i\}}}$$

$$= \prod_{n=1}^{N} \prod_{i=1}^{I+1} P_\theta(T_n \in (\tau_{i-1}, \tau_i])^{\mathbb{1}_{\{z_n = i\}}} = \prod_{n=1}^{N} \prod_{i=1}^{I+1} \left(F_\theta(\tau_i) - F_\theta(\tau_{i-1}) \right)^{\mathbb{1}_{\{z_n = i\}}}.$$

Maximizing the likelihood function $l(\theta)$, we obtain the maximum likelihood estimate $\widehat{\theta}$ of θ:

$$\widehat{\theta} = \mathrm{argmax}_{\theta \in \mathbb{R}^d}\, l(\theta).$$

An asymptotic $(1 - \alpha)$-confidence interval for θ based on the likelihood-ratio test is given by (see, e.g., Bogaerts et al. 2018, p. 17 ff., or Schervish 1995, p. 315 ff., p. 458 ff.)

$$\mathscr{C}(z_1, \ldots, z_N) = \left\{ \theta \in \mathbb{R}^d;\ -2 \ln \left(\frac{l(\theta; z_1, \ldots, z_N)}{l(\widehat{\theta}; z_1, \ldots, z_N)} \right) \leq \chi^2_{d;1-\alpha} \right\},$$

where l is the likelihood function, $\widehat{\theta}$ is the maximum likelihood estimate for θ, and $\chi^2_{d;1-\alpha}$ is the $(1 - \alpha)$-quantile of the chi-square distribution with d degrees of freedom.

As explained in the introduction, for our specific situation with the lifetimes of diamonds, we consider an exponential model, i.e., $T_1, \ldots, T_N \sim \mathrm{Exp}(\lambda)$, where $\lambda > 0$ is unknown. The likelihood function in this case is

$$l(\lambda) = \prod_{n=1}^{N} \prod_{i=1}^{I} \left(e^{-\lambda \tau_{i-1}} - e^{-\lambda \tau_i} \right)^{\mathbb{1}_{\{z_n = i\}}} \left(e^{-\lambda \tau_I} \right)^{\mathbb{1}_{\{z_n = I+1\}}}$$

and the asymptotic $(1 - \alpha)$-confidence interval for λ is

$$\mathscr{C}(z_1, \ldots, z_N) = \left\{ \lambda \in \mathbb{R};\ -2 \ln \left(\frac{l(\lambda; z_1, \ldots, z_N)}{l(\widehat{\lambda}; z_1, \ldots, z_N)} \right) \leq \chi^2_{1;1-\alpha} \right\}. \tag{15.1}$$

15.3.2 Analysis of Lifetimes with Different Interval-Censored Starting Times

Let $D_1, \ldots, D_N$ be independent and identically distributed nonnegative random variables (death times of N objects), but here with different starting times (times of birth) $B_1, \ldots, B_N$ so that lifetime variables $T_1 := D_1 - B_1, \ldots, T_N := D_N - B_N$ are independent identically distributed nonnegative random variables with the cumulative distribution function F_θ, where $\theta \in \mathbb{R}^p$ is an unknown parameter. We assume also that $B_1, \ldots, B_N, T_1, \ldots, T_N$ are independent. However, neither the realizations $d_1, \ldots, d_N$ of $D_1, \ldots, D_N$ nor the realizations $b_1, \ldots, b_N$ of $B_1, \ldots, B_N$ are observed. With given inspection times $0 = \tau_0 < \tau_1 < \ldots < \tau_I < \tau_{I+1} = \infty$, we observe only whether d_n and b_n are lying in $(\tau_{i-1}, \tau_i]$, $i = 1, \ldots, I + 1$, or not. This means that only realizations z_n of $\mathbf{Z}_n$ with

$$\mathbf{Z}_n = (h, i), \quad \text{if} \quad B_n \in (\tau_{h-1}, \tau_h], \quad D_n \in (\tau_{i-1}, \tau_i], \quad h < i, \quad h, i = 1, \ldots, I + 1,$$

are observed for $n = 1, \ldots, N$. Such data are called **doubly interval-censored data**. The goal is, as before, to estimate the unknown parameter θ and to construct corresponding confidence intervals.

The likelihood function for doubly interval-censored data $z_1, \ldots, z_N$ is given by (see Bogaerts et al. 2018, p. 254 ff., Sun 2006, p. 177 ff.)

$$l(\theta) := l(\theta; z_1, \ldots, z_N) = \prod_{n=1}^{N} P_\theta(Z_n = z_n) = \prod_{n=1}^{N} \prod_{\substack{h,i=1 \\ h<i}}^{I+1} P_\theta\big(Z_n = (h, i)\big)^{\mathbb{1}_{\{z_n=(h,i)\}}}$$

$$= \prod_{n=1}^{N} \prod_{i=2}^{I+1} \prod_{h=1}^{i-1} P_\theta\big(B_n \in (\tau_{h-1}, \tau_h], \ D_n \in (\tau_{i-1}, \tau_i]\big)^{\mathbb{1}_{\{z_n=(h,i)\}}}.$$

Since $T_n := D_n - B_n, n = 1, \ldots, N$, we can rewrite

$$P_\theta\big(B_n \in (\tau_{h-1}, \tau_h], \ D_n \in (\tau_{i-1}, \tau_i]\big) = P_\theta\big(B_n \in (\tau_{h-1}, \tau_h], \ B_n + T_n \in (\tau_{i-1}, \tau_i]\big)$$

$$= \iint_{\mathbb{R}^2} \mathbb{1}_{(\tau_{h-1}, \tau_h]}(y_1) \ \mathbb{1}_{(\tau_{i-1}, \tau_i]}(y_2) \ dP_\theta^{(B_n, B_n+T_n)}(y_1, y_2).$$

Using the elementary transformation theorem from the measure theory and then the independence of B_n and T_n, we obtain

$$\iint_{\mathbb{R}^2} \mathbb{1}_{(\tau_{h-1}, \tau_h]}(y_1) \ \mathbb{1}_{(\tau_{i-1}, \tau_i]}(y_2) \ dP_\theta^{(B_n, B_n+T_n)}(y_1, y_2)$$

$$= \iint_{\mathbb{R}^2} \mathbb{1}_{(\tau_{h-1}, \tau_h]}(u) \ \mathbb{1}_{(\tau_{i-1}, \tau_i]}(v + u) \ dP_\theta^{(B_n, T_n)}(u, v)$$

$$= \iint_{\mathbb{R}^2} \mathbb{1}_{(\tau_{h-1}, \tau_h]}(u) \ \mathbb{1}_{(\tau_{i-1}, \tau_i]}(v + u) \ dP_\theta^{T_n}(v) \ dP_\theta^{B_n}(u)$$

$$= \int_{\tau_{h-1}}^{\tau_h} \int_{\tau_{i-1}-u}^{\tau_i-u} dF_\theta(v) \ dG_\theta(u) = \int_{\tau_{h-1}}^{\tau_h} (F_\theta(\tau_i - u) - F_\theta(\tau_{i-1} - u)) \ dG_\theta(u),$$

where F_θ is the distribution function of $T_1, \ldots, T_N$, and G_θ is the distribution function of $B_1, \ldots, B_N$. This implies

$$l(\theta) = \prod_{n=1}^{N} \prod_{i=2}^{I+1} \prod_{h=1}^{i-1} \left(\int_{\tau_{h-1}}^{\tau_h} (F_\theta(\tau_i - u) - F_\theta(\tau_{i-1} - u)) \ dG_\theta(u) \right)^{\mathbb{1}_{\{z_n=(h,i)\}}}. \quad (15.2)$$

Without any further assumptions on the distribution function G_θ, the likelihood function cannot be found explicitly.

For our specific situation with newborn diamonds, we consider the following model: $B_1, \ldots, B_N \sim \text{Exp}(\lambda_0)$ with unknown $\lambda_0 > 0$ and $T_1, \ldots, T_N \sim \text{Exp}(\lambda)$ with unknown $\lambda > 0$, so that $\theta = (\lambda_0, \lambda)$. Using the method described in Sect. 15.3.1—now for interval-censored birth times—we obtain the estimate $\widehat{\lambda}_0$ for λ_0 and assume that $B_1, \ldots, B_N \sim \mathrm{Exp}(\widehat{\lambda}_0)$. Then, a plug-in likelihood function of (15.2) can be calculated as follows:

$$
l(\lambda) = \prod_{n=1}^{N} \prod_{i=2}^{I+1} \prod_{h=1}^{i-1} \left(\frac{\widehat{\lambda}_0}{\lambda - \widehat{\lambda}_0} \left(e^{-\lambda \tau_{i-1}} - e^{-\lambda \tau_i} \right) \left(e^{(\lambda - \widehat{\lambda}_0)\tau_h} - e^{(\lambda - \widehat{\lambda}_0)\tau_{h-1}} \right) \right)^{\mathbb{1}_{\{z_n = (h,i)\}}}.
$$

Using the plug-in likelihood function $l(\lambda)$, we obtain the plug-in maximum likelihood estimate $\widehat{\lambda}$ and a plug-in confidence interval for λ using (15.1). This plug-in confidence interval is an asymptotic confidence interval since $\widehat{\lambda}_0$ is a consistent estimator for λ_0.

15.3.3   Analysis of Lifetimes with Mixed Starting Times

In some situations, a combination of the models presented in Sects. 15.3.1 and 15.3.2 is employed. For example, in drilling experiments, we had some diamonds visible at the beginning and some diamonds, which appeared later during the drilling. For this case, we use the following model.

Let the birth times of the diamonds visible at the beginning be modeled by $B_1 = \cdots = B_{N_0} = 0$ and the birth times of the new diamonds be modeled by the independent and identically distributed random variables $B_{N_0+1}, \ldots, B_N$, $N_0 < N$, with the common distribution function G_θ, where $\theta \in \mathbb{R}^p$ is unknown. Let $T_1, \ldots, T_N$ be the lifetime random variables (also independent and identically distributed) with the common distribution function F_θ. We assume that $B_1, \ldots, B_N, T_1, \ldots, T_N$ are independent. We do not observe the realizations $B_{N_0+1}, \ldots, B_N$ and the realizations of $T_1, \ldots, T_N$. Only realizations z_n of $\mathbf{Z}_n$ with

$$
\mathbf{Z}_n = \begin{cases} (0, i), & \text{if } B_n = 0, \ T_n \in (\tau_{i-1}, \tau_i], \ i = 1, \ldots, I+1, \\ (h, i), & \text{if } B_n \in (\tau_{h-1}, \tau_h], \ B_n + T_n \in (\tau_{i-1}, \tau_i], \ h < i, \ h, i = 1, \ldots, I+1, \end{cases}
$$

are observed for $n = 1, \ldots, N$.

Similar as in Sects. 15.3.1 and 15.3.2, the likelihood function for the data $z_1, \ldots, z_N$ has the following form:

$$
l(\theta) = \prod_{n=1}^{N} \prod_{i=1}^{I+1} P_\theta \big(T_n \in (\tau_{i-1}, \tau_i] \big)^{\mathbb{1}_{\{z_n = (0,i)\}}}.
$$

$$
\cdot \prod_{\substack{h=1 \\ h<i}}^{I+1} P_\theta \big(B_n \in (\tau_{h-1}, \tau_h], \ B_n + T_n \in (\tau_{i-1}, \tau_i] \big)^{\mathbb{1}_{\{z_n = (h,i)\}}}
$$

$$
= \prod_{n=1}^{N} \prod_{i=1}^{I+1} \big(F_\theta(\tau_i) - F_\theta(\tau_{i-1}) \big)^{\mathbb{1}_{\{z_n = (0,i)\}}}.
$$

$$\cdot \prod_{\substack{h=1 \\ h<i}}^{I+1} \left(\int_{\tau_{h-1}}^{\tau_h} (F_\theta(\tau_i - u) - F_\theta(\tau_{i-1} - u)) \, dG_\theta(u) \right)^{\mathbb{1}_{\{z_n=(h,i)\}}}.$$

If we assume that $\theta = (\widehat{\lambda}_0, \lambda)$ with $\widehat{\lambda}_0$ defined in Sect. 15.3.2, $G_\theta = \text{Exp}(\widehat{\lambda}_0)$ and $F_\theta = \text{Exp}(\lambda)$, we obtain

$$l(\lambda) = \prod_{n=1}^{N} \prod_{i=1}^{I+1} \left(e^{-\lambda \tau_{i-1}} - e^{-\lambda \tau_i} \right)^{\mathbb{1}_{\{z_n=(0,i)\}}} \cdot$$

$$\cdot \prod_{\substack{h=1 \\ h<i}}^{I} \left(\frac{\widehat{\lambda}_0}{\lambda - \widehat{\lambda}_0} \left(e^{-\lambda \tau_{i-1}} - e^{-\lambda \tau_i} \right) \left(e^{(\lambda - \widehat{\lambda}_0)\tau_h} - e^{(\lambda - \widehat{\lambda}_0)\tau_{h-1}} \right) \right)^{\mathbb{1}_{\{z_n=(h,i)\}}}.$$

Using the likelihood function $l(\lambda)$, we obtain the maximum likelihood estimate $\widehat{\lambda}$ and an asymptotic confidence interval for λ from (15.1).

15.3.4 Confidence Sets for Related Quantities

Usually, the parameter λ of the exponential distribution itself is not of interest, but rather the following quantities.
(a) The **expected lifetime of each diamond** is

$$q_1(\lambda) := E_\lambda(T_n) = \frac{1}{\lambda},$$

since we assume $T_1, \ldots, T_N \sim \text{Exp}(\lambda)$.

(b) **Expected number of breakouts in N diamonds within $[0, T]$.**
    The probability that one diamond breaks out within $[0, T]$ is

$$p_T := P_\lambda(T_n \leq T) = F_\lambda(T) = 1 - e^{-\lambda T}.$$

Then, the number N_b of breakouts in N diamonds within $[0, T]$ follows the binomial distribution $\text{Bin}(N, p_T)$. Therefore, expected number of breakouts in N diamonds within $[0, T]$ is given by

$$q_2(\lambda) := N p_T = N F_\lambda(T) = N \left(1 - e^{-\lambda T} \right).$$

(c) **Time $T_{L,N,p}$ so that the probability of a breakout of at least L of N diamonds equals p.**

We will consider $p = 0.5$ and denote $T_{L,N} := T_{L,N,0.5}$. Since $N_b \sim \mathrm{Bin}(N, p_T)$ (see (b)), the probability that at least L of N, $L \leq N$, diamonds break out within $[0, T]$ is given by

$$P_\lambda(N_b \geq L) = \sum_{l=L}^{N} \binom{N}{l} \left(1 - e^{-\lambda T}\right)^l e^{-\lambda T(N-l)}.$$

Time $T_{L,N}$ so that the probability of a breakout of at least L of N diamonds is 0.5 can be found by solving the following equation:

$$\sum_{l=L}^{N} \binom{N}{l} \left(1 - e^{-\lambda T_{L,N}}\right)^l e^{-\lambda T_{L,N}(N-l)} = 0.5.$$

Note that the function on the left of the equation is monotone increasing with respect to $\lambda T_{L,N}$ for all L, N with $L \leq N$. This means that there is a unique solution of the equation and it has the following form:

$$q_3(\lambda) := T_{L,N}(\lambda) = \frac{f(L, N)}{\lambda},$$

where $f(L, N)$ is a function of L, N and can be found numerically for each L, N.

All quantities $q_i(\lambda), i = 1, 2, 3$, can be estimated by replacing λ with its estimate $\widehat{\lambda}$, i.e., by $q_i(\widehat{\lambda})$. Since $q_1(\lambda)$ and $q_3(\lambda)$ are decreasing functions with respect to λ, the confidence interval for $q_i(\lambda)$, $i = 1, 3$, is given by

$$[q_i(\widehat{\lambda}_u), q_i(\widehat{\lambda}_l)],$$

where $\mathscr{C}(z_1, \ldots, z_N) = [\widehat{\lambda}_l, \widehat{\lambda}_u]$ is the confidence interval for λ. Similarly, since q_2 is increasing in λ, the confidence interval for $q_2(\lambda)$ is $[q_2(\widehat{\lambda}_l), q_2(\widehat{\lambda}_u)]$.

15.4 Results

The statistical analysis was done in R, see R Development Core Team (2017).

Within the conducted experiments, one or more diamond breakouts were observed for segments B28, B29, and B19. In contrast, segment B18 showed no complete diamond breakout. Hence, a lifetime analysis for single diamonds is not possible here and it was decided to neglect this segment for the analysis. Further experiments need to be conducted for this segment to get enough data for lifetime estimation.

Table 15.2 ML-estimates and 98.4%-confidence intervals for λ based on Sect. 15.3.1 using only the interval-censored data

Experiment	ML-estimate of λ	98.4%-confidence interval
B19	0.00489	[0.00083, 0.01519]
B28	0.03310	[0.01734, 0.05638]
B29	0.00448	[0.00152, 0.00988]

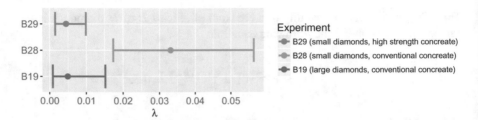

Fig. 15.2 ML-estimates and 98.4%-confidence intervals for λ based on Sect. 15.3.1 using only the interval-censored data

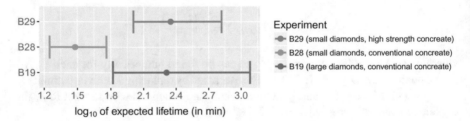

Fig. 15.3 ML-estimates and 98.4%-confidence intervals for expected lifetime $1/\lambda$ based on Sect. 15.3.1 using only the interval-censored data

15.4.1   Using only Diamonds Visible in the Beginning

In this section, we use the statistical method derived in Sects. 15.3.1 and 15.3.4.

The maximum likelihood estimates for λ and the corresponding asymptotic 98.4%-confidence intervals for three experimental setups (B19, B28, B29) are given in Table 15.2 and Fig. 15.2. Note that the level α is chosen so that $(1 - \alpha)^3 \geq 0.95$. This is an adjustment for testing (with level 95%) the hypothesis that the lifetime parameter λ is the same for B19, B28, and B29. Figure 15.2 shows that the confidence intervals have an empty intersection. This allows us to reject the null hypothesis and conclude that there is significant difference between all three experimental setups. At the same time, the chosen level 98.4% for confidence intervals allows to compare the experiments in pairs (with level al least 95%). The null hypotheses about the pairwise equality of the lifetime parameters are rejected when the corresponding confidence intervals are disjoint. So, we see that there is no significant difference between B19 and B29, since the intersection is not empty. However, B28 differs significantly from B19 and B29.

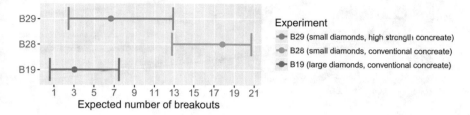

Fig. 15.4 ML-estimates and 98.4%-confidence intervals for the expected number of breakouts of initially visible diamonds within the time interval [0, 50] (in min) based on Sect. 15.3.1 using only the interval-censored data

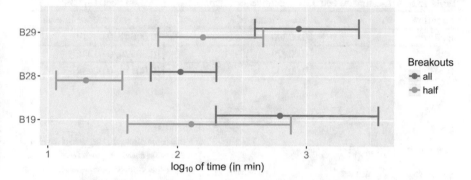

Fig. 15.5 ML-estimates and 98.4%-confidence intervals for the times $T_{L,N}$ such that half ($L = N/2$) of the diamonds and all ($L = N$) diamonds, respectively, are broken out, based on Sect. 15.3.1 using only the interval-censored data

The estimates and the confidence intervals for the quantities represented in Sect. 15.3.4 are given above. Figure 15.3 provides the estimates for the expected lifetime $1/\lambda$ of each diamond. Figure 15.4 gives the estimated expected number of breakouts of initially visible diamonds within the time interval [0, 50] (in min). Figure 15.5 represents $\widehat{T}_{L,N}$ for $L = N/2$ and $L = N$, i.e., the estimated time at which with probability 0.5 at least half (all, respectively) of the initially visible diamonds are broken out. Note that the number N of diamonds visible at the beginning are different for B19, B28, and B29 (14, 22, and 33, respectively) so that the confidence intervals for B28 and B29 in Fig. 15.4 and for B19 and B28 in Fig. 15.5 are not disjunct.

15.4.2 Using All Active Diamonds

In this section, we use the statistical method derived in Sects. 15.3.3 and 15.3.4.

The maximum likelihood estimation for the birth time parameter λ_0 from the model represented in Sect. 15.3.2 yields the following values:

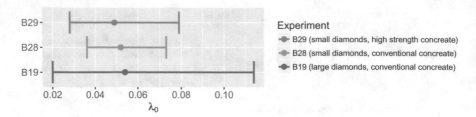

Fig. 15.6 ML-estimates and 98.4%-confidence intervals for λ_0 based on Sect. 15.3.2 using the doubly interval-censored data

Table 15.3 ML-estimates and 98.4%-confidence intervals for λ based on Sect. 15.3.3

Experiment	ML-estimate of λ	98.4%-confidence interval
B19	0.00300	[0.00047, 0.00919]
B28	0.00879	[0.00526, 0.01346]
B29	0.00302	[0.00114, 0.00634]

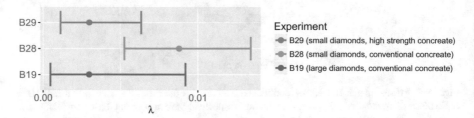

Fig. 15.7 ML-estimates and 98.4%-confidence intervals for λ based on Sect. 15.3.3 using the interval-censored and the doubly interval-censored data

$$\text{B19: } \widehat{\lambda}_0 = 0.054, \quad \text{B28: } \widehat{\lambda}_0 = 0.052, \quad \text{B29: } \widehat{\lambda}_0 = 0.049.$$

These estimates and the corresponding asymptotic 98.4%-confidence intervals for λ_0 are given in Fig. 15.6. Note that the lengths of the intervals are different because of the different number of new diamonds that appeared during the drilling in B19, B28, and B29 (9, 46, and 22, respectively).

The maximum likelihood estimates for the lifetime parameter λ and the corresponding asymptotic 98.4%-confidence intervals for B19, B28, and B29 are given in Table 15.3 and Fig. 15.7. Note again that the level α is chosen so that $(1 - \alpha)^3 \geq 0.95$ (for explanation see Sect. 15.4.1).

Figure 15.7 shows that the confidence intervals are not disjunct anymore in comparison with the confidence intervals from Fig. 15.2. This means that there is no significant difference between the lifetime parameter λ in all three experimental setups. Figure 15.8 provides the estimates for the expected lifetime $1/\lambda$ of each diamond.

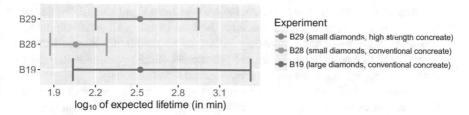

Fig. 15.8 ML-estimates and 98.4%-confidence intervals for expected lifetime $1/\lambda$ based on Sect. 15.3.3 using the interval-censored and the doubly interval-censored data

15.5 Discussion

The wear of diamond-impregnated drilling tools was studied via the lifetime of the single active diamond in three different setups. The three setups concerned small diamonds (grit size 40/50 mesh) applied to conventional concrete (C20/25), small diamonds applied to high-strength concrete (C100/115), and large diamonds (grit size 20/30 mesh) applied to conventional concrete. In order to observe the lifetime of the single diamonds, the drilling process was stopped every minute and the active diamonds were marked via inspection of the surface of the drilling segment. This led to interval-censored lifetimes for the diamonds visible in the beginning and to doubly interval-censored lifetimes for diamonds appearing during the drilling process.

To analyze these interval-censored data, exponential distributions for the lifetimes and the birth times of diamonds appearing during the drilling process were assumed. Moreover, a simple plug-in method for confidence intervals for the lifetimes in the three experimental setups was used. As expected, it turned out that the lifetimes of small diamonds are shorter than those of large diamonds in the tools applied to the conventional concrete. Moreover, the lifetimes of small diamonds in the tools applied to conventional concrete were shorter than those in the tools applied to high-strength concrete. These differences were significant if only the diamonds visible in the beginning were used. However, the significance disappeared when using all active diamonds. A reason might be that, in all setups, the estimated expected lifetimes of all active diamonds were longer than the estimated expected lifetimes of diamonds visible in the beginning so that consequently the variance in a model with exponential distributions should be larger. It is not surprising that the estimated expected lifetimes of the initially visible diamonds were shorter since these diamonds were observed over longer period of time and had more chances to break out within this period.

The significance also disappeared when some related quantities are considered. This holds for the expected number of breakouts of the initially visible diamonds within the time interval up to 50 min and for the time so that the probability of the breakout of all initially visible diamonds is 0.5. However, this is due to the different numbers of diamonds visible in the beginning which were 22 and 33 for the small diamonds and 14 for the large diamonds.

More experiments should be conducted to see whether a significant difference can be observed also for these quantities and for the case of using all active diamonds. Moreover, in future work, the simple plug-in method for confidence intervals should be compared with a method where the two parameters of the two exponential distributions (for lifetimes and birth times) are estimated simultaneously.

The exponential distribution is the simplest failure distribution and therefore widely used. However, a question is whether the exponential distribution is really appropriate and whether other distributions and nonparametric methods will provide similar results. In particular, the study of extensions of the exponential distribution like the Weibull or the gamma distribution is of interest since then a test on the exponential distribution can be done by calculating confidence sets for the parameters as proposed here.

Acknowledgements The authors gratefully acknowledge support from the Collaborative Research Center "Statistical Modelling of Nonlinear Dynamic Processes" (SFB 823, B4) of the German Research Foundation (DFG).

References

Bogaerts, K., Komarek, A., & Lesaffre, E. (2018). *Survival analysis with interval-censored data: A practical approach with examples in R, SAS, and BUGS*. Interdisciplinary Statistics Series Boca Raton: Chapman & Hall/CRC.

Carpinteri, A., Dimastrogiovanni, L., & Pugno, N. (2005). Fractal coupled theory of drilling and wear. *International Journal of Fracture, 131*, 131–142.

Di Ilio, A., & Togna, A. (2003). A theoretical wear model for diamond tools in stone cutting. *International Journal of Machine Tools and Manufacture, 43*, 1171–1177.

Ersoy, A., Buyuksagic, S., & Atici, U. (2005). Wear characteristics of circular diamond saws in the cutting of different hard abrasive rocks. *Wear, 258*, 1422–1436.

Gao, F., Zeng, D., Couper, D., & Lin, D. Y. (2018). Semiparametric regression analysis of multiple right- and interval-censored events. *Journal of the American Statistical Association*. https://doi.org/10.1080/01621459.2018.1482756.

Guillon, O., Gonzalez-Julian, J., Dargatz, B., Kessel, T., Schierning, G., Räthel, J., et al. (2014). Field-assisted sintering technology/spark plasma sintering: mechanisms, materials, and technology developments. *Advanced Engineering Materials, 16*(7), 830–849.

Hu, Y. N., Wang, C. Y., Ding, H. N., & Wang, Z. W. (2006). Wear mechanism of diamond saw blades for dry cutting concrete. *Key Engineering Materials, 304–305*, 315–319.

Ismail, A. A. (2015). Optimum partially accelerated life test plans with progressively Type I interval-censored data. *Sequential Analysis, 34*, 135–147.

Kansteiner, M., Biermann, D., Dagge, M., Müller, C., Ferreira, M., & Tillmann, W. (2017). Statistical evaluation of the wear behaviour of diamond impregnated tools used for the core drilling of concrete. In *Proceedings of the EURO PM 2017*. Milan: EPMA.

Konstanty, J. S., & Tyrala, D. (2013). Wear mechanism of iron-base diamond-impregnated tool composites. *Wear, 303*, 533–540.

Kulldorff, G. (1961). *Contributions to the theory of estimation from grouped and partially grouped samples*. New York: Wiley.

Liao, Y. S., & Luo, S. Y. (1992). Wear characteristics of sintered diamond composite during circular sawing. *Wear, 157*, 325–337.

Malevich, N., & Müller, C. H. (2019). Optimal design of inspection times for interval censoring. *Statistical Papers, 60*(2), 99–114.

Özçelik, Y. (2003). Multivariate statistical analysis of the wear on diamond beads in the cutting of andesitic rocks. *Key Engineering Materials, 250*, 118–130.

R Development Core Team. (2017). R: A language and environment for statistical computing. *R Foundation for Statistical Computing*. Vienna, Austria. http://cran.r-project.org/.

Saleh, A. K. M. E. (1966). Estimation of the parameters of the exponential distribution based on optimum order statistics in censored samples. *Annals of Mathematical Statistics, 37*, 1717–1735.

Schervish, M. J. (1995). *Theory of Statistics*. New York: Springer.

Sun, J. (2006). *The statistical analysis of interval-censored failure time data*. Statistics for biology and health New York: Springer.

Tsai, T.-R., & Lin, C.-W. (2010). Acceptance sampling plans under progressive interval censoring with likelihood ratio. *Statistical Papers, 51*, 259–271.

Wang, S., Wang. C., Wang. P., & Sun, J. (2018). Semiparametric analysis of the additive hazards model with informatively interval-censored failure time data. *Computational Statistics and Data Analysis, 125*, 1–9.

Wu, S.-J., & Huang, S.-R. (2010). Optimal progressive group-censoring plans for exponential distribution in presence of cost constraint. *Statistical Papers, 51*, 431–443.

Yu, Y. Q., Xu, X. P. (2003). Improvement of the performance of diamond segments for rock sawing, Part 1: Effects of segment components. *Key Engineering Materials, 250*, 46–53.

Chapter 16
Detection of Anomalous Sequences in Crack Data of a Bridge Monitoring

Sermad Abbas, Roland Fried, Jens Heinrich, Melanie Horn,
Mirko Jakubzik, Johanna Kohlenbach, Reinhard Maurer,
Anne Michels and Christine H. Müller

Abstract For estimating the remaining lifetime of old prestressed concrete bridges, a monitoring of crack widths can be used. However, the time series of crack widths show a strong variation mainly caused by temperature and traffic. Additionally, sequences with extreme volatility appear where the cause is unknown. They are called anomalous sequences in the following. We present and compare four methods

---

Sermad Abbas and Roland Fried—Responsible for method LCP; Jens Heinrich and Reinhard Maurer—Responsible for experimental setup and bridge monitoring; Melanie Horn—Responsible for data transformation; Mirko Jakubzik and Johanna Kohlenbach—Responsible for method CMAD; Anne Michels—Responsible for methods VCP, MVCP; Christine H. Müller—Responsible for methods VCP, MVCP and the general text.

S. Abbas · R. Fried · M. Horn · M. Jakubzik · J. Kohlenbach · A. Michels · C. H. Müller (✉)
Department of Statistics, TU University Dortmund, Dortmund, Germany
e-mail: cmueller@statistik.tu-dortmund.de

S. Abbas
e-mail: sermad.abbas@uni-dortmund.de

R. Fried
e-mail: biostat@statistik.tu-dortmund.de

M. Horn
e-mail: mhorn@statistik.tu-dortmund.de

M. Jakubzik
e-mail: jakubzik@statistik.tu-dortmund.de

J. Kohlenbach
e-mail: johanna.kohlenbach@tu-dortmund.de

A. Michels
e-mail: anne2.michels@tu-dortmund.de

J. Heinrich · R. Maurer
Department of Architecture and Civil Engineering, TU University Dortmund, Dortmund, Germany
e-mail: jens.heinrich@tu-dortmund.de

R. Maurer
e-mail: reinhard.maurer@tu-dortmund.de

© Springer Nature Switzerland AG 2019
N. Bauer et al. (eds.), *Applications in Statistical Computing*,
Studies in Classification, Data Analysis, and Knowledge Organization,
https://doi.org/10.1007/978-3-030-25147-5_16

which were developed in a pilot study and aim to detect these anomalous sequences in the time series. Volatilities caused by traffic should not be detected.

16.1   Introduction

The accident in Genoa on 14th August 2018 showed that the safety of bridges is a severe problem. This is because many bridges were built in the 50s and 60s of the last century. They were designed and constructed for much lower than today's traffic loads in reality. Furthermore, the principles for designing and detailing have been advanced since then. Hence, an important task is to predict the remaining lifetime of old prestressed concrete bridges. This can be done by monitoring cracks in the bridges. Heeke et al. (2016), Heinrich et al. (2017), Szugat et al. (2016) showed how this can be done in lab experiments with prestressed concrete beams. Crucial in these approaches is to detect the breaking of tension wires. This was mainly done by acoustic measurements. Thereby, it turned out that each break causes a distinct increase in the width of an initial crack.

These acoustic measurements are not possible in bridge monitoring. Only the crack widths can be monitored. However, they are influenced by the traffic and, more heavily, by temperature. The left-hand side of Fig. 16.1 shows crack width measurements together with two temperature measurements. The traffic load of the tram and other heavy vehicles are visible as isolated peaks. In particular, there is a large isolated peak at 6:30 a.m., which was caused by a heavy test load using a mobile crane of 48 tons. Moreover, the measurements of the crack widths are disturbed by

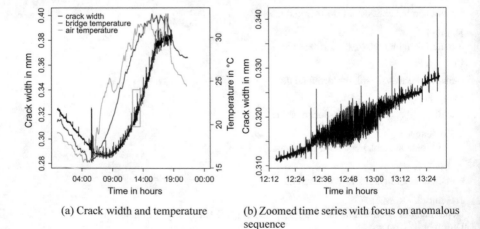

(a) Crack width and temperature

(b) Zoomed time series with focus on anomalous sequence

Fig. 16.1 Left: crack width and temperature measured above (air temperature) and below (bridge temperature) the bridge on 5th June 2016 for sensor WWS4. Right: zoom of the crack width curve given by the green box on the left-hand side

strong anomalous sequences. Such an anomalous sequence is shown on the right-hand side of Fig. 16.1. It provides a zoom around 12:48 p.m. of the left-hand plot.

Anomalous sequences show an extreme volatility. It is unclear whether these phenomena are caused by beat effects or if there are other causes. In any case, they disturb the analysis severely so that a first step is to extract these anomalous sequences. Thereby, it is important that peaks caused by traffic are not extracted.

We present four methods for detecting the anomalous sequences. All of them are based on filtering the time series of crack width data. Three of them use additionally special change point detection methods. A location change point method is used by the LCP (Location Change Point Detection using Squared Residuals) method, in which a two-sample test based on the Hodges–Lehmann estimator for shift is applied on the logarithm of squared residuals in a moving time window, see Abbas and Fried (2017) and Fried (2012). The VCP (Variance Change Point Detection) and MVCP (Modified Variance Change Point Detection) methods use the PELT (Pruned Exact Linear Time) method of Killick et al. (2012) for variance change point detection. The PELT method aims to find the global minimum of a cost function with a penalty for several change points where the number of evaluations is linear in the number of observations. The fourth method, called CMAD (Clustering of MAD filtered data), uses a MAD (Median Absolute Deviation from the median) filter and two times a median filter. It clusters the MAD filtered data by the k-means algorithm.

In Sect. 16.2, the experimental setup and the data are described. The statistical methods are explained in Sect. 16.3. Section 16.4 provides a comparison of the four methods. Finally, in Sect. 16.5 a discussion is given.

16.2 Experimental Setup and Data Description

16.2.1 Bridge Monitoring

16.2.1.1 Motivation

In June 2016, a bridge monitoring on an existing post-tensioned prestressed concrete bridge in Bochum built in 1961 (Fig. 16.2) has been installed and started. The

Fig. 16.2 Bridge view during the installation of the monitoring

reason was an assessment of this bridge by recalculation. It showed a significant deficit of fatigue strength according to the current standards in certain areas of the superstructure. Furthermore, in those critical areas, several bending cracks showing widths of more than 0.5 mm have been detected during a routine bridge inspection. Therefore, it was decided to carry out a crack monitoring with sensors at 16 measurement points to observe a potential increase of crack widths which indicates a damage due to fatigue in the cracked areas.

16.2.1.2   State of the Building

In the course of a design recalculation, deficits with regard to the limit of the state of decompression and fatigue strength were determined. The assessment was based on the *German recalculation guideline for existing buildings* (Bundesministerium et al. 2011). An important aspect of this guideline is the determination of a calculated remaining service life of existing structures under traffic. It was possible to estimate a remaining service life due to fatigue until the year 2019.

Considering this low remaining service life and also the large crack widths, which could already point at a preexisting damage, it has been decided to build a new bridge. Furthermore, additional measures were taken to prevent the collapse of the structure. In addition to the restriction of a maximum weight of 24 tons per vehicle and the limitation to one lane in each direction, a crack monitoring was planned until the demolition of the existing bridge.

16.2.1.3   Objectives of the Monitoring

The primary objective of the monitoring was the early and timely detection of crack width increases, which could indicate the failure of prestressing steel due to fatigue. In case of a successive crack width increase as a result of prestressing steel failure, further emergency measures, such as bridge closure and temporary support, can be initiated.

16.2.1.4   Installation of the Monitoring

During continuous monitoring of the structure, high-frequency crack measurements by means of inductive displacement transducers are used, which were applied at the 16 areas of cracking. These transducers were denoted by WON1 to WON4, WWN1 to WWN4, WOS1 to WOS4, WWS1 to WWS4, where 'O' stands for 'east', 'W' for 'west', 'N' for 'north', and 'S' for 'south'. In addition to the installation of the transducers, three measuring points for temperature recording were installed.

16.2.2   Data Description and Data Transformation

Since 1 June 2016, the bridge monitoring is put into practice. The monitoring is carried out at a total of 16 measuring points for the crack width and three measuring points for the temperature. The measuring frequency is 0.5 Hz, so that the crack width and the temperature is recorded every two seconds. After each day, the collected data are stored in a *Microsoft Access Add-in* data file (.mdt). This results in a total data volume of 43 200 observations per measuring point and day and adds up to a file size of about 1.6 Megabyte per day. As we will analyse the data with R, we have to make the data accessible for it. Converting the MDT files to CSV files for using the standard R-functions is not practical for long time intervals because the data of a single day in CSV format add up to a size of 11.8 Megabyte.

For this reason, we wrote a new R-function to read the data from the MDT file. An MDT file consists of three parts: an ASCII-coded header with information about the data, the measurements stored as signed 2-byte integers, and an ASCII-coded sentence at the end of the file. The header provides useful information, for example, the day and the starting time of the measurements, but it also includes a table with, among other things, the name of the point of measurement and a numerical *offset* and *factor*. The offsets and factors are necessary to transform the 2-byte integers of the crack width and temperature measurements to their real values. Let x be a 2-byte integer. Then, the real value is obtained by $(x - \text{offset}) \cdot \text{factor}$. Thus, in R we can read the table in the header with the help of the function `read.table()` and the measurements with `readBin()` and after transforming the integers to their real values and adding some aesthetics like names, we obtain a `data.frame` with 20 columns (16 crack widths, 3 temperatures and 1 time column) and about 43 200 rows. If we store the obtained data in a `RData`-file, the file has a size of 1.0 Megabyte.

16.3   Statistical Methods

In the following, four methods for finding anomalous sequences are described. Therefore, time series $(X_t)_{t \in T}$ of crack width measurements are examined where $T = \{1, \ldots, N\}$ is the time index set. For simplification, the notation $X = (X_t)_{t \in T}$ is used.

All methods use a filtration in the sense of Brockwell and Davis (2006). It transforms the time series through a time-invariant but not necessarily linear filter $f : \mathbb{R}^{2w+1} \to \mathbb{R}$, $w \in \mathbb{N}$, given by

$$X_t^{(f,w)} = (X_t^{(f,w)})_{t \in T} = f(X_{t-w}, \ldots, X_{t+w}). \tag{16.1}$$

Then $X^{(f,w)}$ is the time series filtered through f with window width $2w + 1$. In this article, the filters $f = med$ (median) and $f = MAD$ (Median Absolute Deviation from the median) are applied.

In particular, all methods use the filter $f = med$ for trend adjustment of X_t. The window width parameter w varies from method to method. The trend-adjusted crack width process is then given by

$$\tilde{X}_t = X_t - X_t^{(med,w)}, \quad t \in T. \tag{16.2}$$

The calculations for all methods were performed with the statistical software R (R Development Core Team 2017).

16.3.1 Location Change Point Detection Using Squared Residuals (LCP)

The Location Change Point Detection using Squared Residuals (LCP) approach consists of several steps, which we will describe in the following. We use Fig. 16.1 to illustrate them on the time series presented in Fig. 16.2.

First, we calculate a trend-adjusted time series as in (16.2). The estimated trend function is depicted in Fig. 16.3b for the parameter $w = 75$. This choice of w is 16 rather intuitive and not the result of a thorough optimization. In general, w should be chosen small enough so that the level of the time series is nearly constant and large enough so that the filter withstands a few large observations and to ensure a satisfying efficiency. In our case, it is only important that no relevant anomalous sequences are smoothed by the filter. The resulting trend-adjusted time series of our example is represented by the black line in Fig. 16.3c, d.

To detect volatility changes, we transform the trend-adjusted time series by computing the logarithm of $\tilde{X}_t^2$. For our example, the result is shown in Fig. 16.3c. The volatility changes are visible as location shifts. In the trend-adjusted time series, about 33% of the values are equal to zero. That is why we add random noise from a uniform distribution to the observations before applying the logarithm. The range of the random numbers is by the factor ten smaller than the scale of the data.

Sequential two-sample tests in a moving time window can now be applied to check for a location shift at each time point t. This approach is, for example, described in Abbas and Fried (2017). At each time point $t \geq k$, we consider a time window $X_t = (X_{t-k+1}, \ldots, X_t, \ldots, X_{t+k})'$ of width $n = 2k + 1$ centred at t. We test for a location shift between t and $t + 1$ by splitting the window into two subwindows

$$\tilde{X}_{t-} = (\tilde{X}_{t-k+1}, \ldots, \tilde{X}_t)' \quad \text{and} \quad \tilde{X}_{t+} = (\tilde{X}_{t+1}, \ldots, \tilde{X}_{t+k})',$$

each of length k. The window width parameter k controls the robustness against outliers and the power of the test. The latter can be improved by using a large window. However, it is crucial that the assumption of a constant underlying signal within the window can be justified. Using this local approach avoids making, possibly unrealistic, global parametric assumptions. Moreover, the methods are able to adapt

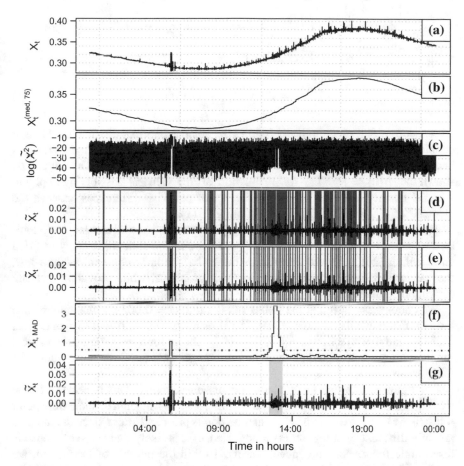

Fig. 16.3 Presentation of LCP method using the crack widths from 5th June 2016 at measuring point WWS4. **a** Original time series, **b** estimated trend after median filtering, **c** logarithm of squared residuals, d) trend-adjusted time series with change point candidates after applying asymptotical HL22-test (annotated in red), **e** estimated change points after using method of Wu and Chu (1993), **f** MAD within candidate intervals on standardized residual time series for anomalous sequences (the red dotted line denotes the threshold $\gamma = 0.5$), g) parts that are detected as anomalous sequences (highlighted in yellow)

to a slowly varying signal and lessen the impact of a possible variation due to the natural behaviour of the process (Fried and Gather 2007) which may not have been completely eliminated by the trend adjustment.

Assuming that the underlying distributions in X_{t+} and X_{t-} differ at most in location allows for a comparison of both subwindows by a two-sample location test.

Generally, any two-sample location test could be used within this framework. In the following, we use an asymptotic test based on the robust and rather efficient two-sample Hodges–Lehmann estimator for shift (HL2 estimator) which is given by

$$\hat{\Delta}_t = \text{med}\{\tilde{X}_{t+j} - \tilde{X}_{t+i} : \ j = 1, \dots, k, \ i = -k+1, \dots, 0\}$$

at time $t = 1, \dots, N$. The test statistic is

$$T_t^{(\text{HL2})} = \sqrt{3} \int g^2(x) \ dx \cdot \sqrt{n} \cdot \hat{\Delta}_t \qquad (16.3)$$

with $n = 2 \cdot k$ and asymptotically follows the standard normal distribution. Here, g denotes the density of the distribution underlying the $\tilde{X}_t$ under $H_0 : \Delta = 0$ (Hodges and Lehmann 1963). Let h be the density of $\tilde{X}_{t+} - \tilde{X}_{t-}$ under H_0. Then, it can be shown that $h(0) = \int f^2(x) \ dx$. Estimation of this value is possible by applying a kernel density estimator applied to the set of pairwise differences between the elements of $\tilde{X}_{t+}$ and $\tilde{X}_{t-}$ (Fried and Dehling 2011).

We use an asymptotic test because some prior studies showed that large windows lead to better results on the herein analysed data. In the simulation study in Fried and Dehling (2011), the HL2-test turns out to be similarly powerful as the t-test, the standard method for two-sample location problems, while providing a strong protection against outliers.

We choose $k = 200$ and the rather small value $\alpha = 0.005$ for the significance level to ensure that the number of false alarms is not too large. About 15% of the test decision lead to a rejection of H_0. As Fig. 16.3d shows for $\alpha = 0.005$ by the vertical red lines, they are scattered all over the time range. This implies a large amount of false alarms.

The figure shows that many alarms occur subsequently. This is because of the moving-window nature of this approach. If an alarm is triggered at some time point t, it is likely that it will also be given at time points prior and after t since the samples are only different by a few observations. Hence, it is likely that the same cause is responsible for several subsequent alarms. To thin the number of alarms out, we use a method proposed by Wu and Chu (1993) to estimate the true change point out of subsequent alarms. Each time point at which H_0 is rejected can be seen as a change point candidate. We calculate the p-values at these change point candidates and determine the time index t^* which belongs to the smallest p-value among the set of candidate points. Within the neighbourhood $\{t^* - k + 1, t^* - k + 2, \dots, t^* - 1, t^* + 1, \dots, t^* + k\}$ of t^*, we remove all other candidate points. This process is repeated until no candidates are left. The remaining time points $t_1^* < t_2^* < \dots < t_m^*$ are estimators for the change points. The result for our example is depicted in Fig. 16.3e, where we now have rejections at approximately 0.02% of the testing time points. However, they are still distributed over the complete time range.

Let now $\{t_1^*, t_1^* + 1, \dots, t_2^*\}, \{t_2^*, t_2^* + 1, \dots, t_3^*\}, \dots, \{t_{m-1}^*, t_{m-1}^* + 1, \dots, t_m^*\}$ be the sets of time points between two estimated change points. To identify potential anomalous sequences, we estimate the variability for the observations in the trend-adjusted time series within each set. Except for t_1^* and t_m^*, each estimated change point can be the starting or the end point of a potential anomalous sequence under the assumption that we did not miss any relevant change point before t_1^* and after t_m^*. Hence, subsequent sets share their first and last element. We estimate the variability within each set by using the MAD on the globally standardized trend-adjusted observations. This means that we first subtract the mean and divide this difference by the empirical standard deviation, both calculated from all observations. We use a threshold approach with which we remove the sets for which the estimated variability is smaller than a value $\gamma > 0$. By standardizing, we want to make sure that a single threshold can be used for different time series. The MAD ensures that single large or small crack widths do not affect the estimation. Figure 16.3f shows the results for the MAD and the threshold (horizontal line) of $\gamma = 0.5$.

In the last step, we combine connected sets for which the threshold is exceeded. By doing so, we assume that they belong to the same anomalous sequence. In Fig. 16.3g, two sequences are identified, highlighted by a yellow background. The first is caused by a test load from a 48-ton mobile crane. The second is an anomalous sequence of interest.

16.3.2   Variance Change Point Detection (VCP)

Variance Change Point Detection (VCP) uses variance change point analysis to detect anomalous sequences. For detection of anomalous sequences, the trend-adjusted time series $\tilde{X}_t$ is used. Therefore, the difference of the original time series and the running median (with $w = 150$) of the time series is calculated (see Fig. 16.4a). On this transformed data set, changes in variance are detected using change point analysis with the method `cpt.var()` from the R package `changepoint` (Killick et al. 2016).

The R method `cpt.var()` uses the Pruned Exact Linear Time (PELT) method of Killick et al. (2012) which is based on the Optimal Partitioning (OP) approach of Jackson et al. (2005). This approach is able to find several change points $1 \leq t_1^* < t_2^* < \ldots < t_M^* < N$ in a time series $(Y_t)_{t \in \{1,\ldots,N\}}$ where the number M of change points is estimated as well. It is based on a cost function $\mathscr{C}$ with the property

$$\mathscr{C}\left(Y_{t_{m-1}:t_m}\right) + \mathscr{C}\left(Y_{(t_m+1):t_{m+1}}\right) = \mathscr{C}\left(Y_{t_{m-1}:t_{m+1}}\right) \tag{16.4}$$

if no change point exists in the interval $[t_{m-1}, t_{m+1}]$ and

$$\mathscr{C}\left(Y_{t_{m-1}:t_m}\right) + \mathscr{C}\left(Y_{(t_m+1):t_{m+1}}\right) < \mathscr{C}\left(Y_{t_{m-1}:t_{m+1}}\right) \tag{16.5}$$

if only one change point exists at $t_m \in [t_{m-1}, t_{m+1})$ so that the change point is lying in the interval $(t_m, t_m + 1]$. Thereby, $Y_{t_k:t_l}$ stands for $(Y_t)_{t \in \{t_k, t_k+1, \ldots, t_l-1, t_l\}}$ for $t_k \leq t_l$, $t_k, t_l \in \{1, \ldots, N\}$.

Here, the adjusted time series $\tilde{X}_t$ is used for Y_t so that we can assume that its mean is zero. Then an appropriate cost function for variance changes satisfying (16.4) and (16.5) is twice the negative log-likelihood function based on the maximum likelihood estimator for the variance with known mean equal to zero. To avoid too many change points, a term $\log(t_m - t_{m-1})$ is added, i.e. the cost function is given by

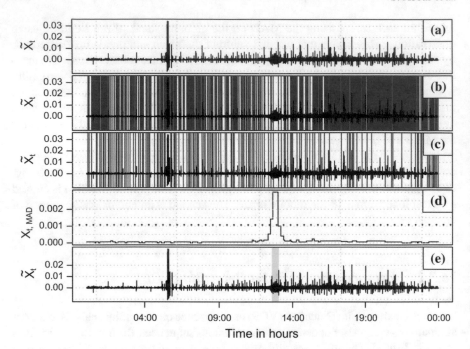

Fig. 16.4 Presentation of VCP method using the crack widths from 5th June 2016 at measuring point WWS4. **a** The trend-adjusted time series $\tilde{X}_t$, **b** $\tilde{X}_t$ with change point detection after applying `cpt.var()` (annotated in red), **c** $\tilde{X}_t$ with change points that have a distance of at least 80 observations to the next change point, **d** MAD between the change points from **c** where the red dotted line marks the threshold between clusters, **e** the part that is detected as anomalous sequence by VCP (highlighted in yellow)

$$\mathscr{C}\left(Y_{(t_{m-1}+1):t_m}\right) \tag{16.6}$$

$$= (t_m - t_{m-1})\left(\log(2\pi) + \log\left(\frac{\sum_{j=t_{m-1}+1}^{t_m}\left(Y_j\right)^2}{t_m - t_{m-1}}\right) + 1\right) + \log\left(t_m - t_{m-1}\right).$$

Given a cost function $\mathscr{C}$, then the change points are determined as the vector $\tau_N^* := \tau_{N,M^*}^* := (t_0^*, t_1^*, t_2^*, \ldots, t_{M^*}^*, t_{M^*+1}^*)$ which solve the following minimization problem

$$\min_{M=0,1,\ldots,N-1} \min_{\tau_M \in \mathscr{T}_{N,M}} \left(M\beta + \sum_{m=1}^{M+1} \mathscr{C}\left(Y_{(t_{m-1}+1):t_m}\right)\right) \tag{16.7}$$

where β is a penalty and

$$\mathscr{T}_{N,M} := \{(t_0, t_1, t_2, \ldots, t_M, t_{M+1}) \in \mathbb{N}^{M+2}; \ 0 = t_0 < t_1 < t_2 < \ldots < t_M < t_{M+1} = N\}.$$

For calculating the solution of (16.7), the OP approach of Jackson et al. (2005) uses the following recursion for $I = 1, \ldots, N$

$$F(I) := \min_{M=0,1,\ldots,I-1} \min_{\tau_{I,M} \in \mathcal{T}_{I,M}} \sum_{m=1}^{M+1} \left(\mathscr{C}\left(Y_{(t_{m-1}+1):t_m}\right) + \beta \right)$$

$$= \min_{J=0,1,\ldots,I-1} \min_{M=0,1,\ldots,J-1} \min_{\tau_{J,M} \in \mathcal{T}_{J,M}} \left[\sum_{m=1}^{M+1} \left(\mathscr{C}\left(Y_{(t_{m-1}+1):t_m}\right) + \beta \right) + \left(\mathscr{C}\left(Y_{(J+1):I}\right) + \beta \right) \right]$$

$$= \min_{J=0,1,\ldots,I-1} \left[F(J) + \left(\mathscr{C}\left(Y_{(J+1):I}\right) + \beta \right) \right]$$

with $F(0) := -\beta$. The idea for the OP algorithm is then to calculate recursively

$$F(I) = \min_{J=0,1,\ldots,I-1} \left[F(J) + \left(\mathscr{C}\left(Y_{(J+1):I}\right) + \beta \right) \right] \tag{16.8}$$

and

$$J(I) := \arg \min_{J=0,1,\ldots,I-1} \left[F(J) + \left(\mathscr{C}\left(Y_{(J+1):I}\right) + \beta \right) \right] \tag{16.9}$$

for $I = 1, \ldots, N$ starting at $I = 1$ and to store all $F(1), \ldots, F(N)$ and $J(1), \ldots, J(N)$. Then backtracking is used to obtain

$$I_0 := N, \quad I_1 := J(N), \quad I_m := J(I_{m-1}) \text{ as long as } I_{m-1} > 0,$$

so that $\tau_N^* = (t_0^*, t_1^*, t_2^*, \ldots, t_{M^*}^*, t_{M^*+1}^*) = (I_{M^*+1}, I_{M^*}, I_{M^*-1}, \ldots, I_1, N)$ with $I_{M^*+1} = 0$ and $I_{M^*} > 0$ is a solution of (16.7). This requires $O(N^2)$ evaluations. The PELT algorithm reduces the number of evaluations to $O(N)$ so that it is linear. This is done by pruning the set $\{0, 1, \ldots, I - 1\}$ for calculating $F(I)$ and $J(I)$ in (16.8) and (16.9), respectively. The pruned set is given by

$$R_I := \{I - 1\} \cup \left\{ i \in R_{I-1} : F(i) + \mathscr{C}\left(Y_{(i+1):(I-1)}\right) < F(I - 1) \right\}$$

with $R_1 = \{0\}$. Hence, $J \in R_I$ is used instead of $J \in \{0, 1, \ldots, I - 1\}$ in (16.8) and (16.9). This pruning is possible since i with $F(i) + \mathscr{C}\left(Y_{(i+1):(I-1)}\right) \geq F(I - 1)$ cannot be the last change point due to (16.4) and (16.5). That this pruning reduces the set $\{0, 1, \ldots, I - 1\}$ drastically is not so easy to see, but the proof is given in Killick et al. (2012).

For the VCP method, cpt.var() is used with its default values. Thereby, the penalty β is the modified Bayes Information Criterion given by $3 \log(N)$. Moreover, a step length of 2 is used as default so that the results for a reverted time series differ from the non-reverted time series although the PELT algorithm should find the global minimum.

With this method, nearly every change in variance of the crack width is labelled as change point (see Fig. 16.4b). Thereby also changes in the variability within an anomalous sequence are labelled as change point. Moreover, changes in crack width triggered by traffic are labelled as change points as well.

To separate changes in crack width triggered by traffic from anomalous sequences and to reduce the change points within anomalous sequences, the MAD is calculated for observations between adjacent change points. All MADs which are based on at least 80 observations are clustered by k-means into two clusters, i.e. in a cluster with high MAD and in a cluster with small MAD. Figure 16.4c shows only those changepoints that are used for clustering and in part d) the corresponding MADs are pictured. If connected sequences of the data are classified to the cluster with high MAD then they are joined to one sequence and this sequence is then a detected anomalous sequence. Then an anomalous sequence is followed by a nonanomalous sequence and this sequence is followed by anomalous sequence and so on. Sequences with less than 80 observations and high MAD are not considered as anomalous sequence here since the high MAD may be caused by the traffic. The anomalous sequence located by this procedure is shown in Fig. 16.4e.

A modification of the above method is to apply the k-means clustering on all MADs, also on those with less than 80 observations. However, then the cluster with high MADs contains mainly very small sequences and these sequences are mainly triggered by traffic. If afterwards the sequences with less than 80 observations are removed then often no anomalous sequences are found. Hence, this modification is not a good alternative.

16.3.3 Modified Variance Change Point Detection (MVCP)

The method cpt.var() from the changepoint-package (Killick et al. 2016) detects changes in variance at different points in the time series if the reversed time series is used. This is due to the dependence on the past in the change point identification process which is caused by the fact that cpt.var() uses a step length of 2 by default so that shifts can appear in the detected anomalous sequences. Therefore, the Modified Variance Change Point Detection (MVCP) method uses the VCP method on the original time series and on the reversed time series. Only those observations which are identified as anomalous sequences in both runs get labelled as anomalous sequences by MVCP.

16.3.4 Clustering of MAD Filtered Data (CMAD)

At first a trend adjustment is conducted on X_t. Therefor the filter $f = med$ with window width parameter $w = 15$ is applied on the time series X_t. The trend-adjusted crack width process is shown in Fig. 16.5a.

For the further process, it is assumed that anomalous sequences certainly exist in adequate large time intervals. This was confirmed in preliminary studies.

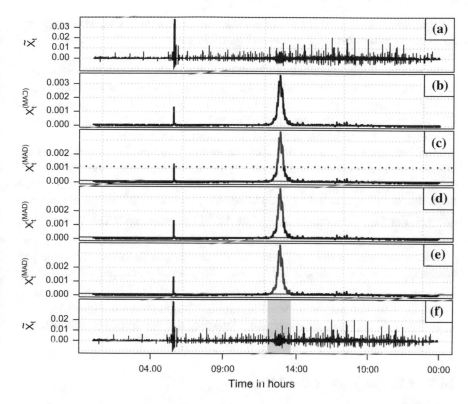

Fig. 16.5 Presentation of CMAD method using the crack widths from 5th June 2016 at measuring point WWS4. **a** The trend-adjusted time series $\tilde{X}_t$, **b** $X_t^{(MAD)}$, the MAD smoothed, trend adjusted time series, **c** $X_t^{(MAD)}$ with red marked candidates where $P_t = 1$, **d** $X_t^{(MAD)}$ with red marked candidates where $P_t^{(med)} = 1$, **e** $X_t^{(MAD)}$ with extension of sequences where $P_t^{(med)} = 1$, **f** $\tilde{X}_t$ with anomalous sequence (highlighted in yellow)

The next step is based on the assumption that anomalous sequences are marked by a locally increased variability. Application of the filter $f = MAD$ with window width parameter $w = 45$ on the trend-adjusted time series $\tilde{X}_t$, i.e.

$$X_t^{(MAD)} = \tilde{X}_t^{(MAD,45)}, \quad t \in T, \tag{16.10}$$

allows the detection of anomalous sequences by finding clusters of higher variability. In Fig. 16.5b, peaks are visible at 06:00 o'clock and 12:00 o'clock.

After that, all observations of the MAD smoothed time series $X_t^{(MAD)}$ are divided into two clusters by the k-means algorithm. Because of the target variable being one dimensional and k being 2, a threshold is the result so that every observation above this threshold is classified as a candidate for an anomalous sequence. A time series $P_t = (P_t)_{t \in T}$ is constructed, whereby $P_t = 1$ means that $X_t^{(MAD)}$ is above the threshold and $P_t = 0$ stands for a value below this threshold. The candidates for anomalous sequences are shown in Fig. 16.5c where $X_t^{(MAD)}$ is shown with red marked candidates for anomalous sequences based on P_t.

To prevent special occurrences like the test load being classified as anomalous sequences, the filter $f = med$ with window width parameter $w = 150$ is applied on P_t. The resulting time series is denoted $P_t^{(med)}$. In Fig. 16.5d, $X_t^{(MAD)}$ is represented with red marked candidates for anomalous sequences where $P_t^{(med)} = 1$. The peak at 06:00 o'clock is no longer classified as an anomalous sequence.

Furthermore, it is apparent that start and end of an anomalous sequence are not included in the cluster. Hence, the boundary points are examined. These points form the transition between a normal and an anomalous sequence. So, the boundary points mark the beginning and ending of an anomalous sequence, but are not classified as one yet. For this, the average $\overline{X_t^{(MAD)}}$ of $X_t^{(MAD)}$ is calculated. The values of the boundary points are compared with this value. If they are larger than $\overline{X_t^{(MAD)}}$, the points are assigned to the candidates for the anomalous sequence. Then the new boundary points are examined iteratively. This process is stopped as soon as the values of the boundary points are smaller or equal to $\overline{X_t^{(MAD)}}$ and hence, the maximal extension of the anomalous sequence is reached. This is shown in Fig. 16.5e.

In Fig. 16.5f, $\tilde{X}_t$ is pictured with anomalous sequences underlaid by yellow boxes.

16.4 Comparison of the Methods

In this section, we investigate the performance of the different approaches on selected time series. The available data sets do not contain information about which sequences fit our definition of anomalous. A categorization was performed manually for a limited number of time series. Thus, a thorough parameter tuning is not possible. Therefore, the following results should only be considered as a descriptive performance study. We use the parameter settings mentioned in the corresponding subsections in Sect. 16.3.

A major difference between the procedures is that, in contrast to VCP, MVCP and CMAD, the LCP method is based on the idea of a sequential online monitoring. It does not use the complete time series for detecting the change points. Moreover, the threshold to identify intervals with a large variability is a fixed value for LCP, whereas the other methods choose it for each data set separately due to clustering on the complete time series.

As illustrated by Figs. 16.3, 16.4, and 16.5 in Sect. 16.3, the approaches do not necessarily identify the same sequences in a time series as being anomalous. For the time series used in Sect. 16.3, only LCP categorizes the short sequence of large variability at around 6:30 a.m. as anomaly which was caused by a heavy test load using a mobile crane. Therefore, we do not consider this short peak caused by traffic as an anomalous sequence of interest. The methods VCP, MVCP and CMAD do not identify this sequence so that, in this case, they are able to distinguish between short

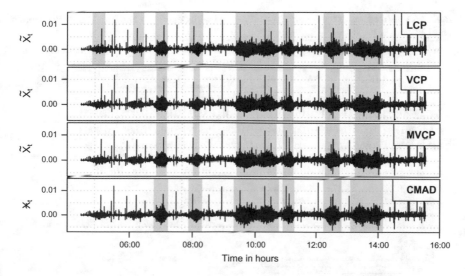

Fig. 16.6 Comparison of the anomalous sequences detected by the four methods (highlighted in yellow) on 12th June 2016 of WOS4. The figure shows the trend-adjusted time series for each method

variance changes caused by traffic from large variance changes where the cause is unknown. The reason for this is that VCP, MVCP and CMAD, unlike LCP, use an assumption on the minimal length of an anomalous sequence.

To point out another difference, we look at the measurements from 12th June 2016 at WOS4. Figure 16.6 shows the trend-adjusted time series for each procedure. VCP, MVCP and CMAD return the same six sequences. The sequences returned by VCP and MVCP are similar and shorter than those returned by CMAD. LCP also returns these six sequences with widths comparable to those of CMAD. In addition, LCP categorizes two sequences, starting at around 4 o'clock and 6 o'clock, as anomalies. A closer inspection of these time intervals shows that the variability of the process is slightly increased compared to the time points before and after. This variability is smaller than for the other returned sequences. Thus, because of the clustering, VCP, MVCP and CMAD are able to adapt to the specific variability of a time series so that they lead to more reasonable results.

The two aforementioned highlighted regions for LCP do not cover the complete sequences with slightly increased variability. This is because the variability in the sequence next to the detected ones is not larger than the threshold. Thus, they are dropped before the combination step.

We now compare the procedures on three other time series with respect to their error rates. The measurements were taken on 5th August 2016 at WOS2, 1st January 2016 at WOS2 and 22nd February 2017 at WWS4. The error rates are computed by comparing the returned sequences to a manually performed categorization. As it is sometimes difficult for the human eye to detect regions with a higher variability, two persons classified the time series into anomalous and nonanomalous sequences.

Table 16.1 Error rates of the four methods compared with counting 1 and 2 and between counting 1 and 2

Day	Error rates between methods and countings								Error rates between counting 1 and 2
	LCP		VCP		MVCP		CMAD		
	1	2	1	2	1	2	1	2	
5th Aug 2016	0.009	0.014	0.010	0.007	0.010	0.007	0.002	0.006	0.0063
1st Jan 2017	0.004	0.005	0.008	0.008	0.008	0.008	0.006	0.006	0.0003
22nd Feb 2017	0.099	0.155	0.070	0.079	0.095	0.048	0.077	0.046	0.1084

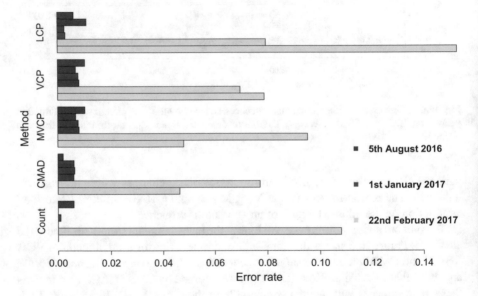

Fig. 16.7 Comparison of the error rates of the four methods for three different days with respect to both countings and the error rate between Counting 1 and 2 (denoted by 'Count'). The upper bar of a day is the result of the comparison with Counting 1 and the lower one is the result of the comparison with Counting 2

Both decided for each time point whether the corresponding observation belongs to an anomalous sequence. The results for each time point are captured in binary vectors called Counting 1 and Counting 2. Similar vectors are obtained by the different methods. Each counting is compared with each result vector. The error rate is defined as the fraction of elements that are different. Table 16.1 shows the rates and Fig. 16.7 illustrates them by a bar plot.

The last column in Table 16.1 quantifies differences between the two manual countings. The large value of about 10% for 22nd February 2017 indicates that the classification in anomalous and nonanomalous sequences can be ambiguous. Here, the time series has, in general, a larger variability which makes it harder to distinguish between normal process behaviour and anomalous sequences.

For 1st January the differences between the procedures are negligble. Some larger differences can be noted for 5th August. Here, the smallest error rate of 0.2% belongs to CMAD on Counting 1 while LCP leads to the largest rate of 1.4%. For the time series from 22nd February, the rates are, compared to those for the other two time series, quite large for all procedures. Differences are mostly visible for Counting 2. LCP has the largest value with 15.5%, whereas MVCP and CMAD perform similarly with about 5%.

There are several reasons for these observations. The countings are, as they were performed by human beings, subjective to a certain degree. This can be seen from the differences between them. Thus, we could not compare our results to the true sequences but only to those that we assume to be plausible. The differences between the countings illustrate that it is not easy to identify anomalous sequences on some time series, for example, due to a generally larger variability in the complete series.

Another reason is that the methods may detect the same sequences, but their lengths are different. This is the case for 5th August and 1st January. Only on 22nd February, LCP discovers more anomalous sequences than the other procedures.

To sum up, in general, the procedures lead to quite similar results when the anomalous sequences express themselves as intervals in which the variability is largely different from the general process behaviour. Differences seem to arise when the variability in the process is generally large so that it might be more difficult to identify the anomalous sequences.

However, the results should be treated with caution. In order to perform a thorough parameter tuning, more of the available data should be used. This was not done up to now because more manually analysed data would be necessary.

Nevertheless, the obtained results look promising and further refinement of the procedures, for example, adding a threshold for the minimal sequence length in the LCP method, might lead to a strong improvement.

16.5   Discussion

A monitoring of cracks in a bridge provided many time series of crack width data. The final aim is to decide whether there is a significant increase in the crack widths. This is not an easy task since the time series show a strongly varying behaviour. Some variation is caused by temperature and traffic. In addition, there are so-called anomalous sequences with much more extreme volatility for which the cause is unknown. Therefore, a further analysis of these time series is only possible if the anomalous sequences are removed. We present four methods to detect them automatically.

The LCP (Location Change Point Detection using Squared Residuals) method is based on applying a two-sample location test in a moving time window to a transformed time series. VCP (Variance Change Point Detection) and MVCP (Modified Variance Change Point Detection) are based on the PELT (Pruned Exact Linear Time) method. CMAD (Clustering of MAD filtered data) uses a classification approach.

We evaluate the procedures on a small set of test time series, all four procedures lead to similar results. Differences occur especially in the length of the sequences. Moreover, VCP, MVCP and CMAD, using an assumption on the minimal length of an anomalous sequence, are able to distinguish between short sequences which may be caused by traffic and long sequences where the cause is unknown. Hence these methods, unlike LCP, do not identify short sequences as anomalous sequences.

When the variability in the time series is generally large, it turns out to be difficult to identify anomalous sequences. Again, LCP turns out to provide more sequences than the other procedures. However, this seems to be primarily due to the chosen parameters. LCP does not use the knowledge on the minimal length of a sequence. Moreover, categorizing sequences into classes, one with large and one with small variability, is done by a fixed threshold by LCP. The other approaches use a clustering algorithm on each time series and thus consider the variability of each time series individually. This may be advantageous since one does not have to specify a threshold value globally. However, it may be prone to errors if there are time series without anomalous sequences. On the other hand, a fixed value needs proper tuning over several time series with a similar variability structure.

The presented analysis concentrates on a few days only. This is mainly because only measurements for three days were visually inspected by two persons on one of the 16 measurement locations on the bridge. Thus, conclusions drawn from our analysis should be treated with caution. However, they show that it is worth to study the procedures further in a larger setting.

In further studies, the methods should be applied to a larger set of time series. This would help to make reliable statements on their advantages and disadvantages. An important aspect is a proper parameter tuning. Moreover, the influence of the method for change point detection should be investigated.

Although we apply the procedures retrospectively, that is under the assumption that all data are available, in practice it may be important to analyse the time series online and detect unusual sequences with only a little time delay. Here, it would be interesting how the procedures can be adapted to this online scenario. For example, LCP is based on sequential two-sample testing that can easily be used in the online analysis.

Acknowledgements The authors gratefully acknowledge support from the Collaborative Research Center 'Statistical Modelling of Nonlinear Dynamic Processes' (SFB 823, B5, C3) of the German Research Foundation (DFG).

References

Abbas, S., & Fried, R. (2017). Control charts for the mean based on robust two-sample tests. *Journal of Statistical Computation and Simulation*, *87*(1), 138–155.

Brockwell, P. J., & Davis, R. A. (2006). *Time series: Theory and methods* (2nd ed.). New York: Springer.

Bundesministerium für Verkehr, Bau und Stadtentwicklung (2011). Richtlinie zur Nachrechnung von Straßenbrücken im Bestand (Nachrechnungsrichtlinie). Berlin: Springer.

Fried, R., & Dehling, H. (2011). Robust nonparametric tests for the two-sample location problem. *Statistical Methods and Applications*, *20*(4), 409–422.

Fried, R., & Gather, U. (2007). On rank tests for shift detection in time series. *Computational Statistics and Data Analysis*, *52*(1), 221–233.

Fried, R. (2012). On the online estimation of local constant volatilities. *Computational Statistics and Data Analysis*, *56*(11), 3080–3090.

Heeke, G., Heinrich, J., Maurer, R., Müller, C.H. (2016). Neue Erkenntnisse zur Ermüdungsfestigkeit und Prognose der Lebensdauer von einbetonierten Spannstählen bei sehr hohen Lastwechselzahlen. In Tagungshandbuch 2016, 2. Brückenkolloquium. Stuttgart: Technische Akademie Esslingen (pp. 529–539).

Heinrich, J., Maurer, R., Hermann, S., Ickstadt, K., Müller, C. (2017). Resistance of prestressed concrete structures to fatigue in domain of endurance limit. In *Proceedings of fib symposium maastricht, high tech concrete: where technology and engineering meet*, 12–14th June 2017.

Hodges, J. L., & Lehmann, E. L. (1963). Estimates of location based on rank tests. *The Annals of Mathematical Statistics*, *34*(2), 598–611.

Jackson, B., Scargle, J. D., Barnes, J., Arabhi, S., Alt, A., Gioumousis, P., et al. (2005). An algorithm for optimal partitioning of data on an interval. *IEEE Signal Processing Letters*, *12*(2), 105–108.

Killick, R., Fearnhead, P., & Eckley, I. A. (2012). Optimal detection of changepoints with a linear computational cost. *Journal of the American Statistical Association*, *107*(500), 1590–1598. https://doi.org/10.1080/01621459.2012.737745.

Killick, R., Heynes, K., & Eckley, I.A. (2016). Changepoint: An R package for changepoint analysis, R package version 2.2.2. https://CRAN.R-project.org/package=changepoint.

R Development Core Team: (2017). R: A language and environment for statistical computing. R Foundation for Statistical Computing. Austria: Vienna. http://cran.r-project.org/.

Szugat, S., Heinrich, J., Maurer, R., & Müller, C. H. (2016). Prediction intervals for the failure time of prestressed concrete beams. *Advances in Materials Science and Engineering*,. https://doi.org/10.1155/2016/9605450.

Wu, J. S., & Chu, C. K. (1993). Kernel-type estimators of jump points and values of a regression function. *The Annals of Statistics*, *21*(3), 1545–1566.

Chapter 17
Optimal Semi-Split-Plot Designs with R

Sebastian Hoffmeister and Andrea Geistanger

Abstract This paper introduces Semi-Split-Plot designs. They are a new class of experimental designs and support factors where only a reduced number of factor settings can be applied inside of one block. An algorithm to generate optimal Semi-Split-Plot designs is presented. A tutorial for the R package **rospd** that implements the algorithm is given. Semi-Split-Plot designs are compared to completely randomized and Split-Plot designs in terms of balance, aliasing and predictive quality.

17.1 Introduction

Split Plot designs are a very well-known quantity in the field of design of experiments (DoE). First introduced by Fisher (1992), today, they are a standard tool in many applications of DoE and well supported by most statistical software packages. Split-Plot designs are the answer to the problem of experimental factors that are hard to change. A factor is hard to change when it is very expensive or time intensive to change the factor's settings from one level to another.

Naturally, Split-Plot designs have been the first thought of the authors when working with a sample preparation robot in the pharmaceutical application of optimizing an assay. Sample preparation robots allow to perform many experiments completely automatically. There is only limited space to place the bottles with solvents on the robot. In the given application, experimenters should compare different solvent solutions. However, there is only limited space to place them on the robot. Whenever a fifth solvent shall be used, the robot has to be stopped and the operator has to manually exchange the bottles of solvents. While this sounds like a typical hard-to-change factor (HTC), there is a major difference: For traditional HTC-factors, we try to minimize the number of changes in factor levels. Instead of minimizing the

S. Hoffmeister (✉) · A. Geistanger
Roche Diagnostics GmbH, Penzberg, Germany
e-mail: sebastian.hoffmeister@roche.com

A. Geistanger
e-mail: andrea.geistanger@roche.com

© Springer Nature Switzerland AG 2019
N. Bauer et al. (eds.), *Applications in Statistical Computing*,
Studies in Classification, Data Analysis, and Knowledge Organization,
https://doi.org/10.1007/978-3-030-25147-5_17

271

overall number of changes, in this scenario, we want to build blocks of experiments in which only a subset of all possible factor levels are used. The problem could be interpreted as generating a blocked design that includes a restriction for each block to use not more than four different solvents.

This paper discusses the problem in more detail. An algorithm to generate optimal designs for this problem is proposed and some resulting designs are evaluated in comparison to completely randomized designs and Split-Plot designs. The experienced DoE-expert might skip Sect. 17.2, which gives a short description of Split-Plot designs and the typical way of analyzing them. Section 17.3 describes the sample preparation robot scenario in greater detail. In Sect. 17.4, we propose an algorithm to generate optimal designs for the described problem. This algorithm is implemented in the rospd-package for the R software (R Core Team 2018). A short tutorial on how to use the package follows in Sect. 17.5. A comparison of multiple design classes in terms of aliasing and predictive quality is done in Sect. 17.6 before we conclude with a summary and an overview of future work (Sect. 17.7).

17.2 Restricted Randomization in Split-Plot-Designs

When applying DoE in the real world, it is often difficult to follow one of the core concepts of DoE: randomization. In a completely randomized design (CRD), treatments (combinations of factor levels) are performed in complete random order to avoid bias in the analysis due to uncontrolled factors. In many applications, this idea is a problem. Often, there are restrictions to the experimental setup that make a complete randomized design extremely expensive or inconvenient at best. Kowalski and Potcner (2003) discusses an example for this problem. When trying to optimize the quality of a printing process (see Fig. 17.1), there are three factors to look at:

- **Blanket Type**: Two possible types of blankets for the printing press.
- **Cylinder Gap**: The distance between the two cylinders of the printing press.
- **Press Speed**: The speed of the printing press.

Fig. 17.1 Printing press example

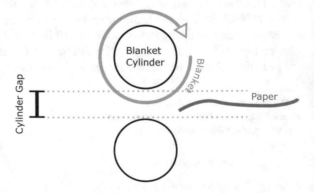

It is fairly simple to change the **Cylinder Gap** or **Press Speed** from one experiment to the next one, as these factors can be reset during the printing process. Exchanging the **Blanket Type** though is much more work. It requires the printing press to be stopped before the old blanket can be removed and the new one can be put on the press. Thus, a completely randomized design, like the following one, is very time intensive as it involves exchanging the blanket quite often.

The reader should be aware that the operator is expected to exchange the blanket after each run even if consecutive runs are using the same type of blanket. The rationale behind that is, that ordinary least squares regression—that is typically used to analyze DoE data—assumes independence of observations. Using the same blanket for multiple runs would violate this assumption as the experiments done with the same blanket would be correlated with each other.

Fisher (1992) first wrote about that problem and gave a solution with Split-Plot designs. A Split-Plot design allows restricted randomization. For the current example, this means that it is no longer necessary to exchange the blanket for each experiment. Instead, the same blanket can be used inside of each whole plot. A whole plot is a group of experiments for which one or more factors are held constant.

In this example, the factor **Blanket Type** would be called an HTC-factor. It is inconvenient to reset this factor for each experiment and thus a restriction to the randomization of the experiment is imposed. The restriction limits the number of resets of the hard-to-change factor to a practical amount.

In contrast to the HTC-factor, all other factors—that can be reset for each experiment without a lot of effort—are called easy-to-change factors (ETC). A Split-Plot design is still randomized inside of the whole plots so that the easy-to-change factors are changing as much as possible. There are many forms of Split-Plot designs and many examples of different types of HTC-factors. Jones and Nachtsheim (2009), Jones and Goos (2011) and Kowalski and Potcner (2003) give a good overview of this topic.

17.2.1 Design Model

The design model of a Split-Plot design is different from the traditional design models of CRD. For the analysis of CRDs, an ordinary least squares regression is applicable:

$$Y = X\beta + \varepsilon,$$

where:

- Y is a $n \times 1$-vector containing the measured responses. Here, n represents the number of runs in the design.
- β is a $p \times 1$-vector containing the model parameters. Here, p is the number of parameters in the model including the intercept.

- X is the $n \times p$-model matrix, where each column represents one model effect and each of the n rows represent one run of the design.
- ε is a $n \times 1$-vector representing the residuals of the model. The residuals are assumed to be identically, independently and normally distributed with mean 0 and standard deviation σ.

As mentioned before, there is a problem with the assumption of independence of residuals when using a Split-Plot design. The design is set up so that some factor settings are not changed for a group of experiments. Thus, all experiments in one whole plot are correlated with each other. This correlation arises because the experimenter is missing the variability that is introduced to the process by changing the hard-to-change factor. To account for the correlation structure in the analysis, Split-Plot designs are usually analyzed by using mixed models of the following structure:

$$Y = X\beta + Z\gamma + \varepsilon.$$

Here, Y, X and β are the same like before.

- Z is a $n \times b$-matrix representing the whole plot structure of the design. b is the number of hard-to-change factors.
- γ is a $b \times 1$-vector containing the random effects of the whole plots. Their estimators represent the variability between different whole plots. It is assumed that $\gamma \sim N(0, \sigma_w^2)$.

The covariance matrix of the response vector Y can be written as (Jones and Nachtsheim 2009)

$$V = \sigma^2 I_n + \sigma_w^2 Z Z'. \tag{17.1}$$

V is a block diagonal matrix, where each block represents one whole plot. σ^2 is the residual variance, while σ_w^2 represents the variance between whole plots. A more in- depth description of the analysis of Split-Plot designs is given at Jones and Nachtsheim (2009) and Næs et al. (2007).

17.2.2 Design Setup

The design of an experiment goes hand in hand with the design model. Thus, the modification of the design model has some implication for the setup of a Split-Plot design. While some programs still provide Split-Plot-variants for classical full- and fractional-factorial designs (Jones and Nachtsheim 2009), optimal designs have proven their value with a much higher flexibility (Jones and Goos 2007; Jones and Goos 2011). When planning optimal Split-Plot designs, there are two major factors to be considered:

1. How many whole plots should be chosen?
2. How to calculate the chosen optimality criterion considering the different type of analysis that is done for Split-Plot designs?

The number of whole plots defines how often the experimenter needs to reset the condition of the corresponding HTC-factor. Thus, this decision is often made with an economical background. At the same time, the number of whole plots is very relevant for the estimation of the random effects part of the model and for the power of the HTC-factor effects (see JMP Help SAS 2018). The minimal number of whole plots is $l + 1$ where l is the number of levels of the HTC-factor. Using only l whole plots would make it impossible to differentiate between the effect of the hard-to-change factor and the whole plot effect (which is the variability that comes from resetting the condition of the hard-to-change factor).

The second aspect that is relevant when working with optimal Split-Plot-designs are the necessary adjustments to optimality criteria. As most optimality criteria depend on the chosen design model ,it is necessary to adjust them to the new design model including the random effects. Jones and Goos (2007, 2012) as well as Hooks et al. (2009) give an overview of how to calculate different optimality criteria in the presence of random effects:

- **D-Optimality**: $D(X) = |X'V^{-1}X|$ with X being the design matrix and V being the covariance matrix of the response as it was defined above (Eq. 17.1). V depends on Z which is the whole plot stuctutre of the design and on the ratio of σ^2 and σ_w^2.
- **I-Optimality**: $I(X) = 2^{-N}tr[(X'V^{-1}X)^{-1}B]$ which corresponds to the average prediction variance, where $B = \int_{x\in[-1,+1]^N} f(x)f'(x)dx$ is called the moments matrix for the experimental region $\chi = [-1, +1]^N$. N is the number of factors.
- **A-Optimality**: $A(X) = tr(X'V^{-1}X)$.

17.3 Less Restrictive Randomization

The following example from an assay development shows that Split-Plot designs using HTC- and ETC-factors are not always flexible enough. A crucial part of assay development is sample preparation. Sample preparation cleans the sample (blood serum, plasma or urine, for example) from undesired components like fats or proteins. Getting rid of those parts of the sample helps in terms of reproducibility of measurements and lowers the minimum concentration of a substance that can be detected in samples. For optimizing the sample preparation workflows, laboratories often use robots. These robots allow to test a lot of different workflows in a reasonable time. In this way, it is possible to test hundreds of different workflows overnight.

Typical DoEs for optimizing sample preparation workflows include factors like **pH** values, **incubation times, incubation temperatures**, usage of different **types of solvents**, etc. The authors were facing a problem with the different **types of solvents**.

Table 17.1 Completely randomized design

Run	Solvent	Run	Solvent	Run	Solvent	Run	Solvent
1	A	6	E	11	E	16	D
2	B	7	A	12	A	17	C
3	A	8	B	13	B	18	A
4	C	9	C	Change Bottles		19	B
5	D	Change Bottles		14	C	20	D
Change Bottles		10	D	15	D	⋮	⋮

As part of the optimization six different types of solvents shall be tested. Solvents are stored in bottles and need to be placed on the robot. The robot will automatically pick the right solvent for any given experiment. But there is only enough space for four different bottles on the robot. It is possible to work with four different solvents and have a completely randomized design.[1] When a fifth solvent is used in the DoE, it becomes more complicated. Now the robot has to be stopped and the operator will exchange one or more bottles before the robot is able to continue its work. This is very inconvenient because it makes it impossible to perform experiments overnight without having an operator supervising the robot at all time. A completely randomized DoE might lead to a workflow like in Table 17.1.

This workflow removes most of the benefits that come from automated experimentation as a lot of user interaction with the robot is required. Treating the factor **type of solvent** as an HTC-factor is an alternative (see left Table 17.2).

Following this approach, we can reduce the work for the operator but there are two problems:

1. **Randomization**: Split-Plot designs are always a nod to practicality but all Split-Plot designs restrict the randomization of the DoE. This makes Split-Plots vulnerable to effects due to noncontrolled factors. For the given use case, it would be possible to do better in terms of randomization compared to a Split-Plot design.
2. **Correlation structure**: One advantage of Split-Plot designs is that a correlation between experiments inside of one whole plot is accepted. For the given use case, this structure does not really fit the problem. Rather than stopping the robot after each whole plot, it would be much more desirable to stop the robot after the fourth whole plot is finished. Thus, the correlation structure is different than it would be assumed by the Split-Plot design setup.

A better structured design for the given problem would look like the right table in Table 17.2. All runs are grouped similarly like in a Split-Plot design. The restriction

---

[1] One might argue that using always the same bottles of solvents is a violation of the assumption of independence. The latter is ignored for the moment as experience shows that the product quality of the solvents is very stable.

Table 17.2 Split-plot design and semi-split-plot design

WP	Solvent	WP	Solvent		WP	Solvent	WP	Solvent
1	A	3	D		1	A	3	A
1	A	3	D		1	B	3	B
1	A	3	D		1	C	3	C
1	A	3	D		1	D	3	E
Change Bottles		Change Bottles			Change Bottles		Change Bottles	
2	C	4	B		2	E	4	F
2	C	4	B		2	F	4	E
2	C	4	B		2	A	4	C
2	C	4	B		2	B	4	A
Change Bottles	:	:			Change Bottles	:	:	

to randomization is different though. Split-Plot designs contain at least one HTC-factor. They are split up into groups of experiments for which the settings of the HTC-factor are not changed. Now, there is one factor—we will call it semi-hard-to-change (SHTC)—that is not fixed to one level inside of each whole plot. Instead, the factor levels can change from one run to the next. But rather than using any of the possible factor levels, the randomization only allows to use four out of the six possible factor levels in each group. We will call designs that include at least one SHTC-factor Semi-Split-Plot designs.

It would not be wrong to think about this setup as a design using random blocks. Each random block contains experiments that can be performed without operator interaction. But here, we introduce an additional restriction that ensures that only a given number of different settings of the SHTC-factor are used inside of each block.

17.4   An Algorithm to Generate Optimal Semi-Split-Plot Designs

There is a wide variety of algorithms to produce optimal designs. The Fedorov algorithm (Miller and Nam-Ky 1994) is probably one of the more popular ones. JMP as an example for a commercial DoE software uses a coordinate-exchange algorithm (Meyer and Nachtsheim 2019). The R package **rospd** uses a modified version of Jones and Goos (2007) candidate-set-free algorithm to create optimal Semi-Split-Plot designs. The following section describes the algorithm and its modifications.

17.4.1 Prerequisites

To use the algorithm, the following information has to be provided:

- **Factors**: The algorithm can handle continuous or categorical ETC-, SHTC- or HTC-factors.
- **Whole Plot Structure**: If there are any SHTC- or HTC-factors, a matrix representing the whole plot structure needs to be defined. This matrix defines the whole plots of the design and therefore when and how often each HTC- or SHTC-factor can be changed.
- **SHTC-Group-Size**: This integer defines how many different settings of an SHTC-factor can be used in one whole plot.
- **Number of runs**: The number of experiments done for the DoE.
- **Constraints**: Possible constraints to the factors space can be defined.
- **Optimality Criterion**: The algorithm works with any optimality criterion that expresses the quality of a design as a numeric value.

17.4.2 Initial Random Design

The algorithm starts by generating a random initial design table. The design table is a data frame containing one combination of factor settings for each run of the design. The initial design uses random factor levels for each cell in the data table. It respects all restrictions like restrictions to the randomization and constraints to the factor space. This includes a possible SHTC-structure.

17.4.3 Updating the Design

The updating begins with the largest whole plot. All cells in that whole plot are updated sequentially beginning with the factors that are the hardest to change and ending with the ETC-factors. The updating works differently depending on the type of the factor:

- **HTC-Factors**: are updated in groups. Each possible value is considered. Instead of updating just one cell, all cells of the current whole plot in the design table are replaced simultaneously with the same value. The value that grants the best optimality for the design is used.
- **SHTC-Factors**: are updated in groups defined by the whole plots. Other than for HTC-factors the values for each cell in one whole plot can be different. All possible combinations are tested and the best one in terms of the chosen optimality is used.
- **ETC-Factors**: are updated one cell at a time. Each possible level for that factor is considered. For continuous factors the user has to specify how many intervals

between minimum and maximum factor level will be considered. If replacing the current value in the design table with any of the other possible values improves the optimality of the design, the best value is used.

As the updating of SHTC-factors can be very time consuming a faster but less comprehensive strategy was implemented additionally. Here, all elements of the SHTC-factor in one whole plot are updated sequentially. This is the same approach that is used for ETC-factors with the exception that not all possible factor levels are used but rather a subset defined by the number of possible settings inside of one whole plot. The sequential updating is done for all possible subsets of factor levels out of all factor levels of the SHTC-factor.

By default, the algorithm will use one random start and update each element of the design table up to 25 times. At the end of each step, the algorithm checks if any cell of the design was updated. If no changes where made, the algorithm converged and stops.

17.5 Using the Rospd-Package

The rospd-package (**R O**ptimal **S**plit **P**lot **D**esigns) is an implementation of the previously described algorithm. This chapter shows how to use the package function to create optimal Semi-Split-Plot designs. The core part of the package are the two classes *doeFactor* and *doeDesign*. The *doeDesign* class collects all information required to generate an optimal design. This includes a list of *doeFactors* that should be investigated in the DoE.

17.5.1 The DoeFactor Class

Objects of class *doeFactor* represent factors in the DoE. The definition of a factor can be done as follows:

```
# solvent - a SHTC categorical
factor solventFactor <- new("doeFactor", name="solvent",
                type="categorical",
                levels=c("A", "B", "C", "D", "E", "F"),
                changes="semi.hard",
                semi.htc.group.size=as.integer(4))
```

doeFactor is a S4 class with the following slots:

- **name**: The name of the factor as a character.
- **type**: Either *'continuous'* or *'categorical'*. Continuous factors are numeric factors that could theoretically take any real number in a given range. Categorical factors use either character or numeric values. They are interpreted as nominal variables.

- **levels**: For continuous factors, the user specifies the range of the factor as a numeric vector containing minimum and maximum. For categorical factors, a vector with all possible factor levels—either numeric or characters—is required.
- **number.levels**: The number of levels is only required for continuous factors. This integer represents how many different levels are used during the design generation. For example, for *number.levels* = 3 and *levels* = $c(100, 300)$, the factor levels that are used during the design updating are 100, 200 and 300. For *number.levels* = 5 and *levels* = $c(100, 300)$, the possible factor levels are 100, 150, 200, 250 and 300. The number of levels has a strong impact on the run-time of the algorithm.
- **changes**: This defines if a factor is *'easy'*-, *'hard'*- or *'semi.hard'*-to-change.
- **semi.htc.group.size**: This is only required for SHTC-factors. This integer defines how many different factor levels can be used inside of one whole plot.

17.5.2 The DoeDesign Class

The *doeDesign*-class is the heart of **rospd**-package. It stores all information required to generate an optimal design and will contain the final design table afterwards as well. In this way, any *doeDesign*-object can serve as a documentation of how the design was created. There is a function **GenerateNewDoEDesign** that should be used to initialize a new *doeDesign*-object. A *doeDesign*-object for the current example might look like this:

```
doeSpecification <- GenerateNewDoeDesign(
  factors = list(solventFactor, phFactor, timeFactor),
  whole.plot.structure = data.frame(solvent=
    rep(1:6, each=10)),
  number.runs = as.integer(60),
  design.model = ~solvent+pH+time,
  optimality.function = DOptimality)
```

doeDesign is a S4-class. As it uses a lot of different slots only the required ones will be discussed here.

- **factors**: A list of all factors that are used in the DoE as objects of class *doeFactor*. This list can contain one or more items.
- **whole.plot.structure**: The whole plot structure is a data.frame representing when changes can be made for all HTC- and SHTC-factor. This data.frame needs to contain one column for each HTC- and SHTC-factor using the factor names as column names. The whole plots of each factor are represented by index numbers starting at 1 for each factor.
- **number.runs**: The number of experiments as *integer*.
- **design.model**: The design model as a *formula*. The formula is supposed to contain only the fixed effects part of the model as the random effects part is already defined by the **whole.plot.structure**.

- **optimality.function**: a *function* that can be used to calculate the optimality of a given design. The algorithm will use the *doeDesign*-object as the only argument of the function. Making use of the slots **design.matrix, design.model** and **variance.ratio** of the class *doeDesign* should allow to calculate most optimality criteria. The **rospd**-package provides a predefined function to calculate D-Optimality. The function is a wrapper of the C-implementation of the **skpr**-package (Morgan-Wall and Khoury 2018).
- **variance.ratio**: The expected ratio of σ^2 (residual variance) and σ_w^2 (between whole plot variance). This is required for the calculation of some optimality criteria. By default a variance ratio of 1 is used.

17.5.3 Generating Optimal Designs

The function **GenerateOptimalDesign** generates an optimal design for the given specification. The following code generates a D-optimal Semi-Split-Plot design as defined in Sect. 17.5.2.

```
optimalDesign <- GenerateOptimalDesign
(doeSpecification)
```

The function uses the following arguments:

- **docSpec**: The DoE specification as an object of class *doeDesign*.
- **random.start**: This argument controls how many different random starting designs are used. The default setting is 1. The function will return a list of one doeDesign for each random start. The designs in this list will be sorted from best to worst in terms of their optimality.
- **max.iter**: The maximum number of iterations. One iteration updates all cells of the design table once.

Table 17.3 shows a subset of the generated design.

17.6 Design Evaluation

This section evaluates the quality of the generated design. For that purpose, we will compare the Semi-Split-Plot design with three alternative designs. Furthermore, multiple Semi-Split-Plot designs with different numbers of whole plots are compared to each other.

Table 17.3 Part of the semi-split-plot design

Whole plot	Solvent	pH	Time		Whole Plot	Solvent	pH	Time
1	B	12	20		⋮	⋮	⋮	⋮
1	B	3	10		5	D	3	10
1	F	3	20		5	E	3	10
1	C	12	10		5	E	12	10
1	D	3	10		5	E	12	20
⋮	⋮	⋮	⋮		6	A	12	20
2	E	12	20		⋮	⋮	⋮	⋮
2	A	12	20		6	B	3	20
2	A	12	10		6	F	12	20
2	E	3	20		6	E	12	10
⋮	⋮	⋮	⋮		6	F	12	10

17.6.1 Sample Preparation Robot Example

The previously generated Semi-Split-Plot design will be compared to the three alternative solutions:

- a **completely randomized design** (CRD).
- a standard **Split-Plot** design using 12 whole plots each of size 5.
- a standard **Split-Plot** design using 8 whole plot. To end up with 60 runs, 4 of the whole plots use size 7 and 4 of the whole plots are of size 8.

The rationale here is to have the CRD as the best case scenario. As the CRD does not include any restrictions on randomization and assumes completely independent runs it will always provide the best statistical properties. As discussed in Sect. 17.3, this design would not be chosen in the real world.

The Split-Plot design with 12 whole plots is not that realistic as well, as it involves much more interactions of the operator with the machine than we would like to do—12 instead of only 6 for the Semi-Split-Plot design. A Split-Plot design with eight whole plots is more feasible but has very limited precision for the whole-plot-to-whole-plot-variance estimator.

Figure 17.2 shows that the CRD, the Semi-Split-Plot and the Split-Plot design using 12 whole plots are perfectly balanced. For the Split-Plot design using eight whole plots, it is not possible to achieve balance for the factor solvent.

The aliasing heat maps in Fig. 17.3 show that there are only minor differences when comparing the designs by their inter-factor correlations. Blue cells in the figure represent low correlations while red cells are representing high correlations. All designs show some correlation between the effects of the different solvents. The

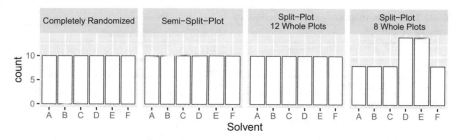

Fig. 17.2 Frequencies of solvent types

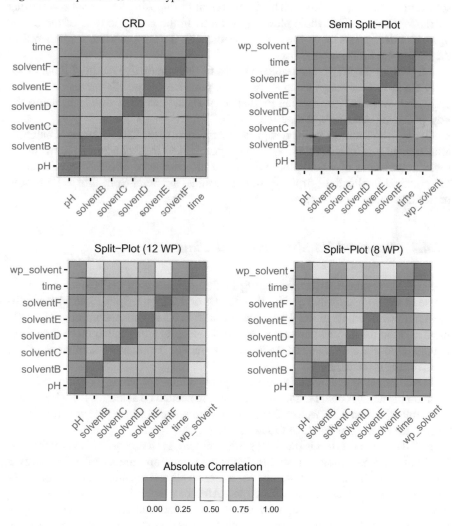

Fig. 17.3 Correlations of factors and whole plots

Table 17.4 Comparison of prediction variances

Design	Median prediction variance
CRD	0.5013
Semi-split-plot	0.5013
Split-plot 8 whole plots	0.5106
Split-plot 12 whole plots	0.5013

Split-Plot design with eight whole plots shows a little less homogeneous correlation structure of the solvent effects. This is the result of imposing very strict restrictions on randomization. The whole plot (**wp_solvent** in the graph) in case of the Semi-Split-Plot design has much lower correlation with the solvent effects compared to the traditional Split-Plot designs.

All designs are very comparable in terms of their predictive quality (Table 17.4). Only the Split-Plot design using eight whole plots has marginally increased median prediction variance. The median prediction variance has been simulated using the **vdg**-package (Schoonees et al. 2016).

Overall, the Semi-Split-Plot design is mostly equal to the CRD in terms of balance, aliasing and predictive quality. When comparing it to the Split-Plot designs, much lower correlations of whole plot and solvent effects can be observed. Considering the practical implications of the different designs, the Semi-Split-Plot is by far the preferred solution for the given problem.

17.6.2 Required Number of Whole Plots

The previous evaluation showed that Split-Plot designs with SHTC-factors are not worse in terms of statistical quality than CRDs or traditional Split-Plots with a sufficient number of whole plots. When dealing with HTC-factors, the number of whole plots is often the critical parameter. As SHTC-factors allow much more flexibility compared to true HTC-factors, we are able to reduce the required number of whole plots for the given example.

In traditional Split-Plot designs, the required number of whole plots depends first of all on the number of HTC-factor levels. There need to be at least $l + 1$ whole plots to be able to estimate the factor effects and the in-between whole plot variation. With l being the number of levels of the HTC-factor.

For SHTC-factors, this is much less of a problem, as changing SHTC-factor levels inside of the whole plots is possible. Thus the minimum number of whole plots is only depending on the overall number of factor levels and the number of factor levels that can be used inside of one whole plot. All factor levels have to be used at least once. At the same time, the correlation of whole plots and all other factors should be minimized to be able to estimate the model effects with high precision and avoid aliasing of whole plot effects and SHTC-factor effects.

In the following, we compare Semi-Split-Plot designs for the previous example with varying number of whole plots. Figure 17.4 shows that the algorithm is not

Fig. 17.4 Frequencies of solvent types for varying numbers of whole plots

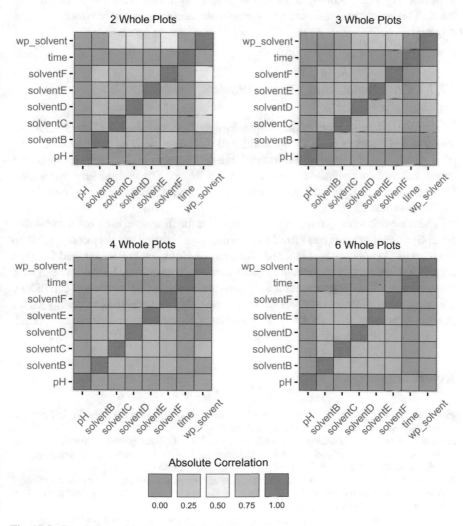

Fig. 17.5 Correlations of factors and whole plots

able to find a solution granting perfect balance in terms of the SHTC-factor when using only two whole plots. This problem is less severe when using three whole plots instead of just two. That statement is only true for the given example. In general, the minimum number of whole plots to achieve perfect balance and minimize correlation of whole plots and the SHTC-factor depends on the size of the whole plots and on the number of possible levels in each whole plot.

The design with two whole plots shows some correlation between solvent effects and the whole plot. This correlation is reduced with every additional whole plot introduced. Going from two whole plots to three whole plots reduces the correlation a lot (see Fig. 17.5). Adding more whole plots does not decrease the correlation by much. This is not a problem as the three whole plot design shows very low correlations anyways.

17.7   Conclusion and Future Work

We introduced the concept of Semi-Split-Plot designs and semi-hard-to-change factors. An algorithm to generate optimal Semi-Split-Plot designs was proposed. The implementation of this algorithm is available on Github https://github.com/neuhier/rospd and will be published on CRAN in the future. The current implementation is done mostly in R and the performance could be improved by changing to a C-implementation.

The comparison of multiple design classes for the described use case showed that Semi-Split-Plot designs can provide statistical qualities that are very comparable to the ones of Completely Randomized Designs. As Completely Randomized Designs are not feasible for this application, the comparison to traditional Split-Plot designs is more relevant. The design evaluation showed that Semi-Split-Plot designs can reduce the correlation of factors and whole plots. Thus, they are preferable over Split-Plot designs with regard to statistical quality and feasibility whenever they can be applied.

References

Fisher, R. A. (1992). *Statistical methods for research workers* (pp. 66–70). New York: Springer. https://doi.org/10.1007/978-1-4612-4380-9_6.

Hooks, T., Marx, D., Kachman, S., & Pedersen, J. (2009). Optimality criteria for models with random effects. *Revista Colombiana de Estadistica 32*, 17 – 31. http://www.scielo.org.co/scielo.php?script=sci_arttext&pid=S0120-17512009000100002&nrm=iso.

Jones, B., & Goos, P. (2007). A candidate-set-free algorithm for generating d-optimal split-plot designs. *Journal of the Royal Statistical Society: Series C (Applied Statistics), 56*(3), 347–364. https://doi.org/10.1111/j.1467-9876.2007.00581.x..

Jones, B., & Goos, P. (2011). *Optimal design of experiments* (pp. 277–282). Wiley-Blackwell. https://doi.org/10.1002/9781119974017.biblio, https://doi.org/10.1002/9781119974017.biblio.

Jones, B., & Goos, P. (2012). I-optimal versus D-optimal split-plot response surface designs. Working Papers 2012002, University of Antwerp, Faculty of Applied Economics. https://ideas.repec. org/p/ant/wpaper/2012002.html.

Jones, B., & Nachtsheim, C. J. (2009). Split-plot designs: What, why, and how. *Journal of Quality Technology, 41*(4), 340–361. https://doi.org/10.1080/00224065.2009.11917790.

Kowalski, S. M., & Potcner, K. J. (2003). *How to recognize a split-plot experiment.* https:// onlinecourses.science.psu.edu/stat503/sites/onlinecourses.science.psu.edu.stat503/files/ lesson14/recognize_split_plot_experiment/index.pdf. Accessed July 29 2018.

Meyer, R. K., & Nachtsheim, C. J. (1995). The coordinate-exchange algorithm for constructing exact optimal experimental designs. *Technometrics 37*(1), 60–69. http://www.jstor.org/stable/ 1269153.

Miller, A.J., & Nam-Ky, N.: A fedorov exchange algorithm for d-optimal design. Journal of the Royal Statistical Society: Series C (Applied Statistics) *43*(4), 669 (1994). http:// search.ebscohost.com/login.aspx?direct=true&db=plh&AN=6119956&site=ehost-live& authtype=sso&custid=s87180.

Morgan-Wall, T., & Khoury, G. (2018). skpr: Design of experiments suite: Generate and evaluate optimal designs. https://CRAN.R-project.org/package=skpr. R package version 0.49.1.

Næs, T., & Aastveit, A., & Sahni, N. (2007). Analysis of split-plot designs: An overview and comparison of methods. *23*, 801–820.

R Core Team. (2018). R: A language and environment for statistical computing. R Foundation for Statistical Computing, Vienna, Austria. https://www.R-project.org/.

SAS. (2018). Split plot designs with different numbers of whole plots. https://www.jmp. com/support/help/14/split-plot-designs-with-different-numbers-of-who.shtml. Accessed July 25 2018.

Schoonees, P., le Roux, N., & Coetzer, R. (2016). Flexible graphical assessment of experimental designs in R: The vdg package. *Journal of Statistical Software 74*(3), 1–22. https://doi.org/10. 18637/jss.v074.i03.

Chapter 18
Continuous Process Monitoring Through Ensemble-Based Anomaly Detection

Jochen Deuse, Mario Wiegand and Kirsten Weisner

Abstract In many production processes, a complete quality inspection of all products is not feasible due to technological and organizational restrictions. In order to ensure zero-defect products, monitoring process parameters in real time and using them to predict product quality by supervised learning methods is a very established approached. However, this approach requires a joining of process parameters and quality features. In order to guarantee high-quality products even in the absence of traceability, a continuous process monitoring approach based on an anomaly detection ensemble method is beneficial.

18.1 Introduction

Due to increasing competition, offering high-quality products is a key success factor to ensure competitive advantages in the long term. Guaranteeing defect-free products through extensive quality testing is therefore essential. However, in many processes, the testing time is a multiple of the cycle time and thus quality testing becomes the bottleneck of production processes (Deuse et al. 2017). In addition, some tests can only be carried out offline under laboratory conditions making quality testing expensive and time-consuming. In order to still realize short throughput and delivery times as well as low testing costs, a lot of companies rely on sample testing instead of testing all products. Procedures such as Statistical Process Control (SPC) are widely known. For evaluating product and process quality SPC is based on random samples taken at periodic intervals, which are used to measure the quality characteristics of interest. Depending on the interval this can lead to delayed detections of product

J. Deuse (✉) · M. Wiegand · K. Weisner
Fakultät Maschinenbau, Technische Universität Dortmund, Dortmund, Germany
e-mail: jochen.deuse@ips.tu-dortmund.de

M. Wiegand
e-mail: mario.wiegand@ips.tu-dortmund.de

K. Weisner
e-mail: kirsten.weisner@ips.tu-dortmund.de

© Springer Nature Switzerland AG 2019
N. Bauer et al. (eds.), *Applications in Statistical Computing*,
Studies in Classification, Data Analysis, and Knowledge Organization,
https://doi.org/10.1007/978-3-030-25147-5_18

defects (Dörmann et al. 2009), so that in the worst case, a large number of defective products are manufactured between two samples.

Modern production processes are increasingly equipped with a large number of different sensors, which enable real-time recording of process parameters. To gain knowledge from these data, the use of so-called soft sensors or virtual sensors is widespread in research and industry. Soft sensors are models of the process allowing to predict the values of difficult-to-measure quality features by easy-to-measure process parameters (Kaneko and Funatsu 2013; Zimek et al. 2014). As a result, a continuous process monitoring can be realized to detect product defects directly after their emergence by analyzing process parameters in real time. To establish a functional relationship between process parameters and quality characteristics, physical models or supervised machine learning methods are used.

In literature, many applications of supervised learning methods for soft sensor construction were proposed in the last three decades. For example, Chen et al. (2007), Yang et al. (2005), and Shi et al. (2004) use Artificial Neural Networks (ANN) for quality prediction in the computer and electronic industry. In metal processing, fuzzy regressions are used to predict wead geometry in robotic welding (Sung et al. 2007). Further publications refer to plastic manufacturing industry (Ozcelik and Erzurumlu 2006; Shen et al. 2007), the chemical process industry (Guessasma et al. 2004), and health care (Austina et al. 2013). In supervised learning, a function is learned to map the inputs to the outputs. This requires a matching between inputs and outputs, at least for a sufficiently large historical dataset. However, a direct matching between process parameters and quality features is not always possible in production processes due to a lack of traceability. In these cases, supervised learning cannot be applied for process monitoring.

Unsupervised learning methods do not need a labeled dataset on the contrary. Particularly, unsupervised methods of anomaly detection provide an opportunity to detect unusual pattern in the process data that can indicate the emergence of product defects. Anomaly detection, synonymously also referred to as outlier recognition, serves to identify observations that deviate greatly from other observations, so that it can be assumed that they are produced by faulty processes (Hawkins 1980). In industrial production, anomaly detection is therefore used to detect faulty states of systems, subsystems, and components. The aim of this, among other things, is to avoid unplanned failures or the production of defective goods. The output of an anomaly detection method is usually an anomaly score indicating the outlierness of the instances.

Concerning quality management literature some applications of unsupervised anomaly detection have been proposed. Most of the work is closely related to SPC, where anomaly detection algorithms are combined with control charts to detect abnormal process states. Sun and Tsung (2003), Kumar et al. (2006) and Kim and Kim (2018) propose variants of the one-class support vector machine (OCSVM) to identify deviations from normal process behavior. Other approaches are based on the k-means algorithm (Gani and Limam 2014), the k-nearest neighbor data descriptor (KNNDD) (Gani and Limam 2013; Liu et al. 2015), or combined methods (Liu et al. 2008). As these approaches are based on unsupervised learning, process anomalies can be detected without a functional mapping from inputs to outputs. Therefore, a labeled dataset for learning this function is not required. However, unsupervised approaches usually assume a direct cause–effect relationship between process anomalies and product defects. Whenever an unusual process state occurs, the emergence of a product defect is presumed.

The work described in the related quality management literature solely focuses on the use of specific anomaly detection methods such as OCSVM or KNNDD. The performance of these learning methods, however, is naturally based on the inherent structure of the dataset. Consequently, it cannot be ensured that the methods function comparably well for different real-world problems. In order to combine the advantages of different methods and to guarantee a robust behavior for different use cases and datasets, the use of so-called ensembles is a popular option in machine learning (Aggarwal 2012). An ensemble is based on the idea that not a single model, but a number of different models are trained and that the results of the models are combined for decision-making.

There are numerous works in literature, in which ensemble methods are proposed. The majority of these works deal with methods of supervised learning (e.g., Caruana et al. 2004; Chen and Zhao 2008; Soares and Araújo 2015). However, in recent years, the principles of ensemble learning have also gained increasing importance in the field of unsupervised learning. The focus of development is on ensemble clustering (e.g., Abbasi et al. 2018; Vega-Pons and Ruiz-Shulcloper 2011). However, there are only a few approaches for unsupervised anomaly detection ensembles (Zimek et al. 2014), whereby no one of these is in the field of quality management.

In this paper, an innovative method for constructing an unsupervised anomaly detection ensemble by transferring the principles of ensemble learning to unsupervised anomaly detection is proposed. This method allows to realize a continuous process monitoring in processes with lacking traceability in order to detect product defects in real time. The paper is structured as follows: Sect. 18.2 presents an approach to train an anomaly detection ensemble with maximum diversity and integrate it into a production process. An application of this approach to a real-world-forming process is provided in Sect. 18.3. Final conclusions are made in Sect. 18.4.

18.2   Process Monitoring with Unsupervised Anomaly Detection Ensembles

In literature, it has been shown that model accuracy and diversity play a major role in ensemble learning (Abbasi et al. 2018; Zimek et al. 2014). On the one hand, the members of the ensemble have to deliver a high overall accuracy of the ensemble. On the other hand, the members have to be diverse to ensure that errors of individual models are compensated by others. In the context of learning theory, increasing models' heterogeneity in an ensemble is synonymous with reducing variance in the logic of bias–variance decomposition (Banfield et al. 2005). The case of unsupervised anomaly detection in production processes quantifying the models' accuracy directly is not possible, since the ground truth (gold standard) is usually not known due to the lack of traceability. Accordingly, only the criterion of diversity explicitly can be considered. For this reason, a number of different models must first be trained in ensemble learning. A subset of these models can then be selected to further increase the variety. Finally, the individual models are combined into an ensemble. After ensemble learning, the entire model must be integrated into the production process to enable real-time process monitoring. These steps are explained in more detail in the following Sections.

18.2.1   Model Training

To achieve diversity, different approaches for model training are proposed in the literature (Zimek et al. 2014):

1. Training on different subsets of attributes.
2. Training on different subsets of objects.
3. Using an algorithm with varying parameter sets.
4. Using different algorithms.

The first two approaches are based on the idea of using different datasets for training individual models. Common ensemble methods like Random Forest or Gradient Boosted Trees follow this idea. In contrast, the last two approaches adapt the learning algorithms directly to induce diversity. Zimek et al. (2014) state that in particular, the use of different, non-congeneric learning methods is promising to improve ensemble performance. For this reason, this approach is adopted here. The authors also claim, that variation of parameter settings of the same learning method leads to homogenous models, especially if it is a more or less stable learning method. Nevertheless, in addition to the application of different learning methods, we also use parameter adjustments to generate a large number of individual models. To avoid combining homogenous models in the ensemble, a methodology for model selection is presented in Sect. 18.2.2.

The first step in our ensemble learning strategy is to define the learning methods and the parameters to be varied as well as the range of variation. As described above, a direct quantification of the model accuracy is not possible in the context of unsupervised anomaly detection. However, in order to implicitly improve the model accuracy, we choose only learning methods for model training that are expected to provide a comparatively high accuracy on average. The most current and largest study on the performance of different unsupervised learning methods for anomaly detection was conducted by Goldstein and Uchida (2016). Based on ten benchmark datasets, they compared different unsupervised anomaly detection methods by using different criteria. The k-nearest neighbor (k-NN) (Ramaswamy et al. 2000), Local Outlier Factor (LOF) (Breunig et al. 2000), Local Outlier Probability (LoOP) (Kriegel et al. 2009), and unweighted Cluster-Based Local Outlier Factor (uCBLOF) (He et al.

2003) methods performed particularly well in terms of accuracy (the ground truth of the datasets was known for evaluation). In addition, k-NN and uCBLOF allow for detecting global outliers, while LOF and LoOP focus on local outliers. For these reasons, these four methods will be used here to train the ensemble model. Afterward, the rough functionality of the methods will be briefly outlined.

As the name implies, the k-NN is a nearest neighbor-based anomaly detection technique. Different variants of the k-NN have been proposed in the last 20 years. Here, we refer to the version of Ramaswamy et al. (2000). In (Ramaswamy et al. 2000), the distance $D^k(p)$ of a point from its kth nearest neighbor is calculated and subsequently all points are ranked according to $D^k(p)$. For calculating $D^k(p)$ any distance measure like the L_1 ("manhattan") or the L_2 ("euclidean"), can be used. Weihs and Szepannek (2009) point out that no distance measure performs equally well for all datasets, but that the distance has to be chosen for every problem individually. The top n points in the resulting ranking are regarded as outliers. The parameters k and n have to be defined by the user.

The LOF is another nearest neighbor-based anomaly detection technique. In contrast to k-NN it is able to detect local outliers that are anomalous compared to its close-by neighborhood (Goldstein and Uchida 2016). To calculate the LOF score, the k-nearestneighbors N_k of a point have to be found and used for calculating the Local Reachability Density (LRD) (Breunig et al. 2000; Goldstein and Uchida 2016):

$$LRD_k(p) = 1 / \left(\frac{\sum_{o \in N_k(p)} d_k(p, o)}{|N_k(p)|} \right), \tag{18.1}$$

Whereas, $d_k(p, o)$ is the reachability distance, which in most cases is the Euclidean distance. Finally, the LOF score is calculated by putting the LRD of point p in relation to the LRD of its k-nearest neighbors (Breunig et al. 2000; Goldstein and Uchida 2016):

$$LOF(p) = \sum_{o \in N_k(p)} \frac{LRD_k(o)}{LRD_k(p)} / |N_k(p)|. \tag{18.2}$$

Therefore, the LOF score can be interpreted as a ratio of local densities. An LOF score near to 1 indicates a normal example with a density comparable to the density of its neighbors, whereas an anomaly will have an LOF score much larger than 1.

The LoOP has a similar principle as the LOF. However, instead of calculating an anomaly score, the LoOP defines an anomaly probability in the range [0, 1]. The local density is estimated on the basis of the so-called probabilistic set distance by assuming a multivariate Gaussian distribution. As with the LOF the density is compared to the local densities of the neighbors and finally, the result is converted to a local outlier probability (Kriegel et al. 2009).

In contrast to the three nearest neighbor-based methods, the uCBLOF is a clustering-based method that utilizes an arbitrary clustering algorithm to find clusters in the dataset. The uCBLOF divides the clusters into small and large clusters and calculates an anomaly score based on the distance between an example and the cluster center. If an example is part of a small cluster, the distance to the closest large cluster is calculated and used for anomaly score calculation (Goldstein and Uchida 2016; He et al. 2003).

In order to induce diversity into the ensemble, all four learning methods for model training are used. Furthermore, the methods' parameter settings are varied to further increase the model set. The appropriate parameters are varied as well as the range of variation depends strongly on the individual dataset. The general suggestion is to vary the distance measure in any case, since all the used methods define outliers on the basis of distance calculations. Besides, the parameter k of the nearest neighbor-based methods seems to be a reasonable choice for variation. Depending on the clustering algorithm selected for uCBLOF, various parameters may be suitable for inducing diversity. Here, the k-means algorithm was chosen due to its popularity and computational efficiency, so that the number of clusters is a further main parameter to be varied. In addition, different n in k-NN were tried.

18.2.2   Model Selection

As mentioned earlier, even though different diversity induction procedures during model training have been used, the models can produce similar outputs. This is, especially, true for models that have been trained with the same learning method and only small differences in the parameter settings. But also different learning methods can lead to similar models, especially if the methods are based on a related learning principle. This applies, for example, to the related methods LoF and LoOP.

Similar models are characterized by the fact that they assign similar outlier scores to the instances of the dataset. As a result, the outlier scores of these models are strongly correlated. In order to ensure a high heterogeneity, correlated models are not included in the ensemble. Instead, only one model of each group of correlated models is kept. We always choose the simplest model according to Occam's razor. Here, the correlation of two models is determined by the Bravais–Pearson correlation coefficient. The coefficients range from -1 to 1, with values close to ± 1 indicating a high negative/positive correlation. Models that show a correlation below/above a defined threshold are therefore eliminated from the model set. The threshold has to be defined manually. In the case study in Sect. 18.4, a coefficient of ± 0.85 as threshold was used.

18.2.3   Model Combination

After model selection, a set of low correlated models remains. As a consequence of using different learning methods for training, the score vectors of the different models differ in scale. The values of the LoOP, for instance, are bound between 0 and 1, whereas the scores of the LOF can take arbitrarily large values. In order to make the scores of the individual models comparable, they must be normalized before combination. For this, there exist several methods, like the z- or range transformation. After normalization, the models can be combined into an ensemble. Various mathematical functions are available for combining the scores. Aggarwal (2012) discusses different possibilities and refers to the maximum and average function as most frequently used. Zimek et al. (2014) take up the discussion and explain why the maximum function can lead to an overestimation of the outlierness and thus to a distortion of the results. In agreement with Zimek et al. (2014), the average function is much better suited to combine the individual models, as errors of individual models are compensated by other models. With the maximum function, on the other hand, the overall ensemble decision is based on one individual model, even if all other models differ significantly from it. Therefore, the models' scores by the average function are combined to an "average anomaly score".

Model combination is the last step in ensemble learning. Before the ensemble can be used in the real-world production process, however, a threshold value has to be defined for the average anomaly score, which distinguishes between anomalous and normal process states. Statistical procedures, domain knowledge or rules of thumbs, such as the three sigma rule or the elbow criterion, can be used to define the decision boundary.

18.2.4 Integration into Production Process

For implementing the ensemble into the production process, different ways exist. One possibility is to transfer the recorded process parameters in real time to an external computer, which calculates the average anomaly score and returns it to the Programmable Logic Controller (PLC). If the score's threshold value is exceeded, the PLC has to eject the product from the running process. Another option is to perform the calculations directly in the PLC. The advantage of this approach is that calculation and control decision are both realized within the PLC, so that transmission times between PLC and an external computer are avoided. This is particularly important, if cycle times are short. However, it must be taken into account that PLCs often only allow simple calculations and are not capable of complex distance calculations for determining the model scores in real time. As opposed to this, simple calculations, e.g., simple if-then statement as in supervised decision tree models, are technically feasible within the PLC. In addition to this, decision trees have, in general, a shorter runtime than distance-based anomaly detection methods, which have to perform a large number of distance calculation within the online phase.

Therefore, it is necessary to transfer the decision structure of the anomaly ensemble into a decision tree. For this purpose, the ensemble is applied to the unmarked training data set and the instances are marked with the classes OK and NOK according to the ensemble decision. This creates a labeled dataset that can be used for training in the decision tree. In order to improve the model accuracy, not only one decision tree has to be trained, but several decision trees have to be trained and combined with a decision tree ensemble. As a rule, only a few products in a process have quality defects, so that the data set is often very unbalanced. To take this into account, the MetaCost method proposed by Domingos (1999) should be used here. The idea behind this method is to assign different costs to different types of classification errors. In addition, the method uses a modification of the bagging approach for learning ensembles (Breiman 1994). In this way, it is possible to train a cost-sensitive decision tree ensemble that punishes false positives harder than false negatives to compensate for class imbalances. Furthermore, the decision structure of the anomaly detection ensemble can be mimicked as closely as possible with a decision tree ensemble to enable direct integration into the PLC.

18.3 Case Study

The case study examined is an assembly process with a cycle time of only a few seconds. Although the process has a significant influence on the final product quality, due to relatively high testing times only a small sample of the products is quality tested every day. That is why quality problems in the process can be only detected with a large time delay. Whenever a quality error occurs during testing, the process is adapted by the operators on the basis of experience-based knowledge, e.g., by parameter adjustments. Various process parameters are currently recorded and stored in a SQL database. However, due to missing traceability, a join of process parameters and quality features is not possible, so that only little is known explicitly about the numerical relationship between parameters and quality defects. To enable an in-process detection of quality problems even so, the aim of the case study was to develop an anomaly detection system that allows for a continuous monitoring of the process, so that quality problems can be detected at the time of their emergence. For this purpose, the existing process data were examined and evaluated with the methodology described in Sect. 18.2. The case study itself is divided into two successive phases: preparation phase and integration phase. The aim of the preparation phase was to train the anomaly detection ensemble in order to find anomalous process states and label the dataset to prepare for supervised learning. In the integration phase, the decision tree ensemble is trained to mimic the decision structure of the anomaly detection ensemble and implement it into the real-world process. Both phases are realized in RapidMiner 9.0 on an Intel Xeon Processor E5-2670 (16 cores, 2.60 GHz, 128 GB RAM).

18.3.1 Offline Phase

Before learning the ensemble, the data had to be collected and cleaned up. Subsequently, the relevance of several process parameters was discussed together with process experts and six parameters could be selected which should be decisive for the product quality. In addition, a sufficiently representative sample of approximately 300,000 instances was drawn. In order to avoid distorted results, noisy instances (instances with unrealistic values in at least one of the process parameters) were deleted and correlated parameters were eliminated according to Pearson's correlation coefficients and based on process knowledge of domain experts, so that five parameters were retained at the end.

After data cleansing, the models for anomaly detection could be trained with the four methods k-NN, LOF, LoOP and uCBLOF and different parameters. Table 18.1 gives an overview of the parameter settings used.

In total, 108 models were trained. To ensure comparability, the model values were normalized by performing a range transformation to the interval [0, 1]. The correlations between the results of the individual models could then be calculated and only uncorrelated models were selected for the combination to induce diversity (correlation coefficient > 0.85). The remaining eleven models were combined to an average anomaly value. In order to define a score's threshold value process, experts decided for the 0.1% highest score as decision boundary, as the elbow criterion did not give satisfying results.

Before the actual integration, the model is to be validated. Although the traceability was not given, in this case at least some offline tests for the artificial simulation of product defects could be carried out and the products could be tracked manually. In this way, the relationship between process parameters and product quality could be maintained. Since the running production process had to be stopped for these

Table 18.1 Parameter setting

Method	Parameter	Range of variation	No. of variation steps	Variation scale
k-NN	k	10–50	10	Logarithmic
	Distance metric	Euclidean, Manhattan, Cosinus, Canberra		
LOF	k	10–50	10	Logarithmic
	Distance metric	Euclidean, Manhattan, Cosinus, Canberra		
LoOP	k	10–50	10	Logarithmic
	Distance metric	Euclidean, Manhattan, Cosinus, Canberra		
uCBLOF	k	10–50	10	Logarithmic
	Distance metric	Euclidean, Manhattan, Cosinus, Canberra		
	α	90–100	10	Linear
	β	1–100	10	Logarithmic

tests, only a few quality defects could be simulated. The ensemble was applied to a data set of approximately 300,000 products including the simulation runs. As a result, all simulation runs had an average anomaly value above the 0.1% limit and were correctly identified as NOK products. In contrast, the results of the individual models were comparatively worse. It is worth mentioning that all process parameters related to the simulated product defects were within their univariate tolerance limits defined on the basis of process knowledge and experience. Thus, the application of the proposed ensemble strategy allows the detection of multivariate anomalies with an unusual combination of in-tolerance parameters.

18.3.2   Online Phase

Until now, the construction of the ensemble and its application for validation purposes was carried out offline. In order to enable an anomaly detection in online mode, the model should run directly in the PLC. Since an PLC only allows comparatively simple calculations, the decision structure of the anomaly detection ensemble was transferred into an ensemble of decision trees. The individual trees were combined into the ensemble by majority vote. The MetaCost was used for model training. In order to consider class imbalance and guarantee the detection of all product defects, the MetaCost was initialized with costs of 1,000 for false positives compared to costs of 1 for false negatives. The confidence matrix of the decision tree ensemble based on a data subset is shown in Table 18.2.

It turns out that the decision tree ensemble can represent the decision structure of the anomaly detection ensemble reasonably well, but not perfectly. On the one hand, this is caused by the different learning principles of the anomaly detection methods and the decision tree method. On the other hand, the number of NOK-instances is comparatively small, so that prediction results are expected to improve as the dataset increases. Particularly, the class precision for the NOK-class is relatively low. However, NOK-predicted parts could be physically tested, in order to confirm or reject the model's decision. The corresponding expense is kept within reasonable limits. Due to its simple structure of if-then statements, the decision tree ensemble can easily be implemented into the PLC. As a result, the model only needs a few milliseconds for classifying a new instance into one of the classes, so that anomalies can be detected in cycle time and real-time control decision can be derived.

Table 18.2 Confusion matrix

		True label		Class precision
		True NOK	True OK	
Prediction	NOK	208	1.276	14.02%
	OK	24	74.064	99.7%
	Class recall	89.66%	98.31%	Accuracy 98.3%

18.4   Conclusion

In this paper, an approach for continuous process monitoring based on an anomaly detection ensemble was presented. Diversity was induced into the ensemble by using different learning methods for model training. Besides, the methods' parameter settings were varied to enlarge the model set. Models with similar outputs were eliminated from the model set based on the results of correlation analysis, in order to further increase diversity. Finally, the models' outputs were normalized and combined into an ensemble by averaging the anomaly scores. For allowing real-time monitoring we mimicked the ensemble's decision structure by transferring it into a decision tree ensemble and implemented it into the machine's PLC. We applied our proposed method on a real-world-forming process and showed its capability to identify process anomalies and detect product defects.

Literature

Abbasi, S., Nejatian, S., Parvin, H., Rezaie, V., & Bagherifard, K. (2018). Clustering ensemble selection considering quality and diversity. *Artificial Intelligence Review*, 1–30.

Aggarwal, C. C. (2012). Outlier ensembles. *SIGKDD Explorations, 14*(2), 49–58.

Austina, P. C., Tua, J. V., Hoe, J. E., Levye, D., & Lee, D. S. (2013). Using methods from the data-mining and machine learning literature for disease classification and prediction: A case study examining classification of heart failure subtypes. *Journal of Clinical Epidemiology, 66*(4), 398–407.

Banfield, R. E., Hall, L. O., Bowyer, K. W., & Kegelmeyer, W. P. (2005). Ensemble diversity measures and their application to thinning. *Information Fusion, 6*, 49–62.

Breiman, L. (1994). *Bagging predictors*. Technical report no. 421. Department of Statistics, University of California.

Breunig, M. M., Kriegel, H.-P., Ng, R. T., & Sander, J. (2000). LOF: Identifying density-based local outliers. In *Proceedings of the ACM SIGMOD International Conference on Management of data* (pp. 93–104).

Caruana, R., Niculescu-Mizil, A., Crew, G., & Ksikes, A. (2004). Ensemble selection from libraries of models. In *Proceedings of the 21st International Conference on Machine Learning*.

Chen, Y., & Zhao, Y. (2008). A novel ensemble of classifiers for microarray data classification. *Applied Soft Computing, 8*, 1664–1669.

Chen, W. C., Lee, A. H. I., Deng, W. J., & Liu, K. Y. (2007). The implementation of neural network for semiconductor PECVD process. *Expert Systems with Applications, 32*(4), 1148–1153.

Deuse, J., Schmitt, J., Stolpe, M., Wiegand, M., & Morik, K. (2017). Qualitätsprognosen zur Engpassentlastung in der Injektorfertigung unter Einsatz von Data Mining. In N. Gronau (Eds.), Industrial Internet of Things in der Arbeits- und Betriebsorganisation. Wissenschaftliche Gesellschaft für Arbeits- und Betriebsorganisation (WGAB) (pp. 47–60).

Domingos, P. (1999). MetaCost: A general method for making classifiers cost-sensitive. In *Proceedings of the Fifth ACM SIGKDD International Conference on Knowledge Discovery and Data Mining* (pp. 155–164).

Dörmann, O., Hans W., Linß, G., Weckenmann, A., & Bettin, V. (2009). Ansatz für ein prozessintegriertes Qualitätsregelungssystem für nicht stabile Prozesse. Ilmenau.

Gani, W., & Limam, M. (2013). Performance evaluation of one-class classification-based control charts through an industrial application. *Journal of Quality and Reliability Engineering International, 29*, 841–854.

Gani, W., & Limam, M. (2014). A one-class classification-based control chart using the K-means data description algorithm. *Journal of Quality and Reliability Engineering, 39*(3), 461–474.

Goldstein, M., & Uchida, S. (2016). A comparative evaluation of unsupervised anomaly detection algorithms for multivariate data. *PLoS ONE, 11*(4).

Guessasma, S., Salhi, Z., Montavon, G., Gougeon, P., & Coddet, C. (2004). Artificial intelligence implementation in the APS process diagnostic. *Materials of Science and Engineering: B, 110*(3), 285–295.

Hawkins, D. M. (1980). *Identification of outliers*. London: Springer.

He, Z., Xu, X., & Deng, S. (2003). Discovering cluster-based local outliers. *Pattern Recognition Letters, 24,* 1641–1650.

Kaneko, H., & Funatsu, K. (2013). Adaptive soft sensor model using online support vector regression with time variable and discussion of appropriate parameter settings. *Procedia Computer Science, 22,* 580–589.

Kim, Y., & Kim, S. B. (2018). Optimal false alarm controlled support vector data description for multivariate process monitoring. *Journal of Process Control, 65,* 1–14.

Kriegel, H.-P., Kröger, P., Schubert, E., & Zimek, A. (2009). LoOP: Local outlier probabilities. In *Proceedings of the 18th ACM Conference on Information and Knowledge Management* (pp. 1649–1652).

Kumar, S., Choudhary, A. K., Kumar, M., Shankar, R., & Tiwari, M. K. (2006). Kernel distance-based robust support vector methods and its application in developing a robust K-chart. *International Journal of Production Research, 44*(1), 77–96.

Liu, X., Xie, L., Kruger, U., Littler, T., & Wang, S. (2008). Statistical-based monitoring of multivariate non-Gaussian systems. *AIChE Journal, 54*(9), 2379–2391.

Liu, Y., Pan, Y., Wang, Q., & Huang, D. (2015). Statistical process monitoring with integration of data projection and one-class classification. *Chemometrics and Intelligent Laboratory Systems, 149,* 1–11.

Ozcelik, B., & Erzurumlu, T. (2006). Comparison of the warpage optimization in the plastic injection molding using ANOVA, neural network model and genetic algorithm. *Journal of Materials Processing Technology, 171*(3), 437–445.

Ramaswamy, S., Rastogi, R., & Shim, K. (2000). Efficient algorithms for mining outliers from large data sets. In *Proceedings of the ACM SIGMOD International Conference on Management of Data* (pp. 427–438).

Shen, C., Wang, L., & Li, Q. (2007). Optimization of injection molding process parameters using combination of artificial neural network and genetic algorithm method. *International Journal of Materials Processing Technology, 183*(2–3), 412–418.

Shi, X., Schillings, P., & Boyd, D. (2004). Applying artificial neural networks and virtual experimental design to quality improvement of two industrial processes. *International Journal of Production Research, 42*(1), 101–118.

Soares, S. G., & Araújo, R. (2015). A dynamic and on-line ensemble regression for changing environments. *Expert Systems with Applications, 42,* 2935–2948.

Sun, R., & Tsung, F. (2003). A kernel-distance-based multivariate control chart using support vector methods. *International Journal of Production Research, 41*(13), 2975–2989.

Sung, B. S., Kim, I. S., Xue, Y., Kim, H. H., & Cha, Y. H. (2007). Fuzzy regression model to predict the bead geometry in the robotic welding process. *Acta Metallurgica Sinica (English Letters), 20*(6), 391–397.

Vega-Pons, S., & Ruiz-Shulcloper, J. (2011). *International Journal of Pattern Recognition and Artificial Intelligence, 25*(3), 337–372.

Weihs, C., & Szepannek, G. (2009). Distances in classification. In *ICDM 2009: Advances in data mining. Applications and theoretical aspects* (pp. 1–12).

Yang, T., Tsai, T., & Yeh, J. (2005). A neural network-based prediction model for fine pitch stencil printing quality in surface mount assembly. *Engineering Applications of Artificial Intelligence, 18*(3), 335–341.

Yao, L., & Ge, Z. (2017). Moving window adaptive soft sensor for state shifting process based on weighted supervised latent factor analysis. *Control Engineering Practice, 61,* 72–80.

Zimek, A., Campello, R. J. G. B., & Sander, J. (2014). Ensembles for unsupervised outlier detection: Challenges and research questions. *SIGKDD Explorations, 15*(1), 15–22.

Part VI
Statistics in Music Applications

Chapter 19
Evaluation of Audio Feature Groups for the Prediction of Arousal and Valence in Music

Igor Vatolkin and Anil Nagathil

Abstract Computer-aided prediction of arousal and valence ratings helps to automatically associate emotions with music pieces, providing new music categorisation and recommendation approaches, and also theoretical analysis of listening habits. The impact of several groups of music properties like timbre, harmony, melody or rhythm on perceived emotions has often been studied in literature. However, only little work has been done to extensively measure the potential of specific feature groups, when they supplement combinations of other possible features already integrated into the regression model. In our experiment, we measure the performance of multiple linear regression applied to combinations of energy, harmony, rhythm and timbre audio features to predict arousal and valence ratings. Each group is represented by a smaller number of dimensions estimated with the help of Minimum Redundancy–Maximum Relevance (MRMR) feature selection. The results show that cepstral timbre features are particularly useful to predict arousal, and rhythm features are the most relevant to predict valence.

19.1 Introduction

Music and listener emotions may impact each other in both directions. Depending on our mood, we prefer to listen to a particular music genre with a characteristic sound, but also listening to a music track with specific sound properties can generate emotions in us. For instance, louder pieces may be associated with joy and quieter pieces with sadness (Scherer 1982); the dynamics properties belonged to the most powerful features for emotion recognition in Saari et al. (2011).

I. Vatolkin (✉)
Department of Computer Science, Technische Universität Dortmund,
Dortmund, Germany
e-mail: igor.vatolkin@tu-dortmund.de

A. Nagathil
Institute of Communication Acoustics, Ruhr-Universität Bochum,
Bochum, Germany
e-mail: anil.nagathil@rub.de

© Springer Nature Switzerland AG 2019
N. Bauer et al. (eds.), *Applications in Statistical Computing*,
Studies in Classification, Data Analysis, and Knowledge Organization,
https://doi.org/10.1007/978-3-030-25147-5_19

305

The identification of emotions characterising given music pieces can improve automatic recommendation systems, either providing users with new music exploration possibilities or helping them to create playlists for different purposes like background music during the working hours, or entertaining pieces for a dance floor.

Models, which have been developed so far to classify the emotional impact of music pieces, assign them either to categorical or numerical values. A prominent example of the categorical methods is the model by Hevner, who has created a circle with eight groups of adjectives for music descriptions (Hevner 1930). Another categorisation into nine clusters was proposed by Schubert (Schubert 2004). The numerical approaches enable easier comparison and ordering of music pieces based on associated emotions. The related emotional adjectives can be then mapped to a point in the two-dimensional space of arousal and valence (Russell 1980). Arousal (the vertical axis) measures the level of activation (from low to high) and valence (the horizontal axis) quantifies the emotional affect (from negative to positive). Based on arousal and valence, emotions can be assigned to one of four quadrants. For example, 'happy' belongs to the quadrant with positive arousal and valence values, 'calm' to the quadrant with negative arousal and positive valence, 'sad' to the quadrant with negative arousal and negative valence and 'annoyed' to the quadrant with positive arousal and negative valence. Other four dimensions were introduced in Tellegen et al. (1999) (levels of positive affect, negative affect, engagement and pleasantness).

In recent years, several studies addressed automatic Music Emotion Recognition (MER), mostly from audio signals, but also using symbolic music data. In the past, the focus was more on the design of handcrafted features (Saari et al. 2011; Grekow 2018). Current works often apply complex methods like Deep Neural Networks (DNNs) (Malik et al. 2017), which perform both feature extraction and classification. However, for both approaches, less work was done on the extended comparison of different feature domains or feature groups. For audio signals, this can be achieved by considering, e.g. the time signal, spectrum, cepstrum, phase domain. Another advantage of differentiating between feature groups is that after measuring their impact on the perceived emotion, music scientists may provide better theoretical explanations of music properties which highly correlate with emotional descriptors.

In this study, we distinguish between five audio feature groups (cepstral timbre, energy, harmony, spectral and phase domain timbre, rhythm) and measure the performance of Multiple Linear Regression (MLR) for the prediction of arousal and valence, applied for all possible combinations of feature groups. In Sect. 19.2, we provide a brief overview of the studies on MER. Section 19.3 introduces features used in our experiments. The study is outlined in Sect. 19.4, and the results are presented and discussed in Sect. 19.5. The final remarks and ideas for future work are summarised in Sect. 19.6.

19.2   Related Work

In the following, we provide a selection of several related works sorted by their publication year. They are restricted to classification and regression methods based on audio features only, because the combination of different feature sources was beyond the scope of our study. However, the latter is also a promising research topic. For instance, (Panda et al. 2013) showed that the combination of audio, lyrics and symbolic (i.e. score-based) features leads to a higher F-measure when classifying music pieces into five emotional clusters. For a general introduction into MER, we refer to Yang and Chen (2011), and for an overview of recent studies, to Yang et al. (2018).

Probably the first work on automatic MER was published in 1988 by Katayose et al. (1988), with the target to predict sentiments based on audio signal analysis. 160 feature dimensions belonging to the groups timbral, loudness, harmonic and rhythmic were used to predict arousal and valence in Huq et al. (2010), however, no combinations of these groups were examined. In Saari et al. (2011), the authors distinguished between six audio feature groups related to music theory (dynamics, rhythm, pitch, harmony, timbre, structure) with 66 features at overall for recognition of four different emotions. With the help of feature selection, it was observed that mostly 'higher-level' features like mode majorness and key clarity were most useful for the classification of emotions. The extension of the 'standard' feature set (458 dimensions) with melodic features (98 dimensions) led to an increase of the regression performance for arousal and valence prediction in Panda et al. (2013), pointing out that the combination of different feature groups make sense. However, the standard set was also constructed with very different features, from low-level signal descriptors to tempo, tonality and key, so that individual contributions of such very different music properties could not be measured. Three groups of 654 features (low-level, rhythm, tonal) were compared in Grekow (2018) for the prediction of arousal and valence with the help of several regression methods, with the outcome, that combination of at least two groups of features led to a higher performance. Panda et al. presented a set of 'novel' features related to musical texture and performance expressive techniques, analysing sets of up to 800 features and coming to the conclusion that the extension of the baseline set (descriptors of rhythm, dynamics, melody, harmony, etc.) significantly improves the performance of classification into four quadrants of the arousal/valence space (Panda et al. 2018).

Some recent approaches apply deep learning for MER, in particular to predict arousal and valence. Schmidt and Kim compared deep belief networks trained with magnitude spectra to standard acoustic feature set (Schmidt and Kim 2011). Malik et al. distinguished between two input sources (baseline and raw) for deep network constructed with convolutional and recurrent layers, which were, however, grouped by the extraction technique and not the relationship to music theory (Malik et al. 2017).

Summarising our literature survey, we may state that even if combinations of different feature groups were examined in several works, either only a few groups were analysed (sometimes without relation to music theory), or the overall dimensionality of the final feature vector was limited to several hundreds.

19.3 Feature Groups

Audio features can be extracted from different signal representations. These include the time, spectral and cepstral domains. These representations are typically computed in a short-time fashion. However, in order to account for the temporal evolution of these short-time representations, for instance, their modulation spectrum can be computed.

Let us denote a discrete-time music signal as $x[n]$ sampled at the sampling frequency f_s with discrete-time index n, which is split into segments $x[\lambda, n] = x[\lambda R + n]$ with λ and R being the segment index and shift, respectively. Then we can obtain a time–frequency representation based on the sliding-window Discrete Fourier Transform (DFT)

$$X[\lambda, k] = \sum_{n=0}^{N-1} x[\lambda, n] \, w[n] \, \exp\left(-\frac{2\pi kn}{N}\right), \qquad (19.1)$$

where k, N and $w[n]$ denote the discrete frequency index, DFT analysis length and a (tapered) window function, respectively. The spectral information in this representation can be decomposed using the cepstral transform

$$X_c[\lambda, q] = \frac{1}{N} \sum_{k=0}^{N-1} \log\left(|X[\lambda, k]|\right) \, \exp\left(\frac{2\pi kq}{N}\right), \qquad (19.2)$$

where q is the cepstral index. This transform maps the log-spectral envelope, which contains timbre information, onto cepstral coefficients with low q values. The spectral fine structure including fundamental frequencies and harmonics is described by coefficients with high values of q. A further sliding-window DFT analysis across the segments of the spectrum or cepstrum results in the spectral or cepstral modulation spectrum (Martin and Nagathil 2009), respectively.

For an illustration of different signal domains Fig. 19.1 shows the spectrogram, cepstrogram and cepstral modulation spectrum of an exemplary music piece, where t, f and f_{mod} denote the time, acoustic frequency and modulation frequency, respectively.

In the following, we briefly outline signal features commonly used in music information retrieval. For the sake of brevity, we refrain from a detailed presentation of all features used in this study but only introduce them descriptively in the context of

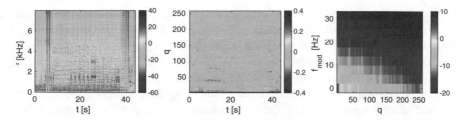

Fig. 19.1 Spectrogram, cepstrogram and cepstral modulation spectrum (from left to right) of an exemplary music signal (B. Whitfield and The Savages—'Bip Bop Bip' from '1000 Songs' collection, Soleymani et al. 2013). The cepstral transform maps the spectral envelope and the spectral fine structure on different coefficients. This allows for a better separation of timbre and pitch/harmony characteristics

feature groups. For a more detailed presentation, the interested reader is referred to, e.g. Nagathil and Martin (2016).

Based on the time, spectral, cepstral and phase domain representations of a music signal different groups of signal features can be extracted, which describe musical properties like energy, dynamics, timbre, pitch, harmony or rhythm.

Typical features in the time domain include the Zero-Crossing (ZC) rate, the Root Mean Square (RMS) value in variable window durations, the low-energy feature or sub-band energy ratios. These features describe the amount of high-frequency content in a signal or short-time signal power dynamics and will we referred to energy features in the following.

Commonly used spectral features are statistical moments such as the spectral centroid, spectral spread, spectral skewness and spectral kurtosis which characterise the spectral distribution in a short-time fashion. Other features describing the coarse spectral shape, i.e. the spectral envelope, are the spectral slope and spectral flatness. A rough measure of spectral dynamics is the spectral flux, which computes the difference between two consecutive time segments of the short-time magnitude spectrum. Note that the flux can also be computed for any other short-time feature to account for temporal changes in a feature series. These features are often used for timbre characterisation.

Phase domain features help to identify the amount of percussive sounds in music signals and were particularly successful to distinguish between pop and classic in Mierswa and Morik (2005).

Signal characteristics describing the pitch and harmonic content are fundamental frequencies, pitch class profiles or chroma vectors (Fujishima 1999), the timbre-invariant extension Chroma DCT-Reduced log Pitch (CRP) (Müller and Ewert 2011) or estimates of the key and its temporal changes, e.g. the Harmonic Change Detection Function (HCDF) features (Mauch and Dixon 2010). There are also spectral features, which describe properties of partials in a complex tone: The tristimulus features measure the relative energy of individual partial tones or groups of partials in relation to the total energy of all partial tones. The inharmonicity feature quantifies the deviation of individual partial tones from an integer multiple of a given fundamental frequency.

Cepstral features comprise another class of signal descriptors. The most frequently used cepstral features are Mel-Frequency Cepstral Coefficients (MFCCs) (Davis and Mermelstein 1980), which are a variant of the linear-frequency cepstral representation in Eq. (19.2). The main difference to the calculation in Eq. (19.2) is that MFCCs are obtained by first grouping the power spectrum into so-called mel-bands and then decorrelating the logarithmised mel-bands using the discrete cosine transform instead of an inverse DFT. In order to take temporal changes into account, it was proposed to compute the MFCC modulation spectrum and extract the energy in specific modulation frequency bands as features (McKinney and Breebaart 2003). Similarly, the cepstral modulation spectrum as shown in Fig. 19.1 was considered for computing ratios of modulation bands which are approximated by polynomial regression. The resulting Cepstral Modulation RAtio REgression (CMRARE) coefficients were proposed as signal features in Martin and Nagathil (2009). An alternative to cepstral features are Octave-Based Spectral Contrast (OBSC) features (Jiang et al. 2002), which are obtained by computing the difference of spectral peaks and valleys per octave followed by a decorrelation step using principal component analysis. Although these are not cepstral features in a strict sense, their computation resembles the steps performed for computing MFCCs. Therefore, in this work, OBSC features are added to the group of cepstral features.

Features describing rhythmic properties are, for instance, note onset times or beats per minute. These features can be estimated in any of the previously mentioned signal domains. For instance, in the case of percussive instruments (e.g. drums or piano), they are characterised by local maxima in the temporal envelope of a signal. Fluctuation patterns are a different feature representation, which are based on the spectral modulation spectrum and describe the rhythmic regularity of a music piece.

Table 19.1 lists all features used in this study along with different sampling frequencies and window lengths used for feature extraction. Note that these settings were heuristic choices. Although the original sampling frequency was $f_s = 44100$ Hz for all music pieces, also other sampling frequencies were considered. To this end, the music pieces were resampled.

Most of these features are available within the AMUSE framework (Vatolkin 2010), MFCC and OBSC features were extracted with librosa (McFee et al. 2015). The features are sorted according to the corresponding feature groups.

To find potentially meaningful features for predicting human arousal and valence ratings of music pieces, in Fig. 19.2 we compare spectrograms of music pieces from the database in Soleymani et al. (2013), which received different combinations of low and high arousal and valence ratings. Obviously, for music pieces with low arousal ratings most of the signal energy concentrates in the lower frequency bands and the spectral representation appears sparser than music pieces with high arousal ratings. This observation leads to the assumption that arousal might be predicted with features describing the spectral envelope such as spectral moments or MFCCs. While this is an interesting observation for the arousal dimension, the spectrograms in Fig. 19.2 do not show any obvious signal properties explaining low and high valence ratings.

On a feature level, Figs. 19.3 and 19.4 show kernel density estimates of selected features for music pieces with low and high values of arousal and valence, respec-

Table 19.1 Feature groups used in this study and extraction settings

Group	Feature name	#Dim.	f_s	Window length
Energy	Zero-crossing rate	1	22050	512, 1024, 2048
	RMS	1	22050	512, 1024, 2048
	Low energy	1	22050	512, 1024, 2048
	RMS peak number	1	22050	66150
	RMS peak number above mean amplitude	1	22050	66150
	Sub-band energy ratio	4	22050	512, 1024, 2048
Harmony	Tristimulus	2	22050	512. 1024, 2048
	Key and its clarity	2	22050	512, 1024, 2048, 4096
	Major/minor alignment	1	22050	512, 1024, 2048, 4096
	Strengths of major/minor keys	24	22050	512, 1024, 2048, 4096
	Tonal centroid vector	6	22050	512, 1024, 2048, 4096
	HCDF	1	22050	512, 1024, 2048, 4096
	Fundamental frequency	1	22050	512, 1024, 2048
	Inharmonicity	1	22050	512, 1024, 2048
	Chroma DCT-reduced log pitch	12	22050	11025
	Number of different chords in 10s	2	22050	220500
	Number of chord changes in 10s	2	22050	220500
	Shares of the most frequent 20, 40 and 60 percents of chords w.r.t. their duration	4	22050	220500
Timbre (cepstral)	MFCCs—best arousal setup after (Kramer 2016)	4	44100	512
	MFCCs—best valence setup after (Kramer 2016)	5	44100	32768
	MFCC modulation features	52	16000	80000, 112000, 160000, 240000, 320000
	OBSC—best arousal setup after (Kramer 2016)	7	44100	16384
	OBSC—best valence setup after (Kramer 2016)	7	44100	32768
	CMRARE (polyn. order = 3)	8	22050	55125, 110250, ..., 441000
	CMRARE (polyn. order = 5)	12	22050	55125, 110250, ..., 441000
	CMRARE (polyn. order = 10)	22	22050	55125, 110250, ..., 441000

(continued)

Table 19.1 (continued)

Group	Feature name	#Dim.	f_s	Window length
	CMRARE (polyn. order = 12)	26	22050	55125, 110250, …, 441000
	CMRARE (polyn. order = 15)	32	22050	55125, 110250, …, 441000
Timbre (spectral and phase)	Spectral centroid	1	22050	512, 1024, 2048
	Spectral irregularity	1	22050	512, 1024, 2048
	Spectral bandwidth	1	22050	512, 1024, 2048
	Spectral skewness	1	22050	512, 1024, 2048
	Spectral kurtosis	1	22050	512, 1024, 2048
	Spectral crest factor	4	22050	512, 1024, 2048
	Spectral flatness measure	4	22050	512, 1024, 2048
	Spectral extent	1	22050	512, 1024, 2048
	Spectral flux	1	22050	512, 1024, 2048
	Spectral brightness	1	22050	512, 1024, 2048
	Sensory roughness	1	22050	512, 1024, 2048
	Spectral slope	1	22050	512, 1024, 2048
	Angles and distances in phase domain	2	22050	512, 1024, 2048
Rhythm	Fluctuation patterns	7	22050	32768
	Rhythmic clarity	1	22050	66150
	Numbers of beats, onsets and tatums per minute	3	22050	229376
	Tempo based on onset times	1	22050	66150
	Five peaks of fluctuation curves	5	22050	229376

tively. These figures indicate that the overlap of densities for pieces with low and high ratings appear to be lower in the case of arousal than for valence. Although this univariate comparison of selected feature densities and the interpretation of the spectrograms in Fig. 19.2 do not allow a conclusion on the predictability of arousal and valence by signal features, they suggest that the estimation of arousal ratings might be more promising.

19.4 Study Outline

The general concept of our study follows (Kramer 2016). However, our target was to provide a more justified comparison between different feature groups. In contrast to Kramer (2016), where each non-cepstral feature group was represented by only three features, we optimise each group similarly, either selecting the top 5 dimensions or using the best set of features identified by means of feature selection (see below).

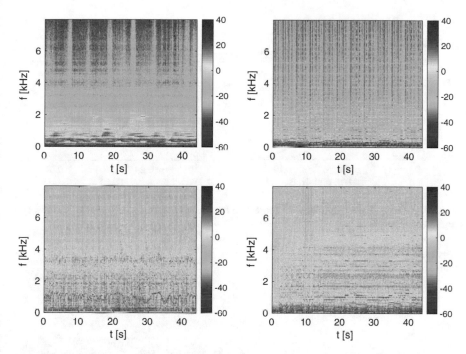

Fig. 19.2 Spectrograms of music pieces with low arousal and low valence (top left), low arousal and high valence (top right), high arousal and low valence (bottom left), high arousal and high valence (bottom right)

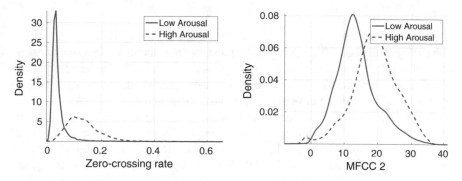

Fig. 19.3 Kernel density estimates for short-time features of music pieces with assigned low and high arousal values

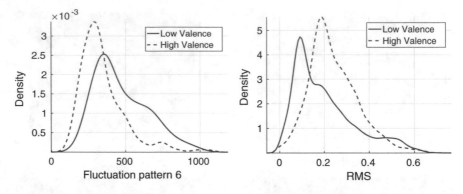
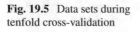

Fig. 19.4 Kernel density estimates for short-time features of music pieces with assigned low and high valence values

Fig. 19.5 Data sets during tenfold cross-validation

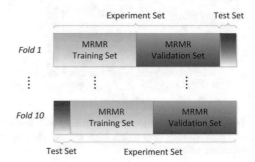

As a database, we use 744 track excerpts of 45s from the '1000 Songs' collection (Soleymani et al. 2013), with arousal and valence annotated by crowdworkers (arousal values in the range [1.6, 8.4] and valence in [1.6, 8.1]). For each feature, we estimate the mean and the standard deviation across all extraction frames which are positioned in the middle between the previously estimated onset events.

Figure 19.5 illustrates the partitioning of tracks during tenfold cross-validation. In each fold, we distinguish between the *experiment set* for the training of a MLR model, and the *test set* for the independent evaluation of the model. Before the training of the MLR model, the MRMR Feature Selection (FS) scheme (Ding and Peng 2005) is applied to reduce the large number of individual feature dimensions and to avoid the overfitting towards the experiment set, which is further divided into training and validation sets (see below). As we keep the original feature values without discretisation (in contrast to Kramer 2016), we apply a simple correlation-based approach as in our previous work (Vatolkin and Rudolph 2018): the relevance is measured by the Pearson correlation coefficient between a feature and a category, and the redundancy by the mean correlation coefficient between a candidate feature and already selected features. The MRMR value which validates feature sets is equal to the relevance subtracted by the redundancy. Starting with an empty feature set, features with the highest MRMR value are added one by one. The experiment set is split into two partitions of equal size: the *MRMR training set* and the *MRMR validation set*. In the first experiment, in which we identify a limited number of the most relevant features, we estimate five-feature dimensions with the highest MRMR values for each feature group, comparable to three- feature dimensions in Kramer (2016). In the second experiment, we use the feature set with the smallest Root Mean Squared Error e (RMSE) on the validation set after MRMR selection on the training set.

It is worth to mention that we restrict our categorisation method to multiple linear regression. Even if it was inferior to support vector regression in Huq et al. (2010), Panda et al. (2013), the main goal of this study was to investigate the individual potential of different feature groups identifying small groups of the most relevant descriptors and do not try to improve the poor feature performance by means of a more complex approach which may benefit from the larger numbers of feature dimensions, requires the optimisation of hyperparameters and may be more prone to overfitting.

19.5   Discussion of Results

Tables 19.2 and 19.3 show mean RMSE values together with standard deviations after tenfold cross-validation for the prediction of arousal and valence. e_{EXP} denotes the RMSE for the experiment set, and e_{TEST} for the test set. Bold font marks the best results among feature sets with the same number of groups, and underlined numbers the overall smallest errors. We distinguished between two feature selection methods: 'Top 5 features' are five features with the highest MRMR values, and 'Best MRMR validation set' is the set of features which achieved the smallest RMSE value estimated for the MRMR validation set, while MLR had been trained on the MRMR training set (cf. Fig. 19.5).

As it can be expected, it is easier to predict arousal rather than valence. For both targets, e_{TEST} is typically comparable to e_{EXP}, which means that no strong overfitting towards the experiment set occurs. The best test RMSE for arousal is achieved when two groups of features are combined (cepstral and rhythm; 'Best MRMR validation set'), or three for valence (harmony, spectral/phase timbre, rhythm; 'Best MRMR validation set'). The higher potential of larger feature groups is also confirmed by the comparison of e_{EXP} columns for both strategies: only in 1 of 62 cases RMSE of 'Best MRMR validation set' is higher than for 'Top 5 features' (harmony features, valence prediction). However, more powerful feature sets can lead to a reduced generalisation ability. For instance, integration of all feature groups leads to slightly smaller e_{TEST} for 'Top 5 features' selection strategy.

The main research question of this study was to measure the relevance of different feature groups for arousal and valence prediction. For this goal, we may compare the performance of two feature sets, when one feature group is added to another one, e.g. measuring the improvements, after cepstral features have been added to all possible combinations of other groups. The results of the statistical comparison of

Table 19.2 RMSE values for arousal prediction. Feature groups: C: cepstral timbre; E: energy; H: harmony; S: spectral and phase timbre; R: rhythm

Groups	Top 5 features		Best MRMR validation set	
	e_{EXP}	e_{TEST}	e_{EXP}	e_{TEST}
C	**0.8189 ± 0.0173**	**0.8251 ± 0.0637**	**0.8103 ± 0.0166**	**0.8405 ± 0.0458**
E	0.9833 ± 0.0125	0.9905 ± 0.1103	0.9243 ± 0.0198	0.9622 ± 0.0922
H	0.9644 ± 0.0122	0.9647 ± 0.1058	0.9551 ± 0.0115	0.9659 ± 0.1028
S	0.9232 ± 0.0160	0.9458 ± 0.1011	0.8796 ± 0.0295	0.9024 ± 0.1105
R	1.1273 ± 0.0184	1.3348 ± 0.6092	1.0641 ± 0.0190	1.0924 ± 0.0860
CE	0.8023 ± 0.0132	0.8154 ± 0.0770	0.7822 ± 0.0079	0.8333 ± 0.0481
CH	**0.8011 ± 0.0147**	**0.8085 ± 0.0778**	0.7917 ± 0.0122	0.8225 ± 0.0482
CS	0.8023 ± 0.0107	0.8167 ± 0.0699	0.7810 ± 0.0172	0.8260 ± 0.0560
CR	0.7963 ± 0.0187	0.8594 ± 0.1200	**0.7645 ± 0.0127**	**0.7931 ± 0.0697**
EH	0.8780 ± 0.0106	0.8823 ± 0.1014	0.8381 ± 0.0164	0.8712 ± 0.0861
ES	0.9017 ± 0.0180	0.9274 ± 0.1151	0.8372 ± 0.0216	0.8788 ± 0.1017
ER	0.9258 ± 0.0122	0.9990 ± 0.1627	0.8506 ± 0.0166	0.9174 ± 0.0727
HS	0.8789 ± 0.0127	0.8973 ± 0.1070	0.8434 ± 0.0244	0.8716 ± 0.1152
HR	0.9004 ± 0.0105	1.0460 ± 0.3437	0.8654 ± 0.0106	0.9279 ± 0.0641
SR	0.8645 ± 0.0240	0.9916 ± 0.2656	0.8032 ± 0.0248	0.8410 ± 0.0618
CEH	0.7882 ± 0.0136	**0.7997 ± 0.0852**	0.7680 ± 0.0069	0.8191 ± 0.0590
CES	0.7879 ± 0.0080	0.8110 ± 0.0801	0.7619 ± 0.0124	0.8218 ± 0.0514
CER	**0.7769 ± 0.0143**	0.8095 ± 0.0560	**0.7396 ± 0.0110**	0.8057 ± 0.0525
CHS	0.7924 ± 0.0123	0.8085 ± 0.0806	0.7672 ± 0.0135	0.8160 ± 0.0630
CHR	0.7801 ± 0.0128	0.8485 ± 0.1164	0.7503 ± 0.0085	0.8105 ± 0.0627
CSR	0.7835 ± 0.0081	0.8525 ± 0.1047	0.7457 ± 0.0149	**0.7941 ± 0.0641**
EHS	0.8634 ± 0.0136	0.8863 ± 0.1150	0.8103 ± 0.0202	0.8591 ± 0.1098
EHR	0.8423 ± 0.0118	0.9213 ± 0.1603	0.7841 ± 0.0159	0.8896 ± 0.1085
ESR	0.8451 ± 0.0280	0.9318 ± 0.1387	0.7794 ± 0.0190	0.8355 ± 0.0576
HSR	0.8357 ± 0.0201	0.9670 ± 0.2661	0.7768 ± 0.0227	0.8423 ± 0.0546
CEHS	0.7792 ± 0.0100	**0.8015 ± 0.0865**	0.7512 ± 0.0114	0.8159 ± 0.0660
CEHR	**0.7661 ± 0.0123**	0.8084 ± 0.0611	**0.7287 ± 0.0109**	0.8334 ± 0.1200
CESR	0.7677 ± 0.0063	0.8201 ± 0.0595	0.7289 ± 0.0129	**0.7992 ± 0.0481**
CHSR	0.7734 ± 0.0094	0.8543 ± 0.1271	0.7297 ± 0.0115	0.8088 ± 0.0547
EHSR	0.8206 ± 0.0212	0.9195 ± 0.1611	0.7579 ± 0.0162	0.8624 ± 0.0963
CEHSR	**0.7595 ± 0.0078**	**0.8207 ± 0.0732**	**0.7172 ± 0.0117**	**0.8266 ± 0.0945**

Table 19.3 RMSE values for valence prediction. Feature groups: C: cepstral timbre; E: energy; H: harmony; S: spectral and phase timbre; R: rhythm

Groups	Top 5 features		Best MRMR validation set	
	e_{EXP}	e_{TEST}	e_{EXP}	e_{TEST}
C	1.0383 ± 0.0165	1.0589 ± 0.0751	0.9972 ± 0.0126	1.0328 ± 0.0842
E	1.1270 ± 0.0161	1.1277 ± 0.0991	1.0768 ± 0.0213	1.1049 ± 0.0861
H	1.1034 ± 0.0101	1.1325 ± 0.0703	1.0386 ± 0.0151	1.0829 ± 0.0549
S	1.1478 ± 0.0085	1.1577 ± 0.0681	1.1231 ± 0.0114	1.1376 ± 0.0691
R	$\mathbf{1.0331 \pm 0.0073}$	$\mathbf{1.0318 \pm 0.0636}$	$\mathbf{0.9888 \pm 0.0114}$	$\mathbf{1.0067 \pm 0.0697}$
CE	1.0225 ± 0.0167	1.0475 ± 0.0880	0.9636 ± 0.0059	1.0239 ± 0.0717
CH	0.9942 ± 0.0170	1.0369 ± 0.0744	0.9191 ± 0.0067	1.0120 ± 0.0842
CS	1.0275 ± 0.0138	1.0492 ± 0.0748	0.9625 ± 0.0168	1.0053 ± 0.0926
CR	0.9698 ± 0.0090	$\mathbf{0.9871 \pm 0.0569}$	0.9154 ± 0.0105	0.9840 ± 0.1115
EH	1.0516 ± 0.0156	1.0742 ± 0.0996	0.9481 ± 0.0246	1.0083 ± 0.0547
ES	1.1040 ± 0.0140	1.1128 ± 0.1016	1.0256 ± 0.0245	1.0719 ± 0.0825
ER	1.0044 ± 0.0075	1.0090 ± 0.0667	0.9413 ± 0.0172	0.9790 ± 0.0880
HS	1.0737 ± 0.0102	1.1040 ± 0.0666	0.9951 ± 0.0200	1.0289 ± 0.0514
HR	$\mathbf{0.9612 \pm 0.0064}$	0.9901 ± 0.0558	$\mathbf{0.8731 \pm 0.0134}$	$\mathbf{0.9170 \pm 0.0544}$
SR	1.0068 ± 0.0099	1.0112 ± 0.0550	$0.9405 + 0.0131$	0.9627 ± 0.0682
CEH	0.9789 ± 0.0169	1.0247 ± 0.0879	0.8875 ± 0.0097	0.9949 ± 0.0491
CES	$1.0155 + 0.0135$	1.0426 ± 0.0822	0.9332 ± 0.0136	1.0196 ± 0.0878
CER	0.9632 ± 0.0087	0.9853 ± 0.0634	0.8948 ± 0.0092	0.9862 ± 0.1190
CHS	0.9852 ± 0.0149	1.0253 ± 0.0712	0.8923 ± 0.0090	0.9520 ± 0.0601
CHR	$\mathbf{0.9313 \pm 0.0073}$	$\mathbf{0.9706 \pm 0.0542}$	$\mathbf{0.8421 \pm 0.0079}$	0.9952 ± 0.2310
CSR	0.9608 ± 0.0069	0.9827 ± 0.0589	0.8978 ± 0.0157	0.9528 ± 0.0760
EHS	1.0402 ± 0.0148	1.0693 ± 0.0964	0.9287 ± 0.0244	0.9965 ± 0.0649
EHR	0.9521 ± 0.0070	0.9839 ± 0.0682	0.8486 ± 0.0175	0.9110 ± 0.0641
ESR	0.9965 ± 0.0096	1.0063 ± 0.0635	0.9108 ± 0.0193	0.9568 ± 0.0791
HSR	0.9567 ± 0.0083	0.9882 ± 0.0527	0.8564 ± 0.0145	$\underline{\mathbf{0.9033 \pm 0.0537}}$
CEHS	0.9732 ± 0.0151	1.0181 ± 0.0826	0.8691 ± 0.0103	0.9605 ± 0.0619
CEHR	0.9265 ± 0.0069	0.9708 ± 0.0601	$\mathbf{0.8280 \pm 0.0092}$	1.0024 ± 0.2389
CESR	0.9586 ± 0.0071	0.9844 ± 0.0611	0.8801 ± 0.0126	0.9798 ± 0.1103
CHSR	$\mathbf{0.9254 \pm 0.0054}$	$\underline{\mathbf{0.9666 \pm 0.0555}}$	0.8307 ± 0.0086	0.9439 ± 0.1114
EHSR	0.9494 ± 0.0087	0.9854 ± 0.0652	0.8382 ± 0.0168	$\mathbf{0.9100 \pm 0.0622}$
CEHSR	$\underline{\mathbf{0.9230 \pm 0.0062}}$	$\mathbf{0.9687 \pm 0.0584}$	$\underline{\mathbf{0.8189 \pm 0.0095}}$	0.9760 ± 0.1785

feature groups are presented in Table 19.4 for arousal and in Table 19.5 for valence prediction.

The tables are organised as follows. The upper halves describe the results with the 'Top 5 features' selection method, the lower halves with 'Best MRMR validation set'. Multi-columns 'Cepstral timbre', 'Energy', etc. correspond to the impact of the examined feature groups. Column 'Group' lists combinations of feature groups, which are extended with the group of the corresponding multi-column. d_e reports the mean reduction of test RMSE with the extended feature set, bold font marks significant reduction. p is the p-value of the paired Wilcoxon test for tenfold. The null hypothesis is, that there is no statistical difference between e_{TEST} values for both feature sets. For example, the entries 'E 0.18 0.004' below 'Cepstral timbre' multi-column header mean, that the extension of energy features with cepstral features leads to the mean reduction of e_{TEST} of 0.18, and this reduction is significant with $p = 0.004$ (we assume the standard level of significance at $p = 0.05$). Even if all feature groups except for cepstral characteristics are extended with the latter group, the error is significantly reduced at mean by 0.10 with $p = 0.006$ (entries 'EHSR 0.10 0.006').

Sometimes the error is increased, when a new feature group is added, in particular, this holds for rhythm features, when they are added to other group combinations for the prediction of arousal: d_e has almost always negative values for the first feature selection strategy, and in several cases for the second one.

For the prediction of arousal, the most relevant group is the cepstral one. Using top 5 MRMR features, p-values are always below 0.05. However, they do not stand the Bonferroni correction: the claim that cepstral features *always* lead to a significant error decrease would require p-values below $0.05/15 = 0.0033$. Also for the second FS strategy, the extension with cepstral features does not lead to a significant reduction of RMSE only for 'EHS' (energy, harmony and spectral/phase features), with $p = 0.064$. The second relevant group is harmony: in 10 of 15 extensions, marked with the bold font, e_{TEST} is reduced significantly with the first FS method, and in 6 of 15 extensions for the second FS method. The impact of energy and spectral/phase is comparable (4/2 significant reductions respectively 3/6). The rhythm group seems to be the least relevant: there is no significant reduction for the first FS method at all (we observe a rather non-significant slight increase of RMSE), and there are only three cases with significant reduction of e_{TEST} for the second FS strategy. However, it is important to mention that 'significant' reductions can be rather marginal, and the extension of feature groups with other characteristics or processing methods may change the order of the 'most relevant' features.

For the prediction of valence (Table 19.5), the importance of feature groups is not the same as for arousal. Here, the rhythm features are the most important (all 15 significant error reductions for the first FS method and 8 for the second one). The second important group is formed with harmony features (13/6), following with cepstral features (10/5). Energy (2/3) and spectral/phase (1/3) are less important.

Finally, we provide lists of the top five features with the highest mean MRMR values from all cross-validation folds in Table 19.6. Some of the features belong to the best five for both arousal and valence: the 6th OBSC coefficient, RMS peak

Table 19.4 Statistical impact of feature groups on arousal prediction. Feature groups: C: cepstral timbre; E: energy; H: harmony; S: spectral and phase timbre; R: rhythm. For details see the text

Top 5 features

Cepstral timbre			Energy			Harmony			Spectral/phase timbre			Rhythm		
Group	d_e	p	Group	d_e	p	Group	d_e	p	Group	d_e	p	Group	d_e	p
E	**0.18**	0.004	C	0.01	0.375	C	**0.02**	0.037	C	0.01	0.160	C	−0.03	0.375
H	**0.16**	0.002	H	**0.08**	0.006	E	**0.11**	0.002	E	**0.06**	0.014	E	−0.01	0.131
S	**0.13**	0.002	S	0.02	0.275	S	**0.05**	0.004	H	**0.07**	0.037	H	−0.08	0.106
R	**0.48**	0.002	R	**0.34**	0.002	R	**0.29**	0.002	R	**0.34**	0.004	S	−0.05	0.084
EH	**0.08**	0.010	CH	0.01	0.375	CE	**0.02**	0.027	CE	0.00	0.193	CE	0.01	0.160
ES	**0.12**	0.004	CS	0.01	0.625	CS	0.01	0.131	CH	−0.00	0.846	CH	−0.04	0.375
ER	**0.19**	0.002	CR	**0.05**	0.049	CR	**0.01**	0.037	CR	0.01	0.375	CS	−0.04	0.625
HS	**0.09**	0.002	HS	0.01	0.193	ES	**0.04**	0.006	EH	−0.00	0.557	EH	−0.04	0.084
HR	**0.20**	0.002	HR	**0.13**	0.004	ER	**0.08**	0.002	ER	0.07	0.020	ES	−0.00	0.106
SR	**0.14**	0.002	SR	0.06	0.084	SR	**0.03**	0.006	HR	0.08	0.065	HS	−0.07	0.131
EHS	**0.09**	0.010	CHS	0.01	0.375	CES	**0.01**	0.037	CEH	−0.00	0.770	CEH	−0.01	0.232
EHR	**0.11**	0.010	CHR	0.04	0.065	CER	0.00	0.557	CER	−0.01	0.846	CES	−0.01	0.275
ESR	**0.11**	0.002	CSR	0.03	0.193	CSR	−0.00	0.695	CHR	−0.01	0.846	CHS	−0.05	0.846
HSR	**0.11**	0.004	HSR	0.05	0.160	ESR	0.01	0.232	EHR	0.00	0.275	EHS	−0.03	0.106
EHSR	**0.10**	0.006	CHSR	0.03	0.131	CESR	−0.00	0.695	CEHR	−0.01	0.770	CEHS	−0.02	0.432

(continued)

Table 19.4 (continued)

Top 5 features

Best MRMR validation set

Cepstral timbre			Energy			Harmony			Spectral/phase timbre			Rhythm		
Group	d_e	p	Group	d_e	p	Group	d_e	p	Group	d_e	p	Group	d_e	p
E	**0.13**	0.002	C	0.01	0.432	C	**0.02**	0.027	C	0.01	0.160	C	**0.05**	0.002
H	**0.14**	0.004	H	**0.09**	0.004	E	**0.09**	0.002	E	**0.08**	0.002	E	0.04	0.064
S	**0.08**	0.006	S	0.02	0.232	S	**0.03**	0.014	H	**0.09**	0.014	H	0.04	0.084
R	**0.30**	0.002	R	**0.18**	0.002	R	**0.16**	0.004	R	**0.25**	0.002	S	**0.06**	0.010
EH	**0.05**	0.037	CH	0.00	0.922	CE	**0.01**	0.049	CE	0.01	0.193	CE	**0.03**	0.020
ES	**0.06**	0.049	CS	0.00	0.846	CS	0.01	0.375	CH	0.01	0.375	CH	0.01	0.232
ER	**0.11**	0.006	CR	−0.01	0.557	CR	−0.02	0.432	CR	−0.00	0.922	CS	0.03	0.105
HS	**0.06**	0.037	HS	0.01	0.432	ES	**0.02**	0.027	EH	0.01	0.492	EH	−0.02	0.432
HR	**0.12**	0.004	HR	0.04	0.105	ER	0.03	0.084	ER	**0.08**	0.002	ES	0.04	0.064
SR	**0.05**	0.002	SR	0.01	0.695	SR	−0.00	0.084	HR	**0.09**	0.002	HS	0.03	0.193
EHS	0.04	0.064	CHS	0.00	1.000	CES	0.01	0.770	CEH	0.00	0.625	CEH	−0.01	0.322
EHR	**0.06**	0.014	CHR	−0.02	0.557	CER	−0.03	0.375	CER	0.01	0.232	CES	0.02	0.105
ESR	**0.04**	0.020	CSR	−0.01	0.922	CSR	−0.01	0.322	CHR	0.00	0.770	CHS	0.01	0.432
HSR	**0.03**	0.010	HSR	−0.02	0.770	ESR	−0.03	0.625	EHR	**0.03**	0.049	EHS	−0.00	0.375
EHSR	**0.04**	0.020	CHSR	−0.02	0.492	CESR	−0.03	0.846	CEHR	0.01	0.375	CEHS	−0.01	0.557

Table 19.5 Statistical impact of feature groups on valence prediction. Feature groups: C: cepstral timbre; E: energy; H: harmony; S: spectral and phase timbre; R: rhythm. For details see the text

Top 5 features

Cepstral timbre			Energy			Harmony			Spectral/phase timbre			Rhythm		
Group	d_e	p	Group	d_e	p	Group	d_e	p	Group	d_e	p	Group	d_e	p
E	**0.08**	0.006	C	0.01	0.084	C	0.02	0.064	C	0.01	0.064	C	**0.07**	0.002
H	**0.10**	0.002	H	**0.06**	0.002	E	**0.05**	0.006	E	0.01	0.160	E	**0.12**	0.002
S	**0.11**	0.002	S	**0.04**	0.014	S	**0.05**	0.010	H	**0.03**	0.014	H	**0.14**	0.002
R	**0.04**	0.020	R	0.02	0.275	R	**0.04**	0.037	R	0.02	0.064	S	**0.15**	0.002
EH	**0.05**	0.006	CH	0.01	0.193	CE	**0.02**	0.027	CE	0.00	0.275	CE	**0.06**	0.002
ES	**0.07**	0.006	CS	0.01	0.375	CS	**0.02**	0.037	CH	0.01	0.105	CH	**0.07**	0.002
ER	0.02	0.084	CR	0.00	0.922	CR	**0.02**	0.027	CR	0.00	0.375	CS	**0.07**	0.004
HS	**0.08**	0.002	HS	0.03	0.105	ES	**0.04**	0.014	EH	0.00	0.695	EH	**0.09**	0.002
HR	0.02	0.160	HR	0.01	0.492	ER	**0.03**	0.014	ER	0.00	0.492	ES	**0.11**	0.002
SR	**0.03**	0.027	SR	0.00	0.492	SR	0.02	0.064	HR	0.00	0.770	HS	**0.12**	0.002
EHS	**0.05**	0.004	CHS	0.01	1.000	CES	**0.02**	0.027	CEH	0.01	0.160	CEH	**0.05**	0.002
EHR	0.01	0.232	CHR	-0.00	0.770	CER	**0.01**	0.037	CER	0.00	0.557	CES	**0.06**	0.006
ESR	**0.02**	0.049	CSR	-0.00	0.922	CSR	**0.02**	0.020	CHR	0.00	0.375	CHS	**0.06**	0.004
HSR	0.02	0.064	HSR	0.00	0.770	ESR	**0.02**	0.014	EHR	-0.00	0.695	EHS	**0.08**	0.002
EHSR	0.02	0.131	CHSR	-0.00	0.770	CESR	**0.02**	0.020	CEHR	0.00	0.432	CEHS	**0.05**	0.004

(continued)

Table 19.5 (continued)

Top 5 features

Cepstral timbre			Energy			Harmony			Spectral/phase timbre			Rhythm		
Group	d_e	p	Group	d_e	p	Group	d_e	p	Group	d_e	p	Group	d_e	p
Best MRMR validation set														
E	**0.08**	0.010	C	0.01	0.492	C	0.02	0.432	C	0.03	0.084	C	0.05	0.064
H	0.07	0.064	H	**0.07**	0.002	E	**0.10**	0.004	E	0.03	0.064	E	**0.13**	0.002
S	**0.13**	0.002	S	**0.07**	0.002	S	**0.11**	0.002	H	**0.05**	0.004	H	**0.17**	0.002
R	0.02	0.275	R	0.03	0.322	R	**0.09**	0.002	R	**0.04**	0.004	S	**0.17**	0.002
EH	0.01	0.322	CH	0.02	0.432	CE	0.03	0.275	CE	0.00	0.492	CE	0.04	0.105
ES	**0.05**	0.027	CS	−0.01	0.193	CS	0.05	0.064	CH	**0.06**	0.006	CH	0.02	0.084
ER	−0.01	0.625	CR	−0.00	0.922	CR	−0.01	0.375	CR	0.03	0.084	CS	**0.05**	0.020
HS	**0.08**	0.002	HS	**0.03**	0.027	ES	**0.08**	0.010	EH	0.01	0.432	EH	**0.10**	0.002
HR	−0.08	0.432	HR	0.01	0.432	ER	**0.07**	0.027	ER	0.02	0.232	ES	**0.12**	0.004
SR	0.01	0.557	SR	0.01	0.557	SR	**0.06**	0.014	HR	0.01	0.322	HS	**0.13**	0.002
EHS	**0.04**	0.020	CHS	−0.01	0.322	CES	0.06	0.084	CEH	0.03	0.131	CEH	−0.01	0.105
EHR	−0.09	0.232	CHR	−0.01	0.557	CER	−0.02	0.557	CER	0.01	0.375	CES	0.04	0.084
ESR	−0.02	0.625	CSR	−0.03	0.322	CSR	0.01	0.492	CHR	0.05	0.322	CHS	0.01	0.105
HSR	−0.04	0.322	HSR	−0.01	1.000	ESR	0.05	0.064	EHR	0.00	0.695	EHS	**0.09**	0.004
EHSR	−0.07	0.375	CHSR	−0.03	0.232	CESR	0.00	0.557	CEHR	0.03	0.375	CEHS	−0.02	0.105

Table 19.6 Features with highest MRMR values after cross-validation. p denotes the polynomial degree for CMRARE characteristics, and t is the extraction frame size. M($\cdot$) are mean values, S($\cdot$) standard deviation values. For multi-dimensional features, the number after the feature name corresponds to the dimension. HCDF: harmonic change detection function; MFCC: mel-frequency cepstral coefficients; OBSC: octave-based spectral contrast; RMS: root mean square (of the time signal); ZC: zero crossings

	Arousal		Valence	
	Name	MRMR	Name	MRMR
C	M(MFCC 1 - best valence setup)	0.5167	M(OBSC 6 - best valence setup)	0.4674
	M(OBSC 5 - best arousal setup)	0.3621	M(MFCC modulation 27, $t = 7s$)	0.1866
	M(OBSC 6 - best valence setup)	0.1425	S(CMRARE 31, $p = 15$, $t = 55125$)	0.1741
	S(CMRARE 27, $p = 15$, $t = 55125$)	0.1314	M(OBSC 7 - best arousal setup)	0.1678
	M(MFCC modulation 1, $t = 20s$)	0.0975	M(CMRARE 4, $p = 5$, $t = 330750$)	0.1391
E	M(ZC rate, $t = 2048$)	0.4542	M(RMS peak number in 3s)	0.3219
	M(RMS peak number in 3s)	0.2898	M(ZC rate, $t = 2048$)	0.1603
	M(ZC rate, $t = 1024$)	0.1442	S(RMS, $t = 2048$)	0.1215
	M(RMS, $t = 512$)	0.1194	M(Low energy, $t = 2048$)	0.0434
	M(RMS, $t = 2048$)	0.0547	M(ZC rate, $t = 1024$)	0.0162
H	M(Tristimulus 2, $t = 2048$)	0.6181	M(HCDF, $t = 4096$)	0.4160
	M(HCDF, $t = 4096$)	0.1698	S(Major/minor alignment, $t = 1024$)	0.1516
	M(Inharmonicity, $t = 512$)	0.1438	M(Fundamental frequency, $t = 512$)	0.1006
	M(Tristimulus 1, $t = 512$)	0.1230	M(Major/minor alignment, $t = 4096$)	0.0807
	M(HCDF, $t = 2048$)	0.0986	S(Tristimulus 2, $t = 2048$)	0.0594
T	M(Phase domain distances, $t = 2048$)	0.3214	M(Spectral bandwidth, $t = 2048$)	0.2648
	M(Spectral brightness, $t = 2048$)	0.3005	M(Spectral skewness, $t = 2048$)	0.1143
	M(Spectral flatness, $t = 2048$)	0.1436	M(Spectral flux, $t = 2048$)	0.0703
	M(Spectral skewness, $t = 512$)	0.1109	S(Spectral flatness 2, $t = 2048$)	0.0250
	M(Spectral slope, $t = 2048$)	0.0919	S(Phase domain angles, $t = 512$)	0.0124
R	M(Onsets per minute)	0.3871	M(Fluctuation patterns 6)	0.4745
	M(Beats per minute)	0.1854	M(Fluctuation patterns 5)	0.1257
	M(Fluctuation patterns 6)	0.0964	M(first peak of fluct.curves)	0.1239
	M(Fluctuation patterns 2)	0.0937	M(Beats per minute)	0.1034
	M(Fluctuation patterns 5)	0.0657	M(second peak of fluct.curves)	0.0929

number in 3s, RMS, ZC rate, HCDF, spectral skewness, spectral flatness and beats per minute. In case it is desired to have a rather small feature set, which has a good performance for general emotion recognition in arousal/valence space, these features can be considered.

19.6 Conclusions

In this study, we have analysed the importance of five audio feature groups for predicting arousal and valence in music based on multiple linear regression: cepstral timbre, energy, harmony, spectral/phase timbre and rhythm descriptors. Each feature group was represented by the most relevant dimensions identified with the help of a minimal redundancy–maximal relevance feature selection scheme. In the first approach, we have stored for each group five dimensions with the highest correlation to the target (arousal or valence) and the lowest correlation between each other. In the second approach, we have selected more feature dimensions which contributed to the feature set with the smallest validation error. This procedure was repeated in line with tenfold cross-validation.

The analysis of the test root mean squared errors has shown that cepstral features were the most relevant for the prediction of arousal, followed by harmony descriptors. For valence prediction, the most important features were rhythm and harmony descriptors. The smallest test errors were achieved for the combinations of two to four feature groups, depending also on the feature selection strategy. The combination of all tested feature groups led to the smallest training error in three of four cases, but in none of the cases to the smallest test error, which confirms empirically that the uncontrolled extension of feature dimensions leads to a higher danger of overfitting towards the training set and a reduced generalisation performance.

In future, we plan to analyse other feature categories, grouping them with respect to music theory, extraction domain and their extraction properties. Furthermore, the experiments can be repeated with other regression methods.

Acknowledgements We thank Philipp Kramer for providing the code and explanations of experiments from his bachelor's thesis, in particular for the extraction of MFCC and OBSC features.

References

Davis, S. B., & Mermelstein, P. (1980). Comparison of parametric representations for monosyllabic word recognition in continuously spoken sentences. *IEEE Transactions on Acoustics, Speech, and Signal Processing, 28*(4), 357–366.

Ding, C. H. Q., & Peng, H. (2005). Minimum redundancy feature selection from microarray gene expression data. *Journal on Bioinformatics and Computational Biology, 3*(2), 185–206.

Fujishima, T. (1999). Realtime chord recognition of musical sound: a system using common lisp music. In *Proceedings of the international computer music conference (ICMC)* (pp. 464–467).

Grekow, J. (2018). Audio features dedicated to the detection and tracking of arousal and valence in musical compositions. *Journal of Information and Telecommunication, 2*(3), 322–333. https://doi.org/10.1080/24751839.2018.1463749.

Hevner, K. (1930). Tests for aesthetic appreciation in the field of music. *Journal of Applied Psychology, 14*, 470–477.

Huq, A., Pablo Bello, J., & Rowe, R. (2010). Automated music emotion recognition: A systematic evaluation. *Journal of New Music Research, 39*(3), 227–244.

Jiang, D. N., Lu, L., Zhang, H. J., Tao, J. H., & Cai, L. H. (2002). Music type classification by spectral contrast feature. In *Proceedings IEEE international conference on multimedia and expo (ICME)* (vol. 1, pp. 113–116). IEEE.

Katayose, H., Imai, M., & Inokuchi, S. (1988). Sentiment extraction in music. In *Proceedings of the 9th international conference on pattern recognition (ICPR)* (pp. 1083–1087). IEEE.

Kramer, P. (2016). Relevanz cepstraler Merkmale für Vorhersagen im Arousal-Valence Modell auf Musiksignaldaten. Bachelor's thesis. TU Dortmund: Department of Computer Science.

Malik, M., Adavanne, S., Drossos, K., Virtanen, T., Ticha, D., & Jarina, R. (2017). Stacked convolutional and recurrent neural networks for music emotion recognition. *CoRR.* arXiv:abs/1706.02292. (2017)

Martin, R., & Nagathil, A. (2009). Cepstral modulation ratio regression (CMRARE) parameters for audio signal analysis and classification. In *Proceedings of the IEEE international conference on acoustics, speech and signal processing (ICASSP)*.

Mauch, M., & Dixon, S. (2010). Approximate note transcription for the improved identification of difficult chords. In J. S. Downie, & R. C. Veltkamp (Eds.), *Proceedings of the 11th international society for music information retrieval conference (ISMIR)* (pp. 135–140).

McFee, B., Raffel, C., Liang, D., Ellis, D., McVicar, M., & Battenberg, E. (2015). Librosa: Audio and music signal analysis in python. In *Proceedings of the 14th Python in science conference* (pp. 1–7).

McKinney, M. F., & Breebaart, J. (2003). Features for audio and music classification. In *Proceedings of international society of music information retrieval conference (ISMIR)* (vol. 3, pp. 151–158).

Mierswa, I., & Morik, K. (2005). Automatic feature extraction for classifying audio data. *Machine Learning Journal, 58*(2–3), 127–149.

Müller, M., & Ewert, S. (2011). Chroma toolbox: MATLAB implementations for extracting variants of chroma-based audio features. In: A. Klapuri, & C. Leider (Eds.), *Proceedings of the 12th international conference on music information retrieval (ISMIR)* (pp. 215–220). University of Miami.

Nagathil, A., & Martin, R. (2016). Signal-level features. In C. Weihs, D. Jannach, I. Vatolkin, & G. Rudolph (Eds.), *Music data analysis: foundations and applications* (pp. 145–164). CRC Press.

Panda, R., Malheiro, R., Rocha, B., Oliveira, A., & Paiva, R. P. (2013). Multi-modal music emotion recognition: A new dataset, methodology and comparative analysis. In *Proceedings of the 10th international symposium on computer music multidisciplinary research (CMMR)*. Berlin: Springer.

Panda, R., Malheiro, R. M., & Paiva, R. P. (2018). Novel audio features for music emotion recognition. *IEEE Transactions on Affective Computing*, 1–1. https://doi.org/10.1109/TAFFC.2018.2820691

Panda, R., Rocha, B., & Paiva, R. P. (2013). Dimensional music emotion recognition: Combining standard and melodic audio features. In *Proceedings of the 10th international symposium on computer music multidisciplinary research (CMMR)*. Berlin: Springer.

Russell, J. A. (1980). A circumplex model of affect. *Journal of Personality and Social Psychology, 39*(6), 1161–1178.

Saari, P., Eerola, T., & Lartillot, O. (2011). Generalizability and simplicity as criteria in feature selection: Application to mood classification in music. *IEEE Transactions on Audio, Speech, and Language Processing, 19*(6), 1802–1812.

Scherer, K. R. (1982). *Vokale Kommunikation: Nonverbale Aspekte des Sprachverhaltens.* Weinheim/Basel: Beltz.

Schmidt, E.M., & Kim, Y. E. (2011). Learning emotion-based acoustic features with deep belief networks. In *2011 IEEE workshop on applications of signal processing to audio and acoustics (WASPAA)* (pp. 65–68). https://doi.org/10.1109/ASPAA.2011.6082328.

Schubert, E. (2004). Modeling perceived emotion with continuous musical features. *Music Perception*, *21*(4), 561–585.

Soleymani, M., Caro, M. N., Schmidt, E. M., Sha, C. Y., & Yang, Y. H. (2013). 1000 songs for emotional analysis of music. In *Proceedings of the 2nd ACM international workshop on crowdsourcing for multimedia* (pp. 1–6). USA: CrowdMM 13. https://doi.org/10.1145/2506364.2506365.

Tellegen, A., Watson, D., & Clark, L. A. (1999). On the dimensional and hierarchical structure of affect. *Psychological Science*, *10*(4), 297–303.

Vatolkin, I., Theimer, W., & Botteck, M. (2010). AMUSE (Advanced MUSic Explorer)—a multitool framework for music data analysis. In: J. S. Downie, & R. C. Veltkamp (Eds.), *Proceedings of the 11th international society on music information retrieval conference (ISMIR)* (pp. 33–38).

Vatolkin, I., & Rudolph, G. (2018). Comparison of audio features for recognition of western and ethnic instruments in polyphonic mixtures. In *Proceedings of the 19th International Society for Music Information Retrieval Conference, ISMIR 2018* (pp. 554–560). Paris, France.

Yang, X., Dong, Y., & Li, J. (2018). Review of data features-based music emotion recognition methods. *Multimedia Systems*, *24*(4), 365–389. https://doi.org/10.1007/s00530-017-0559-4.

Yang, Y. H., & Chen, H. H. (2011). *Music emotion recognition*. CRC Press.

Chapter 20
The Psychological Foundations of Classification

Martin Ebeling and Günther Rötter

Abstract Phenomenology assumes that human thought is decisively shaped by the conditions of human perception. Therefore, all human concepts have paragons in perception. Accordingly, classification is based on elementary psychic functions, and here, we demonstrate that the same psychic functions are also effective in hearing and music perception. Thus, we deduce the psychological prerequisites of classification from music perception and elementary music cognition. Classification is a certain case of categorization, a mental property which determines all thinking (Sect. 20.1). The principles of categorization were discussed since ancient times, i.e., by Aristotle (Sect. 20.2). Categorization requires the ability to compare objects according to a certain criterion. Carl Stumpf (1848–1936) demonstrated that all comparisons are psychologically based on fundamental relations (German: *Grundverhältnisse*) which also determine pitch perception (Sect. 20.3). These fundamental relations govern the perception of Gestalt, which follows certain Gestalt laws (Sect. 20.4). Intuitive classification criterions result from differences concerning Gestalt. The importance of the Gestalt laws is demonstrated by some recent studies in current musicology which all refer to the concept of pitch (Sect. 20.5). Remarkably, the pitch space and the number space have the same topological structure. Discrete and continuous parameters in statistic have analogies in discrete and continuous concepts of pitch (Sect. 20.6). Particularly noteworthy is the analogy between periodic data and the cyclic organization of tone chroma by the classification of pitch "modulo octave" due to the sound phenomenon of octave identification (Sect. 20.7).

M. Ebeling (✉)
Institute of Music and Musicology, TU Dortmund University, Mönchengladbach, Germany
e-mail: martin.ebeling@tu-dortmund.de

G. Rötter
Institute of Music and Musicology, TU Dortmund University, Dortmund, Germany
e-mail: guenther.roetter@tu-dortmund.de

© Springer Nature Switzerland AG 2019　　327
N. Bauer et al. (eds.), *Applications in Statistical Computing*,
Studies in Classification, Data Analysis, and Knowledge Organization,
https://doi.org/10.1007/978-3-030-25147-5_20

20.1 Categorical Thinking

Man is constantly confronted with an overwhelming multitude of objects. The objects are compared in the mind and are grouped together as to similarities or separated according to differences. Elementary perception already exhibits this organizing effort, which is described in psychology as categorical perception and categorical thinking. All human action is based on organizational structures that have been experienced and thought. In the various sciences, different orders for the objects of research are used for which terms such as taxonomy or typology have been coined.

The organizational structures can be very different. The objects can be divided into disjoint classes, such as an equivalence relation in mathematics. Then every object is in exactly one class. The number of objects in each equivalence class can be counted (or measured), a favorable condition for statistical data collection and evaluation.

Sets of objects can be uniquely (bijectively) mapped to the sets of numbers. There is a fundamental difference between the countable sets, which can be bijectively mapped to the natural numbers and to the rational numbers, and the incountable sets that can be mapped to the irrational numbers and at the end to the real numbers. The latter form continua with the "$<$"-relation ($a < b$ means: a is $less$ $than$ b), which is transferred from the real numbers to the continuum. If a set can bijectively be mapped to the natural numbers, it contains discrete objects that can be arranged with the "$<$"-relation of the natural numbers. If a set can bijectively be mapped to the rational numbers, its objects are not discrete but countable.

The prerequisite for categorical information is the observation of characteristics. Every observation is based on sensory perception, which is here aimed at a characteristic as a classification criterion. The assignment to the classes is made by comparing the observation with the classification criterion. The classification criterion may be elemental perception, e.g., the categorization for a color. But far more common are classifications of patterns for which the psychological concept of Gestalt has been coined.

The division of observations or data into classes is an integral part of human thinking. Philosophers of all times have thought about the basics of categorical thinking of man. The elementary concepts of being have also been categorized by them, as the following examples show.

20.2 Category Theory According to Aristotle

The philosopher and scientist Aristotle is considered the founder of the concept of "categories". He examined how statements about being can be made and established the so-called Aristotelian categories, which mean concepts that have no common generic concepts. For Aristotle, the question arose about the "highest class of predicates, i.e., of the categories in the sense in which he introduced the term

Table 20.1 The 10 categories by Aristole

	Category	Meaning	Example
1.	Substance	Essence	Socrates is a man
2.	Quantity	How much	Socrates is 1.80 m tall
3.	Quality	Of what kind	Socrates is pale
4.	Relation	In relation to	Socrates is taller than Alkibiades
5.	Place	Where (somewhere)	Socrates is in the house
6.	Time	When (sometime)	Socrates is here now
7.	Position	To be in a position	Socrates is standing on the marketplace
8.	State	To have	Socrates is armed
9.	Action	To make, do	Socrates is leaving
10.	Affection	To suffer, undergo	Socrates is being hit (by a rock)

https://www.vaticarsten.de/theologie/philosophie/aristoteles-kategorie.pdf accessed on November 26, 2018

http://www.philolex.de/aristote.htm accessed on November 26, 2018

into philosophy" (Brentano 1933, p. 110). The categories thus serve to "distinguish and determine logical and metaphysical concepts, to determine general concepts" (Trendelenburg 1963, p. 216).

What is special about Aristotle's writing on category is that the categories no longer comprise only predicates or types of predication, but also nouns. These categories are substance, quantity, quality, relation, place, time, position, state, action, and affection. Later, however, he reduced the list to eight categories by deleting the last two categories (Table 20.1).

However, the concept of category has changed over time, which was initiated by Immanuel Kant, among others. He defines categories as deduced by reasoning (a priori) and views them as tools of thought and judgment. As a result, they have no existence, but are only present in the human mind. Kant, through his new approach to categories, is regarded as a renewer of the category concept, and he defines twelve "pure concepts of the understanding" which can be divided into four basic functions of judgment: quantity, quality, relation, and modality.[1]

20.3   Carl Stumpf: The Origin of Categories in Perception

The philosopher and cofounder of empirical perceptual psychology Carl Stumpf (1848–1936), from 1900 is first Director of the Psychological Institute of the University of Berlin, sees the cause for the formation of categories in the laws of perception and not, as Kant, in a priori categories of reason. He traces the formation of concepts back to perception:

---

[1] https://korpora.zim.uni-duisburg-essen.de/Kant/aa03/093.html accessed on November 26, 2018.

> If in this way we show the origin of a concept in and from perceptions, we are at the same time explaining its original meaning. With simple terms there is no other way than to teach a person who does not know the meaning of the name in question. One just has to refer to examples (Stumpf 1939/40, p. 9).

The derivation of concepts, and in particular, the basic concepts (categories) from perception, is at the same time an essential step from philosophy to psychology: the investigation of sensory "phenomena" as the basis of psychical activity becomes the subject of empirical investigations. In addition to the sense of sight, hearing is perceived as an equal sense. In addition to external perception, which is conveyed by the senses, "*internal perception or psychological perception*" (Stumpf 1939/40, p. 11) is also important:

> Colors, tones, tastes are perceived through the senses, but thinking, feeling, desiring are only perceived through self-contemplation. ... Therefore also the concepts of psychical states or activities must not be derived from sensory perception, but from this so-called 'inner sense' (Stumpf 1939/40, p. 12).

The ordering functions of the mind are based on Gestalt perception. Gestalts are mental formations, abstractions of the same or similarly shaped objects of perception, which—in modern terms—enable the arrangement of objects in object structures according to features of their Gestalt.

20.3.1 Comparison: Fundamental Relationships According to Stumpf

The comparison is based on elementary psychical functions that are directly linked to sensory perception:

> First of all, all absolute contents, especially sensory contents, belong to these evaluable materials. The affirmation or notice (grasping, setting, acknowledgment) of such a content we call its perception. After all, it can be a deception and differs from cognition. But what we are to be concerned with now is not the noticing of absolute contents, but certain relations (relationships) that are found on or 'between' them. Which and how many basic conditions are there, may be left open here. We will deal essentially with four: plurality, increase, similarity, fusion (Stumpf 1883, p. 96).

Other factors, especially attention, individual disposition, etc., do not determine the essence, but do determine the depth of analysis and comparison (cf. Stumpf 1883, p. 67).

The basic conditions can easily be explained by listening to tones. Each tone simultaneously has different properties which are called tone parameters: pitch, volume, duration, timbre. The same applies to all senses: Each sensation is a unit (a whole) in which different sensory moments (cf. Stumpf 1890, p. 65) or attributes (cf. Stumpf 1939/40, p. 22) can be recognized. The comparison can refer to any moment of sensation. For tones it is the tone parameters, for sight it is the shape and color of a surface.

Sensory moments have two important properties in perception:

- they are constitutive of the sensation,
- they are independent in perception.

Constitutive means that the sensation cannot be without the particular moment of sensation. There is no tone without a pitch or without a volume, there is no color without a surface and no surface without a color (Stumpf 1890, p. 65).

Independent in perception means that the moment of sensation can be changed without changing the other moments of sensation. The pitch of a tone can be changed without changing the volume, conversely, the volume can be changed at the same pitch. Or, if the color is the same, the color-bearing surface can be changed, or the same surface can be given a different color.

These two characteristics of sensory moments are psychological conditions: This is how it is experienced. Here the hiatus between the sensation as something elementary psychical and the external stimulus as the physical cause of the sensation is clear: Psychoacoustics, for example, proves that the volume depends not only on the sound pressure but also on the frequency. Nevertheless, the sensory moments of volume and pitch are experienced and thought to be independent, because the frequency dependence of volume does not immediately become conscious. Only the measurement and comparison between the sensory moments in the psychoacoustic experiment reveal this connection (Zwicker and Fastl 1999, pp. 119 and 204).

In a plurality of tones, the question is whether one or more tones are heard. This concept may seem unproblematic at first, but it is not trivial from the point of view of perceptual psychology: If, for example, several synchronous sine tones are in integer oscillation relationships with one another, they fuse and are perceived as one tone with a certain pitch and timbre. Multiple tones are not heard, but only one single tone. Under special circumstances, however, e.g., with louder partial tones or slight detuning of individual partial tones (Moore et al. 1984; Moore and Ohgushi 1993) or asynchronous tone onsets (Bregman 1990, p. 222) or with special disposition of the listener (Schneider et al. 2005), it is however possible to discern individual sine tones and thus perceive a plurality of tones.

If multiple tones are recognized, it will also be heard which tone is higher or louder than the other. This is the ratio of the increase. Both tones can also be similar in one or more tone parameters. Similar loudness means that both tones are perceived as having almost the same volume, similar pitch indicates that they are perceived as having almost the same pitch.

The holistic approach of analyzing and progressing from perception of the whole to perceptions of parts differs fundamentally from the concept of "grouping" used in modern music psychology, as in the ASA (auditory scene analysis, Bregman 1990). In this case, it is assumed that the figures simply emerge synthetically from individual perceptual elements according to certain laws. This difference can be traced back to historical reasons: The doctrine of the Berlin Gestalt psychology school—Carl Stumpf and his students—and their epistemological foundations were not fully taken note of by American psychology. At the neural level, correlates of Gestalt perception can be identified. These suggest that the boundaries of shapes are

Fig. 20.1 Isophones. On the abscissa, the frequencies are scaled logarithmically. The ordinate shows the sound pressure level in dB. Different tones on the same curve are perceived as equally loud

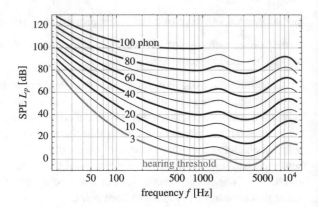

detected neuronally faster than the details of the shape (Griffiths et al. 2012). That would speak for Stumpf's holistic view.

The basic ratio of the increase is elementary for listening to music. It makes it possible to decide which of two tones is higher. Increase and similarity are the prerequisites for forming scales. In psychophysics, they allow to scale perceptual content. The isophones (equal-loudness contour, cf. Zwicker and Fastl 1999, p. 203) shown in Fig. 20.1 are based on the similarity—"Equality of sensory phenomena is nothing but extreme similarity" (Stumpf 1883, p. 111)—of two different tones with respect to the perceived loudness. The sone scale can be derived by estimating half or double the perceived loudness.

20.3.2 The Fundamental Relationship of Fusion—Already Unity or Rather Plurality?

The concept of fusion is derived from the auditory perception of simultaneously perceived or imagined tones and is closely related to the sense of consonance (cf. Ebeling 2008a). Although Stumpf defines fusion as an abstract fundamental relationship (cf. Stumpf 1890, p. 65), he describes and examines it only for pitch hearing.

> At the center of the theory of fusion is the assertion that fusion, i.e. the unity of impression in octaves, fifths, and thirds decreases in this order. For since the intervals mentioned also stand in the same order with regard to their consonance […], and as the whole scale results from these intervals with the help of the concept of indirect relationship, that fact contains […] an indication […] that the essential characteristic of consonance and thus the basic concept of the whole theory of music is to be found in tone fusion (Stumpf 1897, p. 281).

By showing *"that we have no reason at all to seek the cause for the tone fusion in the psychological field"*, Stumpf concludes: *"The cause of the tone fusion is physiological"* (Stumpf 1890, p. 211). *"It must therefore correspond to the differences of the degrees of fusion of certain differences of the last processes in the hearing center as physical correlation or as cause […]"* (Stumpf 1890, p. 213). Modern neuroacoustics can show by electrophysiological means that pitch and timbre sensations are based on a periodicity analysis between auditory neurons in the cochlear nucleus (first auditory pathway station in the brain stem) and the inferior colliculus (second auditory pathway station in the midbrain) (Langner 1997, 2015). The mathematization of this periodicity analysis shows that the tonal fusion of simultaneous tones has its cause in the structure of this neuronal periodicity detection (Ebeling 2008b).

20.3.3   Perceptional Complexes and the Concept of Gestalt

Against the background of the co-perceived relationships, the contents of perception are *complexes* of sensations consisting of absolute contents, e.g., pitch and relationships, e.g., intervals. In other words: the sensations are always embedded in a structure in the perception. The structure, however, is determined solely by the totality of the relationships; it can be abstracted from the current, absolute contents of the sensation. When listening to the C major triad, e.g., with the notes c'- e'- g', the ratios or intervals of the lower major third c'- e', the upper minor third e'- g', and the pure fifth c'- g' as frame interval are also given. If one abstracts from the concrete tones as the absolute contents, the structure from the intervals remains: major third below, minor third above, pure fifth as frame interval, and an abstract structure determined solely by the relationships, which is called triad.

If one thus abstracts its absolute contents from a perceptual complex, the relationships from the complex remain. The totality of these relationships, which determine the structure of the complex of perception, is its Gestalt. According to Stumpf's definition:

> We designate a whole of sensory contents (phenomena) as a complex, but a whole of relationships between sensory contents as a Gestalt (Stumpf 1939/40, pp. 229/230).

The relationships defining a Gestalt are invariant under a uniform displacement, a *translation*. If the absolute sensory contents of a complex are changed in the same degree and in the same direction, we obtain new absolute sensory contents, but the relations between the sensory contents are preserved. The perceptual complex has been transposed, the Gestalt remains intact.

A melody can be played one or more notes higher or lower, it remains the same melody, the C major triad, the G major triad ,and the F major triad all have the Gestalt of the major triad, stand only on different fundamental notes and can be transposed or reversed into each other. Tone Gestalts are, therefore, invariant under the transposition or reversal as a musical translation. According to Stumpf's formulation *"Gestalts must in principle be transferable, if they are nothing other than the quintessence of relationships"* (Stumpf 1939/40, p. 230). The invariance of relationships is the prerequisite for the cognition and recognition of Gestalts in an otherwise altered material.

20.4 The Six Gestalt Laws of Max Wertheimer

The political conditions in Germany during the Third Reich led to the emigration of influential Gestalt psychologists. As a result, Gestalt psychology research in America was continued almost exclusively in the field of visual perception. A former assistant to Carl Stumpf, the psychologist Max Carl Wertheimer, summarized the results of his studies, at first with regard to visual stimuli, in **six Gestalt laws** or principles:

(1) *The Gestalt law of similarity* means the perceived togetherness of similar elements, even if they are at a certain distance from each other. These stand in comparison to other elements, which are identical in distance, but show a lesser similarity to these.
(2) Contrary to this, elements that are close to each other are perceived to be more similar than elements that are similar but farther away from each other. This *law of proximity* can also be responsible for things being perceived as more similar to each other than they actually are.
(3) If we see an unfinished figure or an unfinished form, we often perceive it as completed. Wertheimer speaks here of the *Gestalt law of closure*.
(4) Also superimposed elements, which continue preceding elements, appear to belong together according to the *law of continuity*. According to Wertheimer, a distinction is also made between the *law of the common movement* and the *law of the inversion of figure and ground*.
(5) The *law of common movement* means the perceived togetherness of elements that either seem to move in the same direction or perform the same movement.
(6) The *law of the inversion of figure and ground*, on the other hand, states on the one hand that a perception of several figure–ground relationships of such a conceived image is possible. On the other hand, it can be stated that simple and regular forms are preferred (cf. Zimbardo 1978, pp. 221/222).

The Gestalt laws of Max Wertheimer also apply to the musical field. Under certain conditions, individual tones are formed into a Gestalt in the sense of a coherent melody. Wertheimer explains "*I hear a melody (17 notes!) with its accompaniment (32 notes!). I hear the melody and accompaniment, not just "49" or at least certainly not normally or ad libitum 20 plus 29*" (Wertheimer 1923, p. 301). In general, it can be said that in all people there is an "*expected kind of summarization [sic] and separateness and, moreover, only sometimes, rarely, under certain circumstances, can other things result or artificially, through special measures, only this also - is more difficult to produce*" (ibid., p. 302).

20.5   Gestalt Perception in Modern Musicology: Segregation and Integration

In modern psychology of perception and music, one speaks in the same sense of integration and segregation, which play an important role especially in the perception of hearing and music (see below and cf. Bregman 1990). In 1940, Charles Warren Fox showed in his essay *Modern Counterpoint—A Phenomenological Approach* that Wertheimer's Gestalt laws were fundamental both to classical counterpoint and to the music that was modern at the time.

Also, in more recent times, perception phenomena were explained by Gestalt laws. In this context, for example, a phenomenon is referred to as the "trill threshold". A tone sequence with alternating ascending and descending tones, depending on the size of the interval and tempo, will be perceived either as one "jumping" melody or as two melodies, one ascending and the other descending. The distance of the frequency of the successive notes must be greater than 15%. This corresponds to an interval between a major second and a minor third. The second condition is tempo. While at a slow tempo one can still perceive the tone sequence as such, at a tempo of 10 notes per second one begins to separate the high notes from the low notes and thus perceive an ascending and a descending melody (cf. Butler 1992, p. 105) (Fig. 20.2).

For such hearing phenomena, Bregman uses the terms *stream integration* and *stream segregation*: Depending on the tempo and the range of the jumps, the ear follows the jumps, and only one melody is heard, or the notes are grouped in the ear into different simultaneous melody layers. These grouping processes are fundamental to the perception of music and are analyzed in detail by Bregman (1990).

W. J. Dowling investigated Gestalt formation with the help of "interleaved melodies" (Fig. 20.3). He combined two melodies so that one tone of each melody

Fig. 20.2 From a certain tempo, an ascending and a descending melody are perceived in this example

Fig. 20.3 Interleaved melodies

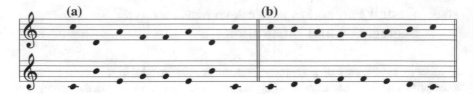

Fig. 20.4 Another study on the Gestalt laws of proximity and the good continuation

was heard alternately. If the notes of each melody are in different registers, both melodies are easy to identify. If, however, the two melodies move at a similar pitch, it takes much longer to assign the notes to the melodies. It is noticeable that it is easier to name the second melody as soon as the first one has been recognized. Butler (1992) explains this with the "cocktail party effect" (see p. 107). The cocktail party effect was described as such by E. Colin Cherry in 1953. This effect describes the phenomenon of being able to follow a conversation even though several other conversations take place nearby. This phenomenon, which in Cherry's study referred to language, can thus be applied equally to the reception of melodies (cf. Cherry, "Some Experiments on the Recognition of Speech, with One and Two Ears").

Another phenomenon was investigated by Deutsch in 1972. If you play a song with the right tones, but spread it over three octaves, the listeners have clear problems identifying the songs. However, as soon as the listeners knew which song it was, they were able to follow the actual melody line without any problems (cf. Butler 1992, p. 108).

In 1975, Deutsch carried out the following study: Two different tone sequences were played to the participants via stereo headphones. The tones were arranged so that the first tone of an ascending scale could be heard on one ear, followed by the second tone of a descending scale and so on (see Fig. 20.4). Almost all listeners perceived the tone sequences as shown in the diagram. The tones were, therefore, grouped according to their linear relationship or their proximity to each other instead of the melodies played being assigned to each ear (cf. Butler 1992, p. 108).

20.6 The Continuum of Pitch

The pitches form a continuum, where the perception of and the thinking in continuum can be clarified. Two tones can be used to decide whether they are the same or different in pitch. If they are different, it is usually recognized which of the tones is higher. The pitch perception is also transitive. Transitive means with a "<"-relation: $a < b$ and $b < c \Rightarrow a < c$ (cf. Stumpf 1883, p. 141).

Stumpf states: *"First, of every three unequal tones, one is always a middle one between the other two"* (Stumpf 1883, p. 142). *"Second, the domain of sounds has only one dimension. This means that of every three tones, only one can be the middle one under any circumstances"* (Stumpf 1883, p. 144).

The sensations of the pitches are thus perceived and presented in the perception as an ordered and unbranched set: *"thus, under the aforementioned condition the perception of the sum of the tones as a one-dimensional series is also given"* (Stumpf 1883, p. 141).

There are continuous transitions between pitch sensations: *"One can ascribe external as well as internal infinity to the tone series. External, i.e., the possibility of ever deeper and higher tones, internal, i.e., the possibility of ever shorter distances"* (Stumpf 1883, p. 176).

The pitches form a continuum that shows the same topology as the number line. Thinking in number range has a paragon in pitch perception (Stumpf 1890, p. 275).

Because the real numbers also belong to the one-dimensional continua, each one-dimensional continuum can be scaled, and measurability is created by constructing a unambiguous mapping between the one-dimensional continuum and the real numbers. The respective one-dimensional continuum becomes a metrically scaled variable, and after the determination of a fixed reference value, an absolute metrical scaling is created.

> Every continuum is a unit, not a mere aggregate. But it is possible to distinguish parts that border on each other in points, lines or areas. But the boundaries themselves are not parts (Stumpf 1890, p. 374).

20.6.1 Continuum Perception and Metric Variables

From Carl Stumpf's observations, it becomes clear how the classification in the case of metric variables has its psychological origin in the perception of one-dimensional continua, such as pitches. The scalability of the continuum and the differentiability of parts with their boundaries make the terms class boundary, class center, and frequency density of a class plausible in statistics. The possibility of a bijection between the one-dimensional continuum and the (real) numbers provides the prerequisite for the mathematical–statistical description of a metric variable.

20.6.2 Pitch Is a Cyclic Continuum

The classification of pitch, however, is still being pursued. The continuum of pitches can be understood as a straight line. In music, however, the pitches also have a cyclic structure, which is the result of the strong tonal fusion of the octave. Tones at octave intervals are identified in almost all tone systems. In all music theories, only a few discrete pitches per octave are used, which are usually distributed relatively evenly over the octave. In western music, the diatonic scale has only 7 notes per octave, the chromatic scale 12 notes per octave. The cyclicity is shown by the fact that the scale repeats itself in the same way in each octave. This categorially divides the infinite tone series into 7 or 12 discrete, i.e., clearly separated pitches, which makes the listening reproduction of music much easier. In modern music theory these 12 tones are called pitch classes. These 12 pitch classes are isomorphic to the residue class ring modulo 12. In mathematical music theory, algebraic properties of this residue class ring have been extensively analyzed with regard to music (cf. Mazzola 2002; Tymoczko 2011).

A similar order can also be found in the continuum of time. Every year the similar climatic processes, the almost equally repeating astronomical processes and the classification perceived by man based on the lunar cycles show a rhythm of 12 months. In time series analysis the statistician can regard the "month" e.g., as category.

20.7 Outlook

In research on musical perception, Gestalt laws are still used today as an explanatory approach. As an example, on a melodic level, the works of Mungan and Kaya (2017) and Lee et al. (2015) should be referred to. On the harmonic level, Gestalt formations were last investigated by Slana et al. (2016). But who actually explains the Gestalt laws? Terms such as "law of gravity", "Gestalt law", or "Gestalt principles" suggest that these phenomena are simply given and do not require further explanation. This is not the case. The Gestalt laws are purely descriptive and do not provide any explanation as to how Gestalt formations actually function. There is a huge need for research here (the same also applies, by the way, to the "law of gravity").

Approaches on the neurological level can be found at Griffiths et al. (2012), Lai et al. (2013) and Ono et al. (2015).

20.8 Conclusion

From the psychological point of view, classification is a special case of categorial thinking. It is based on psychic functions rooted in perception as can be shown at pitch perception and elementary musical structures. Particularly imported are grouping processes and pattern recognition which are based on comparisons and Gestalt perception. Quite remarkable are the analogies between pitch systems in music and mathematical number spaces.

Classification has paragons in perception, especially in hearing.

References

Bregman, A. S. (1990). *Auditory scene analysis*. Cambridge: MIT Press.
Brentano, F. (1933). Kategorienlehre. Meiner, Darmstadt (1933, unchanged reprint 1968).

Butler, D. (1992). *The Muscian's guide to perception and cognition*. New York: Schirmer.

Colin, C. E. (1953). Some experiments on the recognition of speech, with one and with two ears. *JASA, 25*, 975–979. https://doi.org/10.1121/1.1907229.

Deutsch, D. (1972). Octave generalization and tune recognition. *Perception and Psychophysics, 11*, 411–412. https://doi.org/10.3758/BF03206280.

Deutsch, D. (1975). Two-channel listening to musical scales. *JASA, 57*, 1156–1160. https://doi.org/10.1121/1.380573.

Dowling, W. J. (1973). The perception of interleaved melodies. *Cognitive Psychology, 5*, 322–337. https://doi.org/10.1016/0010-0285(73)90040-6.

Ebeling, M. (2008a). Zum Wesen der Konsonanz. Neuronale Koinzidenz, Verschmelzung und Rauhigkeit. In W. Auhagen (Ed.), *Jahrbuch der Deutschen Gesellschaft für Musikpsychologie* (Vol. 20, pp. 71–93).

Ebeling, M. (2008b). Neuronal periodicity detection as a basis for the perception of consonance: A mathematical model of tonal fusion. *JASA, 124*(4), 2320–2329. https://doi.org/10.1121/1.2968688.

Fox, C. W. (1948). Modern counterpoint: A phenomenological approach. *Notes for the Music Library Association Dec 1, 6*, 46–57. https://doi.org/10.2307/891495.

Griffiths, T. D., Micheyl, C., & Overath, T. (2012). Auditory object analysis. In D. Poeppel, T. Overath, A. N. Popper, & R. R. Fey (Eds.), *The human auditory cortex*. Springer: New York.

Lai, H., Xu, M., Sony, Y., & Liu, J. (2013). The neural mechanism underlying music perception: A meta-analysis off MRI studies. *Acta Psychologica Sinica, 45*(5), 491–507.

Langner, G. (1997). Temporal processing of pitch in the auditory system. *Journal of New Music Research, 26*, 116–132. https://doi.org/10.1080/09298219708570721.

Langner, G. (2015). *The Neural code of pitch and harmony*. Cambridge: CUP.

Lee, Y.-S., Janata, P., Frost, C., Martinez, Z., & Granger, R. (2015). Melody recognition revisited: Influence of melodic Gestalt on the encoding of relational pitch information. *Psychonomic Bulletin & Review, 22*(1), 163–169. https://doi.org/10.3758/s13423-014-0653-y.

Mazzola, G. (2002). *The Topos of music. Geometric logic of concepts, theory, and performance*. Basel: Birkhäuser.

Moore, B. C. J., & Ohgushi, K. (1993). Audibility of partials in inharmonic complex tones. *JASA, 93*(1), 452–461. https://doi.org/10.1121/1.405625.

Moore, B. C. J., Glasbert, B. R., & Shailer, M. J. (1984). Frequency and intensity difference limens for harmonics within complex tones. *JASA, 75*(2), 550–561. https://doi.org/10.1121/1.390527.

Mungan, E., & Kaya, M. U. (2017). Perceiving boundaries in unfamiliar Turkish makam music: Evidence for Gestalt universals? *Music Perception, 34*(3), 267–290. https://doi.org/10.1525/mp.2017.34.3.267.

Ono, Z., Altmann, C. F., Matsuhashi, M., Mima, T., & Fukuyama, H. (2015). Neural correlates of perceptual grouping effects in the processing of sound omission by musicians and nonmusicians. *Hearing Research, 319*, 25–31. https://doi.org/10.1016/j.heares.2014.10.013.

Schneider, P., Sluming, V., Roberts, N., Scherg, M., Goebel, R., Specht, H. J., et al. (2005). Structural and functional asymmetry of lateral Heschl's gyrus reflects pitch perception preference. *Nature Neuroscience, 8*, 1241–1247. https://doi.org/10.1038/nn1530.

Slana, A., Repovš, G., Fitch, W. T., & Gingras, B. (2016). Harmonic context influences pitch class equivalence judgments through Gestalt and congruency effects. *Acta Psychologica, 166*, 54–63. https://doi.org/10.1016/j.actpsy.2016.03.006.

Stumpf, C. (1883/1890). Tonpsychologie. Band 1/II. Leipzig: Barth.

Stumpf, C. (1897). Neues zur Tonverschmelzung. *Zeitschrift für Psychologie und Physiologie der Sinnesorgane, 15*, 280–303.

Stumpf, C. (2011). Erkenntnislehre. Barth, Leipzig (1939/1940). Reprint with an introduction by Margret Kaiser-el-Safti. Lengerich: Papst.

Trendelenburg, A. (1963). *Geschichte der Kategorienlehre*. Hildesheim: Georg Olms Verlagsbuchhandlung.

Tymoczko, D. (2011). *A geometry of music. Harmony and counterpoint in the extended common practice*. New York: Oxford University Press.

Wertheimer, M. (1923). Untersuchungen zur Lehre von der Gestalt. *Psychologische Forschung: Zeitschrift für Psychologie und ihre Grenzwissenschaften 4*, 301–350.

Zimbardo, P. G., & Ruch, F. L. (1978). *Lehrbuch der Psychologie*. Berlin: Springer, 221/22.

Zwicker, E., & Fastl, H. (1999). *Psychoacoustics. Facts and models*. Berlin: Springer.

Printed in the United States
By Bookmasters